Two Manuals Combined into One Book

Physical Security

And

Military Intelligence

Department of Defense

Pentagon
Publishing

*FM 3-19.30

Field Manual
No. 3-19.30

Headquarters
Department of the Army
Washington, DC, 8 January 2001

PHYSICAL SECURITY

Contents

Distribution Restriction: Approved for public release; distribution is unlimited.

*This publication supersedes FM 19-30, 1 March 1979.

Preface

This field manual (FM) sets forth guidance for all personnel responsible for physical security. It is the basic reference for training security personnel. It is intended to be a "one-stop" physical-security source for the Department of Defense (DOD), the Department of the Army (DA), and other proponents and agencies of physical security.

Prevention and protection are the two primary concerns of physical security. Both serve the security interests of people, equipment, and property. These interests must be supported at all staff and command levels; and this support must be unified in joint, multinational, and interagency operations.

Support to joint, multinational, and interagency operations relies on the fact that the Army will not conduct operations alone. Additionally, force-projection operations conducted by the military will involve the integration of war-fighting capabilities with stability and support operations. This manual's primary focus is the articulation of a balanced understanding of physical security for joint, multinational, and interagency operations throughout the environments of peacetime, conflict, and war (whether in the continental United States [CONUS] or outside the continental United States [OCONUS]).

Physical security must integrate the various capabilities of joint, multinational, and interagency operations in pursuit of a seamless connection between the strategic, operational, and tactical levels of war. Physical security must also address an expanded range of threats that embraces not only traditional threat components of war, but also nontraditional threats generated by guerrillas, terrorists, criminals, and natural or man-made disasters. In addition, physical security must address the concept of Homeland Defense due to the aforementioned threats.

Homeland Defense is the military's role in the United States (US) government's principal task of protecting its territory and citizens. This is accomplished by joint, interagency, and multijurisdictional organizations. Homeland Defense includes—

- Supporting domestic authorities for crisis and consequence management with regard to weapons of mass destruction (WMD).
- Protecting national-security assets (such as installations) and deploying forces and ensuring the availability, integrity, and adequacy of other critical assets.
- Deterring and defending against strategic attacks while maintaining freedom of action through antiterrorism and force-protection operations.

With this in mind, it is essential to address the five pillars of force protection—combating terrorism, physical security, personal security, law enforcement, and operations security (OPSEC). Physical security is a central component of force protection and provides an integrated venue to express support for operations. Physical security is a primary-leader task and an inherent part of all operations to protect soldiers, family members, civilians, and resources. This function directly supports the Army's universal task list.

While the effects of these changes (when viewed individually) appear revolutionary, the basic activities remain relatively unchanged, though executed under different conditions and standards. Another component that remains unchanged is our reliance upon quality soldiers and leaders well versed in physical-security fundamentals. Leaders will be challenged to ensure that they are functionally proficient; possess an understanding of physical-security operations; are educated in joint, multinational, and interagency operations; and have the ability to perform physical-security functions in support of full-dimension operations.

Appendix A contains an English-to-metric measurement conversion chart. Appendix B is a sample installation crime-prevention handbook. This handbook is designed to assist commanders in developing crime-prevention programs for their installation and units.

The proponent of this publication is HQ TRADOC. Send comments and recommendations on DA Form 2028 directly to Commandant, US Army Military Police School (USAMPS), ATTN: ATSJ-MP-TD, Directorate of Training, 401 Engineer Loop, Suite 2060, Fort Leonard Wood, Missouri 65473-8926.

Unless this publication states otherwise, masculine nouns and pronouns do not refer exclusively to men.

Chapter 1

Physical-Security Challenges

Physical security is defined as that part of security concerned with physical measures designed to safeguard personnel; to prevent unauthorized access to equipment, installations, material, and documents; and to safeguard against espionage, sabotage, damage, and theft. As such, all military operations face new and complex physical-security challenges across the full spectrum of operations. Challenges relative to physical security include the control of populations, information dominance, multinational and interagency connectivity, antiterrorism, and the use of physical-security assets as a versatile force multiplier.

OVERVIEW

1-1. Reductions in manpower and funding are critical challenges to physical security. Manpower for supporting physical-security activities is reduced through deployments and cutbacks. The rapid evolution of physical-security-equipment technology also lends to physical-security challenges, which are exponentially multiplied by the introduction of the information age.

1-2. Physical-security challenges must be understood, and measures must be taken to minimize them to enhance force protection. Leaders must create order when coming upon a situation; and when they depart, some semblance of that order must remain. They must be aware of the human-dimension factors and ensure that their soldiers do not become complacent. It was human error rather than modern technology that took lives in the bombings of the African embassy. Warning was given, but not heeded. Complacency became a physical-security challenge.

AUTOMATED INFORMATION SYSTEMS

1-3. Success on past battlefields has resulted not so much from technological advances, but from innovative ways of considering and combining available and new technologies as they apply to war fighting. Some of these technologies dealt with disseminating and processing information. For example, the telegraph, the telephone, the radio, and now the computer have redefined the fire-support paradigm.

1-4. As the armed forces move into the technological age, a greater need for physical-security measures is required. The risks associated with automated information systems (AISs) are widespread because computers are used for everything. Army Regulation (AR) 380-19 outlines the requirements that commanders and managers need for processing unclassified and classified information and for securing media, software, hardware, and different systems.

1-5. The threat to AISs and information systems security (ISS) involves deliberate, overt, and covert acts. This includes the physical threat to tangible property, such as the theft or destruction of computer hardware. Also included is the threat of electronic, electromagnetic-pulse, radio-frequency (RF), or computer-based attacks on the information or communications components that control or make up critical Army command and control (C^2) infrastructures. In most cases, the threat's target is the information itself rather than the system that transmits it. The threat comes from a range of sources, including the following:

- Unauthorized users (such as hackers) are the main source of today's attacks, primarily against computer-based systems. The threat they pose to AIS networks and mainframe computers is growing.
- Insiders are those individuals with legitimate access to an AIS. They pose the most difficult threat to defend against. Whether recruited or self-motivated, the AIS insider has access to systems normally protected by ISS against an attack.
- Terrorists once had to operate in the immediate vicinity of a target to gain access to or collect intelligence on that target. The proximity to the target risked exposure and detection. Today, a terrorist can accomplish most target selection, intelligence collection, and preoperational planning by gaining access through a computer network. He can increase his probability of success by using computer systems to reduce his "time on target." Terrorist access to an AIS also increases the threat of critical-data destruction or manipulation. Although his presence would be virtual, the potential for damage to Army C^2 systems could be equal to or greater than that achieved by physical intrusion, especially when used as a force multiplier in conjunction with a traditional terrorist attack. Therefore, while traditional preventive measures are still needed to protect unwanted access to information, the information age has added additional concerns for the commander and new opportunities for those with hostile intent.
- Non-state- and state-sponsored groups provide additional challenges. In many cases, it is difficult to confirm state sponsorship of threat activity against an AIS, no matter how apparent the affiliation might seem. Activists of all persuasions are increasingly taking advantage of information-age technology. Neither AISs nor ISS are immune from an adversary's interest in exploiting US military information systems or disrupting communication infrastructures. The availability of low-cost technology and the proliferation of an AIS increase the risk to the Army by potential adversaries.
- Foreign-intelligence services (FIS), both civil and military, are continually active and are another source of contention concerning information systems. In peacetime, they are increasingly targeted against US commercial and scientific interests, rather than military information. With little effort, this peacetime intrusiveness could easily be refocused on AISs and ISS using a wide range of information operations tactics.

- Political and religious groups are other potential adversaries to AISs and ISS. The world's political climate is diverse and complicated. It embraces traditional mainstream political values, as well as radical religious fundamentalism and political extremism. When political or religious viewpoints also incorporate anti-US sentiment, US information infrastructures (including AISs) are increasingly at risk of penetration or exploitation by these potential adversaries.

1-6. When considering an AIS, physical security is more than just safeguarding the equipment. It includes the following elements:

- Software is marked for each system and secured when not in use.
- Initial logon is password-protected (at a minimum).
- Passwords are a minimum of eight characters, using a mixture of letters and numerals.
- Access to an AIS is allowed only to authorized and cleared personnel (per AR 380-19).

1-7. Classified material is entered and transmitted only on approved devices with the following considerations:

- Approved classified devices are operated in a secured environment.
- Classified devices are secured in appropriate containers when not in use.
- Secure telephone unit–III (STU-III) keys are secured in an appropriate safe when not in use (as outlined in AR 380-19).

1-8. Additional information regarding AISs can be found in ARs 380-5 and 380-19. Required training of personnel working with an AIS is located in AR 380-19.

OPSEC AND THE THREAT

1-9. OPSEC is a process of identifying critical information and subsequently analyzing friendly actions attendant to military operations and other activities. The threat is identified using the factors of mission, enemy, terrain, troops, time available, and civilian considerations (METT-TC). The threat defines the physical-security challenges. Implementing physical-security measures supports OPSEC. Providing soundproof rooms for conducting briefings is a simple but invaluable measure.

1-10. Another issue to consider when evaluating physical-security challenges is what actions to take in case of political implications interfering with physical-security measures. In the devastating event at Khobar Towers, a warning was given but not everyone received it. It took too long to evacuate the building after the warning was issued because a cohesive plan was not in place.

1-11. Commanders can minimize the challenges to physical security through proactive measures. They should periodically change the physical-security posture of their area of responsibility to throw off perpetrators.

Chapter 2

The Systems Approach

Commanders must ensure that appropriate physical-security measures are taken to minimize the loss of personnel, supplies, equipment, and material through both human and natural threats. Commanders commonly exercise those protective responsibilities through the provost marshal (PM) and/or physical-security officer and the force-protection officer. The force-protection officer must coordinate with several different agencies to complete his mission. For example, the Army's Intelligence and Counterintelligence Program (see Appendix C) provides information that will be used to complete the unit's crisis-management plan (see Appendix D).

PROTECTIVE SYSTEMS

2-1. The approach to developing protective measures for assets should be based on a systematic process resulting in an integrated protective system. The protective system focuses on protecting specific assets against well-defined threats to acceptable levels of protection. The system is organized in-depth and contains mutually supporting elements coordinated to prevent gaps or overlaps in responsibilities and performance.

2-2. Effective protective systems integrate the following mutually supporting elements:

- Physical protective measures, including barriers, lighting, and electronic security systems (ESSs).
- Procedural security measures, including procedures in place before an incident and those employed in response to an incident. (These include procedures employed by asset owners and those applied by and governing the actions of guards.)
- Terrorism counteraction measures that protect assets against terrorist attacks.

2-3. The following determinations are made when considering system-development procedures:

- The resources available.
- The assets to be protected.
- The threat to those assets.
- The risk levels applicable to those assets.
- The applicable regulatory requirements for protecting the assets.
- The applicable level of protection for those assets against the threat.
- Additional vulnerabilities to the assets (based on the threat).

SYSTEMS DEVELOPMENT

2-4. AR 190-51, DA Pamphlet (Pam) 190-51, and Technical Manual (TM) 5-853-1 are useful tools for developing protective systems using the systems approach. The key to applying these tools successfully is to use a team approach. A team may include physical-security, intelligence, and operations personnel; the installation engineers; and the user of the assets. It may also include representatives from the multinational, host-nation (HN), and local police as well as the regional security office from the embassy.

ASSETS

2-5. Protective systems should always be developed for specific assets. The goal of security is to protect facilities and buildings and the assets contained inside. The risk-analysis procedure in DA Pam 190-51 is used to identify assets. This procedure is applied to all mission-essential or vulnerable areas (MEVAs) according to AR 190-13. It represents the majority of assets with which DOD is commonly concerned. These assets include—

- Aircraft and components at aviation facilities.
- Vehicle and carriage-mounted or -towed weapons systems and components at motor pools.
- Petroleum, oil, and lubricants (POL).
- Controlled medical substances and other medically sensitive items.
- Communication and electronics equipment; test, measurement, and diagnostic equipment (TMDE); night-vision devices (NVDs); and other high-value precision equipment and tool kits.
- Organizational clothing and individual equipment stored at central-issue facilities.
- Subsistence items at commissaries, commissary warehouses, and troop-issue facilities.
- Repair parts at installation-level supply activities and direct-support (DS) units with authorized stockage lists.
- Facilities-engineering supplies and construction materials.
- Audiovisual equipment, training devices, and subcaliber devices.
- Miscellaneous pilferable assets (not included above) and money.
- Mission-critical or high-risk personnel.
- General military and civilian populations.
- Industrial and utility equipment.
- Controlled cryptographic items.
- Sensitive information (included in TM 5-853-1, but not included in DA Pam 190-51).
- Arms, ammunition, and explosives (AA&E).
- Installation banks and finance offices.

RISK LEVELS

2-6. DA Pam 190-51 provides a procedure for determining risk levels—assessing the value of the assets to their users and the likelihood of compromise. These factors are assessed by answering a series of questions leading to value and likelihood ratings.

2-7. Asset value is determined by considering the following three elements:

- The criticality of the asset for its user and the Army as a whole.
- How easily the asset can be replaced.
- Some measure of the asset's relative value.

2-8. The relative value differs for each asset. For some assets, the relative value is measured in terms of monetary cost.

2-9. The likelihood of the threat is assessed for each applicable aggressor category by considering the asset's value to the aggressor, the history of or potential for aggressors attempting to compromise the asset, and the vulnerability of the asset based on existing or planned protective measures.

REGULATORY REQUIREMENTS

2-10. The risk level is the basis for determining the required protective measures for assets covered in AR 190-51. For each asset type, there may be physical protective measures, procedural security measures, and terrorism counteraction measures. These measures are specified by risk level. The measures identified in AR 190-51 are the minimum regulatory measures that must be applied for the identified threat level. The minimum regulatory measures for AA&E are based on the risk category established in AR 190-11.

ANTITERRORISM/FORCE-PROTECTION CONSTRUCTION STANDARDS

2-11. In accordance with DOD Instruction 2000.16, the commanders in chief (CINCs) have developed standards for new construction and existing facilities to counter terrorism threat capabilities within the area of responsibility. These construction standards have specific requirements for such measures as standoff distance, perimeter barriers, building construction, and parking. The DOD construction standard provides for minimum standards that must be incorporated into all inhabited DOD structures regardless of the identified threat. These standards provide a degree of protection that will not preclude the direct effects of blast but will minimize collateral damage for buildings and people and will limit the progressive collapse of structures. These standards add relatively little cost, may facilitate future upgrades, and may deter acts of aggression. (All services have adopted common criteria and minimum standards to counter antiterrorism/force-protection [AT/FP] vulnerabilities and terrorism threats.) Protection to identified threat levels is described in the following paragraphs. Physical-security personnel must be familiar with the CINC and DOD AT/FP construction standards because these standards may affect elements of physical-security plans and how individual facilities are secured.

THREAT IDENTIFICATION

2-12. The threat must be described in specific terms to help determine the assets' vulnerabilities or to establish protective measures. This description should include the tactics that aggressors will use to compromise the asset (weapons, tools, and explosives are likely to be used in an attempt). For example, the threat might be described as a moving vehicle bomb consisting of a 4,000-pound vehicle containing a 500-pound explosive. Another example would be a forced-entry threat using specific hand, power, or thermal tools. These types of threat descriptions (called the design-basis threat) can be used to design detailed protective systems to mitigate the attacks. TM 5-853-1 and DA Pam 190-51 contain procedures for establishing design-basis threat descriptions in the format described above. These procedures can be used together or separately. Threats listed in the TM will be summarized later in this chapter. When using the TM as a lone source or in conjunction with DA Pam 190-51, the following actions occur:

- When the TM process is used alone, the user goes through an identical process to that in DA Pam 190-51 up to the point where the risk level would be determined. In TM 5-853-1, the value and likelihood ratings are used differently than in DA Pam 190-51. The likelihood rating is used to determine the weapons, tools, and explosives that will be used by a particular aggressor in carrying out a specific tactic. In this procedure, higher likelihood ratings result in more severe mixes of weapons, tools, and explosives. The assumption is that the more likely the attack, the more resources the aggressor is likely to use in carrying out the attack.
- When the procedure in TM 5-853-1 is used in conjunction with the results of the DA Pam 190-51 risk analysis, the likelihood rating is taken directly from the risk analysis and applied as described above.

LEVEL OF PROTECTION

2-13. The level of protection applies to the design of a protective system against a specified threat (for example, a bomb, breaking and entering, pilfering, and so forth). The level of protection is based on the asset's value rating from either DA Pam 190-51 or TM 5-853-1. The level increases as the asset's value rating increases. There are separate levels of protection for each tactic. TM 5-853-1 provides detailed guidance on how to achieve the levels of protection, and Chapter 3 of this manual provides a summary of the levels of protection as they apply to various tactics.

VULNERABILITIES

2-14. Vulnerabilities are gaps in the assets' protection. They are identified by considering the tactics associated with the threat and the levels of protection that are associated with those tactics. Some vulnerabilities can be identified by considering the general design strategies for each tactic described in TM 5-853-1 and as summarized in Chapter 3 of this manual. The general design strategies identify the basic approach to protecting assets against specific tactics. For example, the general design strategy for forced entry is to provide a way to detect attempted intrusion and to provide barriers to delay the aggressors until a response force arrives. Vulnerabilities may involve inadequacies in intrusion-detection systems (IDSs) and barriers. Similarly, the general design strategy for a moving vehicle bomb is to keep the vehicle as far from the facility as possible and to harden the facility to resist the explosive at that distance. Vulnerabilities may involve limited standoff distances, inadequate barriers, and building construction that cannot resist explosive effects at the applicable standoff distance.

PROTECTIVE MEASURES

2-15. Where vulnerabilities have been identified, protective measures must be identified to mitigate them. AR 190-13, AR 190-51, DA Pam 190-51, and TM 5-853-1 are effective tools for developing protective measures. The key to effective development of protective systems is a partnership between physical-security personnel and the installation engineers. Appendix E of this manual discusses information for office security, which should be listed in the physical-security plan (see Appendix F). Appendix G discusses personal-protection measures.

THE INTEGRATED PROTECTIVE SYSTEM

2-16. Protective systems integrate physical protective measures and security procedures to protect assets against a design-basis threat. The characteristics of integrated systems include deterrence, detection, defense, and defeat.

DETERRENCE

2-17. A potential aggressor who perceives a risk of being caught may be deterred from attacking an asset. The effectiveness of deterrence varies with the aggressor's sophistication, the asset's attractiveness, and the aggressor's objective. Although deterrence is not considered a direct design objective, it may be a result of the design.

DETECTION

2-18. A detection measure senses an act of aggression, assesses the validity of the detection, and communicates the appropriate information to a response force. A detection system must provide all three of these capabilities to be effective.

2-19. Detection measures may detect an aggressor's movement via an IDS, or they may detect weapons and tools via X-ray machines or metal and explosive detectors. Detection measures may also include access-control elements that assess the validity of identification (ID) credentials. These control elements may provide a programmed response (admission or denial), or they may relay information to a response force. Guards serve as detection elements, detecting intrusions and controlling access.

2-20. Nuclear, biological, and chemical (NBC) detection systems must be used to measure and validate acts of aggression involving WMD. NBC detection systems should also be used to communicate a warning.

DEFENSE

2-21. Defensive measures protect an asset from aggression by delaying or preventing an aggressor's movement toward the asset or by shielding the asset from weapons and explosives. Defensive measures—

- Delay aggressors from gaining access by using tools in a forced entry. These measures include barriers along with a response force.
- Prevent an aggressor's movement toward an asset. These measures provide barriers to movement and obscure lines of sight (LOSs) to assets.
- Protect the asset from the effects of tools, weapons, and explosives.

2-22. Defensive measures may be active or passive. Active defensive measures are manually or automatically activated in response to acts of aggression. Passive defensive measures do not depend on detection or a response. They include such measures as blast-resistant building components and fences. Guards may also be considered as a defensive measure.

DEFEAT

2-23. Most protective systems depend on response personnel to defeat an aggressor. Although defeat is not a design objective, defensive and detection systems must be designed to accommodate (or at least not interfere with) response-force activities.

SECURITY THREATS

2-24. Security threats are acts or conditions that may result in the compromise of sensitive information; loss of life; damage, loss, or destruction of property; or disruption of mission. Physical-security personnel and design teams must understand the threat to the assets they are to protect in order to develop effective security programs or design security systems. Historical patterns and trends in aggressor activity indicate general categories of aggressors and the common tactics they use against military assets. Aggressor tactics and their associated tools, weapons, and explosives are the basis for the threat to assets.

THREAT SOURCES

2-25. There are many potential sources of threat information. Threat assessment is normally a military-intelligence (MI) responsibility. MI personnel commonly focus on such security threats as terrorists and military forces. Within the US and its territories, the Federal Bureau of Investigation (FBI) has primary responsibility for both foreign and domestic terrorists. The FBI, the US Army Criminal Investigation Command (USACIDC [CID]), and local law-enforcement agencies are good sources for physical-security personnel to obtain criminal threat information. Coordinating with these elements on a regular basis is essential to maintaining an effective security program.

THREAT CATEGORIES

2-26. Security threats are classified as either human or natural. Human threats are carried out by a wide range of aggressors who may have one or more objectives toward assets such as equipment, personnel, and operations. Aggressors can be categorized and their objectives can be generalized as described below. (See DA Pam 190-51 and TM 5-853-1 for more information.)

Aggressor Objectives

2-27. Four major objectives describe an aggressor's behavior. Any one of the first three objectives can be used to realize the fourth. These objectives include—

- Inflicting injury or death on people.
- Destroying or damaging facilities, property, equipment, or resources.
- Stealing equipment, materiel, or information.
- Creating adverse publicity.

Aggressor Categories

2-28. Aggressors are grouped into five broad categories—criminals, vandals and activists, extremists, protest groups, and terrorists. Hostile acts performed by these aggressors range from crimes (such as burglary) to low-intensity conflict threats (such as unconventional warfare). Each of these categories describes predictable aggressors who pose threats to military assets and who share common objectives and tactics.

- Criminals can be characterized based on their degree of sophistication. They are classified as unsophisticated criminals, sophisticated criminals, and organized criminal groups. Their common objective is the theft of assets; however, the assets they target, the quantities they seek, their relative efficiency, and the sophistication of their actions vary significantly. Vandals and activists may also be included under this category.
- Vandals and activists are groups of protesters who are politically or issue oriented. They act out of frustration, discontent, or anger against the actions of other social or political groups. Their primary objectives commonly include destruction and publicity. Their selection of targets will vary based on the risk associated with attacking them. The degree of damage they seek to cause will vary with their sophistication.
- Extremists are radical in their political beliefs and may take extreme, violent actions to gain support for their beliefs or cause.
- Protesters are considered a threat only if they are violent. Lawful protesters have to be considered, but significant protective measures and procedures are not normally needed to control their actions. The presence of extremists or vandals/activists at a peaceful protest increases the chance of the protest becoming violent.
- Terrorists are ideologically, politically, or issue oriented. They commonly work in small, well-organized groups or cells. They are sophisticated, are skilled with tools and weapons, and possess an efficient planning capability. There are three types of terrorists—CONUS, OCONUS, and paramilitary OCONUS.
  - CONUS terrorists are typically right- or left-wing extremists operating in distinct areas of the US.
  - OCONUS terrorists generally are more organized than CONUS terrorists. They usually include ethnically or religiously oriented groups.

- Paramilitary OCONUS terrorist groups show some military capability with a broad range of military and improvised weapons. Attacks by OCONUS terrorists are typically more severe.

2-29. Natural threats are usually the consequence of natural phenomena. They are not preventable by physical-security measures, but they are likely to have significant effects on security systems and operations. They may require an increase in protective measures either to address new situations or to compensate for the loss of existing security measures. They may reduce the effectiveness of existing security measures by such occurrences as collapsed perimeter fences and barriers, inoperable protective lighting, damaged patrol vehicles, and poor visibility. Natural threats and their effects relative to security include the following:

- Floods may result in property damage, destruction of perimeter fences, and damage to IDSs. Heavy rains or snowfalls may have similar effects even if they do not result in flooding.
- Storms, tornadoes, high winds, or rain may cause nuisance alarms to activate and cause damage to IDSs. They may limit the visibility of security personnel and may affect closed-circuit television (CCTV) systems. Winds may also disrupt power or communication lines and cause safety hazards from flying debris.
- Earthquakes may cause nuisance alarms to activate or may disrupt IDSs. They may also cause broken water or gas mains, fallen electrical or communication lines, and weakened or collapsed buildings.
- Snow and ice can make travel on patrol roads difficult, may delay responses to alarms, may impede the performance of IDSs, and may freeze locks and alarm mechanisms. Heavy ice may also damage power and communication lines.
- Fires may damage or destroy perimeter barriers and buildings, possibly leaving assets susceptible to damage or theft.
- Fog can reduce the visibility of security forces, thereby requiring additional security personnel. It may also increase the response time to alarms and reduce the effectiveness of security equipment such as CCTV systems.

Aggressor Tactics

2-30. Aggressors have historically used a wide range of offensive strategies reflecting their capabilities and objectives. These offensive strategies are categorized into 15 tactics that are specific methods of achieving aggressor goals (see TM 5-853-1). Separating these tactics into categories allows facility planners and physical-security personnel to define threats in standardized terms usable as a basis for facility and security-system design. Common aggressor tactics include—

- **Moving vehicle bomb.** An aggressor drives an explosive-laden car or truck into a facility and detonates the explosives. His goal is to damage or destroy the facility or to kill people. This is a suicide attack.
- **Stationary vehicle bomb.** An aggressor covertly parks an explosive-laden car or truck near a facility. He then detonates the explosives either by time delay or remote control. His goal in this tactic is the

same as for the moving vehicle bomb with the additional goal of destroying assets within the blast area. This is commonly not a suicide attack. It is the most frequent application of vehicle bombings.

- **Exterior attack.** An aggressor attacks a facility's exterior or an exposed asset at close range. He uses weapons such as rocks, clubs, improvised incendiary or explosive devices, and hand grenades. Weapons (such as small arms) are not included in this tactic, but are considered in subsequent tactics. His goal is to damage the facility, to injure or kill its occupants, or to damage or destroy assets.
- **Standoff weapons.** An aggressor fires military weapons or improvised versions of military weapons at a facility from a significant distance. These weapons include direct (such as antitank [AT] weapons) and indirect LOS weapons (such as mortars). His goal is to damage the facility, to injure or kill its occupants, or to damage or destroy assets.
- **Ballistics.** The aggressor fires various small arms (such as pistols, submachine guns, shotguns, and rifles) from a distance. His goal is to injure or kill facility occupants or to damage or destroy assets.
- **Forced entry.** The aggressor forcibly enters a facility using forced-entry tools (such as hand, power, and thermal tools) and explosives. He uses the tools to create a man-passable opening or to operate a device in the facility's walls, doors, roof, windows, or utility openings. He may also use small arms to overpower guards. His goal is to steal or destroy assets, compromise information, injure or kill facility occupants, or disrupt operations.
- **Covert entry.** The aggressor attempts to enter a facility or a portion of a facility by using false credentials or stealth. He may try to carry weapons or explosives into the facility. His goals include those listed for forced entry.
- **Insider compromise.** A person authorized access to a facility (an insider) attempts to compromise assets by taking advantage of that accessibility. The aggressor may also try to carry weapons or explosives into the facility in this tactic. His goals are the same as those listed for forced entry.
- **Visual surveillance.** The aggressor uses ocular and photographic devices (such as binoculars and cameras with telephoto lenses) to monitor facility or installation operations or to see assets. His goal is to compromise information. As a precursor, he uses this tactic to determine information about the asset of interest.
- **Acoustic eavesdropping.** The aggressor uses listening devices to monitor voice communications or other audibly transmitted information. His goal is to compromise information.
- **Electronic-emanations eavesdropping.** The aggressor uses electronic-emanation surveillance equipment from outside a facility or its restricted area to monitor electronic emanations from computers, communications, and related equipment. His goal is to compromise information.

- **Mail-bomb delivery.** The aggressor delivers bombs or incendiary devices to the target in letters or packages. The bomb sizes involved are relatively small. His goal is to kill or injure people.
- **Supplies-bomb delivery.** The aggressor conceals bombs in various containers and delivers them to supply- and material-handling points such as loading docks. The bomb sizes in this tactic can be significantly larger that those in mail bombs. His goal is to damage the facility, kill or injure its occupants, or damage or destroy assets. Appendix H addresses the actions to take when a bomb is suspected.
- **Airborne contamination.** An aggressor contaminates a facility's air supply by introducing chemical or biological agents into it. His goal is to kill or injure people.
- **Waterborne contamination.** An aggressor contaminates a facility's water supply by introducing chemical, biological, or radiological agents into it. These agents can be introduced into the system at any location with varying effects, depending on the quantity of water and the contaminant involved. His goal is to kill or injure people.

2-31. The aforementioned tactics are typical threats to fixed facilities for which designers and physical-security personnel can provide protective measures. However, some common terrorist acts are beyond the protection that facility designers can provide. They cannot control kidnappings, hijackings, and assassinations that take place away from facilities or during travel between facilities. Protection against these threats is provided through operational security and personal measures (see Appendices G and I), which are covered in doctrine relative to those activities and are under the general responsibility of the CID.

TACTICAL ENVIRONMENT CONSIDERATIONS

2-32. When determining the assets and threats, the same considerations should be given to the systems approach in the tactical environment as when in the cantonment area. The same process of determining the assets, their risk level, and any regulatory guidance apply. Identifying potential threats and the level of protection required for the assets are necessary. Commanders and leaders must also identify additional vulnerabilities and other required protective measures. Commanders are not expected to have the same physical protective measures due to the impact of resources, budget, location, and situations.

2-33. Commanders must consider the various tactics used by aggressors and use their soldiers' abilities to counteract these tactics. Considerations for specific assets (such as military-working-dog [MWD] and explosive-ordnance-disposal [EOD] teams and their abilities to detect and disassemble a bomb) must be identified. Units must have the ability to improvise in a tactical environment. Their training and resourcefulness will compensate for shortcomings in the field.

2-34. The systems approach to security provides focus and integration of resources. Protective systems are mutually supporting and systematically developed to negate the threat. Commanders conduct an intelligence preparation of the battlefield (IPB) and vulnerability assessments (VAs) to determine risks. Security resources and measures are applied to mitigate risks and to deter, detect, defend, and defeat the threat.

Chapter 3
Design Approach

Developing protective systems to protect assets depends on an effective partnership between engineers and physical-security personnel. Physical-security personnel need to understand the basic approaches the engineers will take in laying out protective systems. Engineers must understand the issues involved with ensuring that anything they lay out is compatible with security operations and the operations of the asset users. The best way to ensure a viable design is through teamwork. This chapter provides a summary of the basic approaches to protecting assets against threats (the design strategies). Understanding these strategies is critical to being an effective team member in developing protective systems.

DESIGN STRATEGIES

3-1. There are separate design strategies for protecting assets from each tactic described in Chapter 2. There are two types of strategies associated with each tactic—the general-design and specific-design strategies. The general-design strategy is the general approach to protecting assets against tactics. The specific-design strategy refines the general-design strategy to focus the performance of the protective system on a particular level of protection. (See TM 5-853-1 for more information.)

PROTECTIVE MEASURES

3-2. Protective measures are developed as a result of the general- and specific-design strategies. These protective measures commonly take the form of site-work, building, detection, and procedural elements.

- Site-work elements include the area surrounding a facility or an asset. Technically, they are associated with everything beyond 5 feet from a building. They can include perimeter barriers, landforms, and standoff distances.
- Building elements are protective measures directly associated with buildings. These elements include walls, doors, windows, and roofs.
- Detection elements detect such things as intruders, weapons, or explosives. They include IDSs, CCTV systems used to assess intrusion alarms, and weapon and explosive detectors. These elements can also include the guards used to support this equipment or to perform similar functions.
- Procedural elements are the protective measures required by regulations, TMs, and standing operating procedures (SOPs). These elements provide the foundation for developing the other three elements.

VEHICLE BOMBS

3-3. Vehicle-bomb tactics include both moving and stationary vehicle bombs. In the case of a moving vehicle bomb, the aggressor drives the vehicle into the target. This is commonly known as a suicide attack. In a stationary vehicle bomb, he parks the vehicle and detonates the bomb remotely or on a timed delay.

GENERAL-DESIGN STRATEGY

3-4. Blast pressures near an exploding vehicle bomb are very high, but they decrease rapidly with distance from the explosion. The design strategy for these tactics is to maintain as much standoff distance as possible between the vehicle bomb and the facility and then, if necessary, to harden the facility for the resulting blast pressures. Barriers on the perimeter of the resulting standoff zone maintain the required standoff distance. The difference between moving and stationary vehicle-bomb tactics is that the aggressor using the moving vehicle bomb will attempt to crash through the vehicle barriers; the aggressor using the stationary vehicle bomb will not. Therefore, vehicle barriers for the moving vehicle bomb must be capable of stopping a moving vehicle at the perimeter of the standoff zone. For a stationary vehicle bomb, vehicle barriers must mark the perimeter of the standoff zone, but they are not required to stop the moving vehicle. They only need to make it obvious if an aggressor attempts to breach the perimeter.

LEVELS OF PROTECTION

3-5. There are three levels of protection for vehicle bombs—low, medium, and high. The primary differences between the levels are the degree of damage allowed to the facility protecting the assets and the resulting degree of damage or injury to the assets.

- **Low.** The facility or the protected space will sustain a high degree of damage but will not collapse. It may not be economically repairable. Although collapse is prevented, injuries may occur and assets may be damaged.
- **Medium.** The facility or the protected space will sustain a significant degree of damage, but the structure will be reusable. Occupants and other assets may sustain minor injuries or damage.
- **High.** The facility or the protected space will sustain only superficial damage. Occupants and other assets will also incur only superficial injury or damage.

SITE-WORK ELEMENTS

3-6. The two primary types of site-work elements for vehicle bombs are the standoff distance and vehicle barriers. The vehicle's speed must also be taken into consideration.

Standoff Distance

3-7. The standoff distance is the maintained distance between where a vehicle bomb is allowed and the target. The initial goal should be to make that distance as far from the target facility as practical. Figure 3-1 shows the distances required to limit building damage to particular levels (including the levels of protection described above) for a range of bomb weights. All bomb weights are given in terms of equivalent pounds of trinitrotoluene (TNT), which is a standard way of identifying all explosives regardless of their composition. The example in Figure 3-1 is a building of conventional construction (common, unhardened construction). Buildings built without any special construction at these standoff distances will probably withstand the explosive effects. Conventionally constructed buildings at standoff distances of less than those shown in Figure 3-1 will not adequately withstand blast effects. (Refer to TM 5-853-1 for information on hardening buildings to resist a blast.) Do not allow vehicles to park within the established standoff distances. Recognize that this restriction can result in significant operational and land-use problems.

3-8. **Exclusive Standoff Zone.** When an exclusive standoff zone is established, do not allow vehicles within the perimeter unless they have been searched or cleared for access. The zone's perimeter is established at the distance necessary to protect the facility against the highest threat explosive. All vehicles should be parked outside the exclusive standoff zone; only

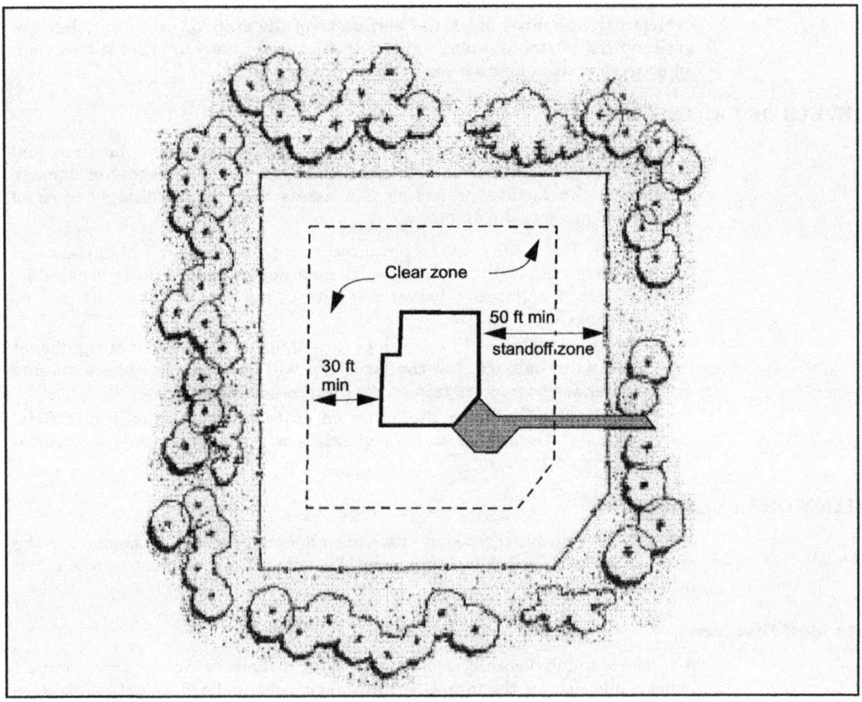

Figure 3-1. Standoff Distance

maintenance, emergency, and delivery vehicles should be allowed within the zone after being searched. Figure 3-2 shows an exclusive standoff zone.

3-9. **Nonexclusive Standoff Zone.** A nonexclusive standoff zone is established in a location having a mixture of cars and trucks (with relatively few trucks). A nonexclusive standoff zone takes advantage of aggressors being able to conceal a smaller quantity of explosives in a car than they can in a truck. Therefore, a nonexclusive standoff zone includes inner and outer perimeters. The inner perimeter is set at a distance corresponding to the weight of explosives that can be concealed in cars. The outer perimeter is set at a distance associated with the weight that can be placed in trucks (refer to TM 5-853-1). With these two perimeters, cars can enter the outer perimeter without being searched but they cannot enter the inner perimeter. Trucks cannot enter the outer perimeter, since it is established based on what they can carry. Figure 3-3 shows a nonexclusive standoff zone. The nonexclusive standoff zone provides the advantages of allowing better use of the parking areas and limiting the number of vehicles that need to be searched at the outer perimeter.

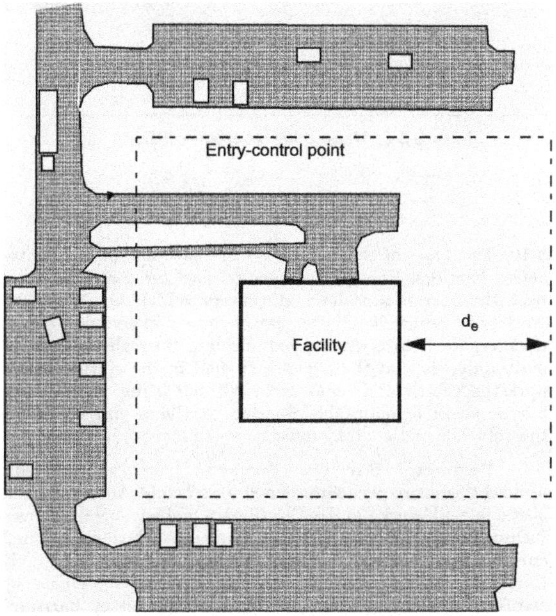

d_e = exclusive standoff-zone distance

Figure 3-2. Exclusive Standoff Zone

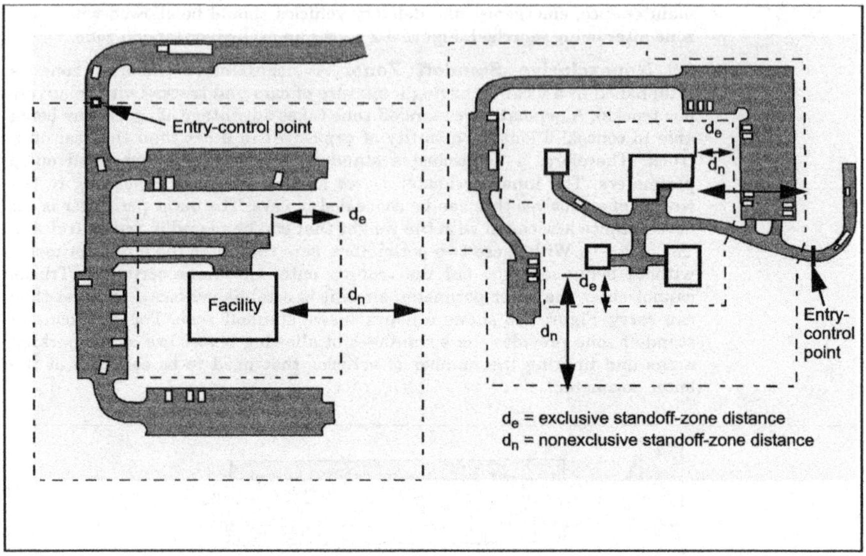

Figure 3-3. Nonexclusive Standoff Zone

Vehicle Barriers

3-10. Two types of vehicle barriers are used for vehicle bombs—perimeter and active barriers. The type of barrier used for a moving vehicle bomb differs from the barrier used for a stationary vehicle bomb. The barrier used for a stationary vehicle bomb does not have to stop a vehicle's motion. The goal for that barrier is to make anybody driving through the barrier noticeable. The assumption is that the aggressor's goal in the stationary vehicle bomb is to park the vehicle and sneak away without being noticed. Crashing through a barrier would be noticeable. Barriers for the moving vehicle bomb need to stop the vehicle's motion; they must be much more substantial.

3-11. **Perimeter Barriers.** Perimeter barriers are fixed barriers placed around the entire perimeter of a standoff zone. Anything that presents a fixed obstacle will work for the stationary vehicle bomb. Common applications include chain-link fences, hedges made of low bushes, and high (over 8 inches) curbs. Aggressors driving through such barriers are likely to be noticed. Barriers capable of stopping moving vehicles include chain-link fences reinforced with cable, reinforced concrete "Jersey barriers", pipe bollards, planters, ditches, and berms. When barriers such as the Jersey barriers and planters are used to stop moving vehicles, they must be anchored into the ground to be effective. The cables in the reinforced fence also have to be anchored into the ground or partially buried. Spaces between barriers should be no greater than 4 feet. Figure 3-4 shows common perimeter barriers for stationary or moving vehicle bombs. Refer also to TM 5-853-1.

3-12. **Active Barriers.** Active barriers are placed at openings in perimeters where vehicles need to enter or exit. These barriers must be able to be raised and lowered or moved aside. For the stationary vehicle bomb, barriers can be as simple as chain-link, pipe, or wooden gates that can be raised and lowered. Aggressors crashing through any of these or similar obstructions will likely draw attention. For the moving vehicle bomb, the barriers are heavy structures and have many construction and operations considerations

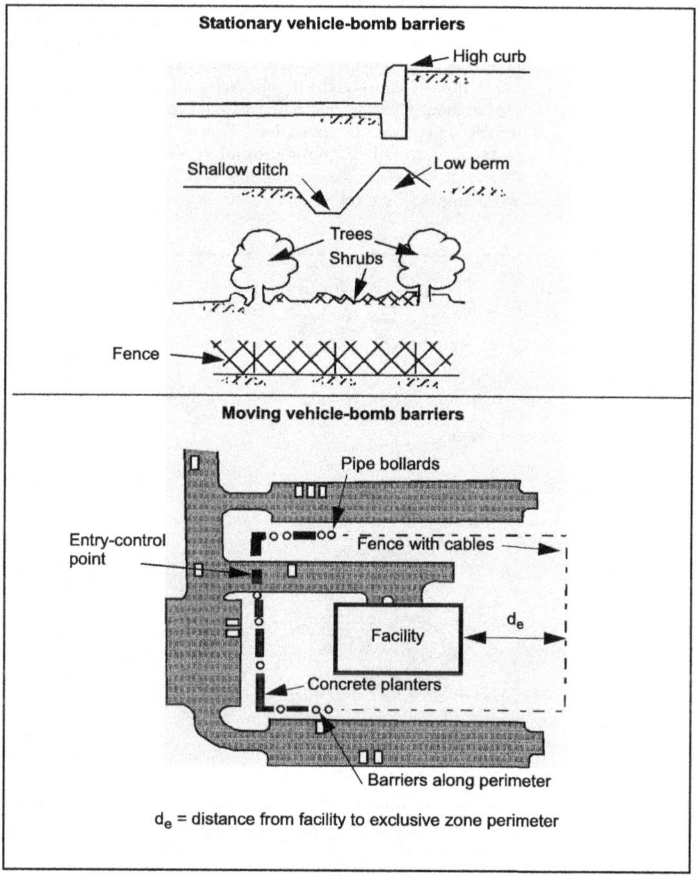

Figure 3-4. Perimeter-Barrier Application

associated with them. These barriers may stop vehicles weighing up to 15,000 pounds and travelling 50 miles per hour. They commonly cost tens of thousands of dollars (refer to TM 5-853-1). Some common active vehicle barriers are shown in Figure 3-5. For temporary or deployed conditions, park a vehicle across an opening and move it aside to grant access.

Speed Control

3-13. It is important to control the speed of a vehicle approaching a barrier used for a moving vehicle bomb. The energy from a vehicle that a barrier must stop increases as its speed increases. The energy also increases with more weight, but the effect of speed is much greater. Therefore, decreasing the vehicle's speed results in smaller and less costly barriers. The best way to limit a vehicle's approach speed to perimeter barriers is to place or retain obstacles in potential approach paths. The vehicles are forced to reduce speed when going around these obstacles. The same principle applies for road approaches. Placing obstacles in a serpentine pattern on the road forces a vehicle to reduce its speed (see Figure 3-6, page 3-8). If the vehicle hits the obstacles instead of going around them, they are still slowed down. Other means to slow vehicles include forcing them to make sharp turns and installing traffic circles.

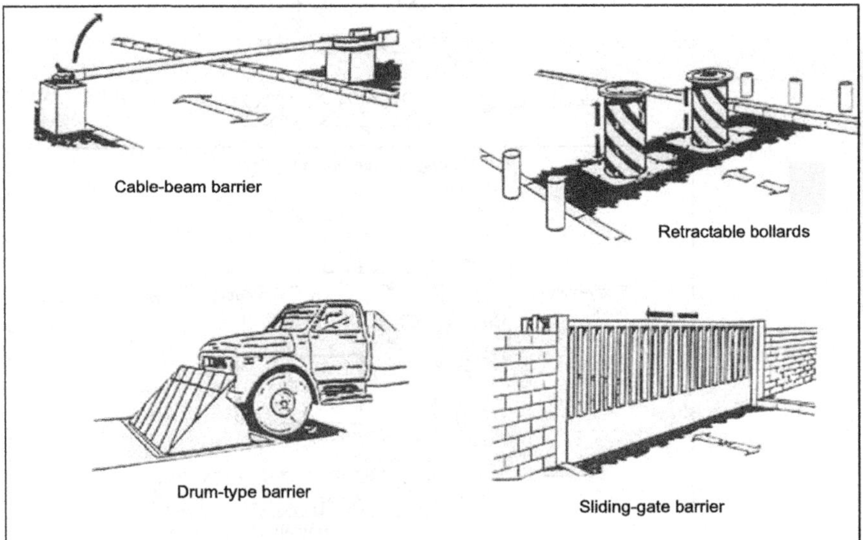

Figure 3-5. Active Vehicle Barriers

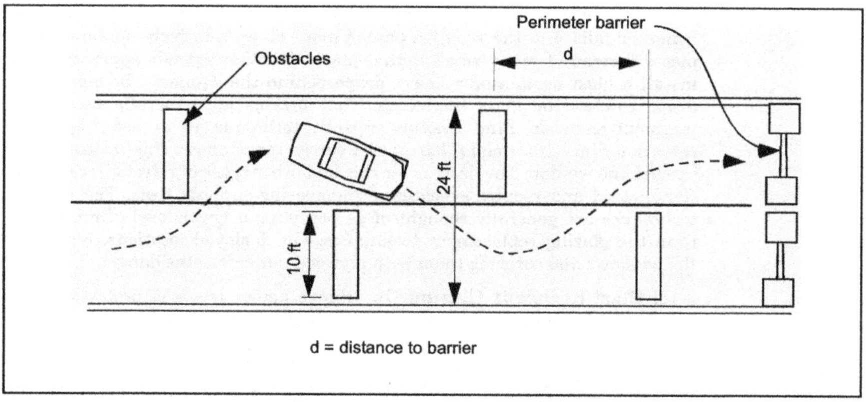

Figure 3-6. Serpentine Pattern

BUILDING ELEMENTS

3-14. Once the standoff distance is established and the site has been laid out, the designers can select the building components necessary to protect the assets against the threat explosives at the standoff distance. The building components include the walls, roofs, doors, and windows. Detailed design issues related to these building elements are covered in TM 5-853-1.

Walls and Roofs

3-15. If the distances shown for the desired damage levels in Figure 3-1, page 3-3, cannot be enforced, the building's walls and roofs will need to be strengthened. This can be achieved in new construction by using reinforced masonry or reinforced concrete in the walls and reinforced concrete in the roof. When the standoff distance is not available for existing construction, a more detailed analysis may be required to determine what the explosion's impact will be on the structure. When the construction is inadequate, more standoff distance should be investigated or the engineers should apply specialized techniques for retrofitting the construction to increase its strength.

Windows

3-16. Historically, glass fragments have caused about 85 percent of injuries and deaths in bomb blasts. There are two basic approaches to mitigating the effects of bomb blasts on glass—retrofitting the windows with film or curtains and using blast-resistant glazing.

3-17. **Retrofitting Windows.** One of the most common means of decreasing the hazards from broken glass is to install fragment-retention film on the glass. The film is a plastic (polyester) sheet that adheres to the window glass with a special adhesive. The film does not strengthen the glass; but when the glass breaks, it keeps the fragments from spreading throughout the room. The glass fragments stick to the film, and the film either stays in the window frame or falls into the room in one or more large, relatively nonhazardous pieces instead of many small, lethal pieces. Another retrofit approach is to install a blast curtain or a heavy drape behind the window. The curtain or drape catches the glass fragments. The curtains are generally used with fragment-retention film. Another retrofit technique is to use fragment-retention film with a metal bar placed across the window. This "catcher bar" catches the window. The designs for this and other types of retrofit devices are complicated and require specialized engineering-analysis tools. The retrofit techniques are generally thought of as providing a lower level of protection than the glazing replacement techniques. For deployed locations, removing the windows and covering them with plywood minimizes the danger.

3-18. **Blast-Resistant Glazing.** To achieve higher levels of protection, the window glass must be replaced and the window frame should be reinforced. Because of its expense, this procedure is generally limited to new construction and major renovations. Special blast-resistant glazing and frames are available that use either tempered glass or a plastic glazing (such as polycarbonate). Another promising type of blast-resistant glazing is laminated glass, in which several layers of common glass are adhered together with a special interlayer. The resulting laminated construction is usually stronger than common glass while retaining the same thickness. The interlayer acts similarly to fragment-retention film. For deployed locations, a means of minimizing the danger of windows is to remove them and replace them with plywood.

Doors

3-19. Doors are another building component particularly vulnerable to an explosive blast. Common metal and wood doors provide little resistance to a blast. The two ways to address the problem of doors is to install them in foyers or to replace them. Glass doors or doors containing windows should be avoided.

Foyers

3-20. Door hazards can be reduced by installing doors in foyers during construction or by adding foyers to existing buildings. When a door is located in a foyer and the outer door fails, the outer door flies into a wall instead of the building's interior (see Figure 3-7, page 3-10). The inner door then has a greater chance of remaining intact. This option generally provides a low level of protection.

3-21. Another option is to replace the doors with specially constructed blast-resistant doors and frames. These doors are commercially available and can provide a high level of protection, but they are very expensive and heavy. The doorframe must be made of the same type of material and provide the same level of protection as the door.

DETECTION ELEMENTS

3-22. Detection elements for vehicle bombs are limited to the use of guards to control access into standoff zones. The guards search vehicles seeking entry

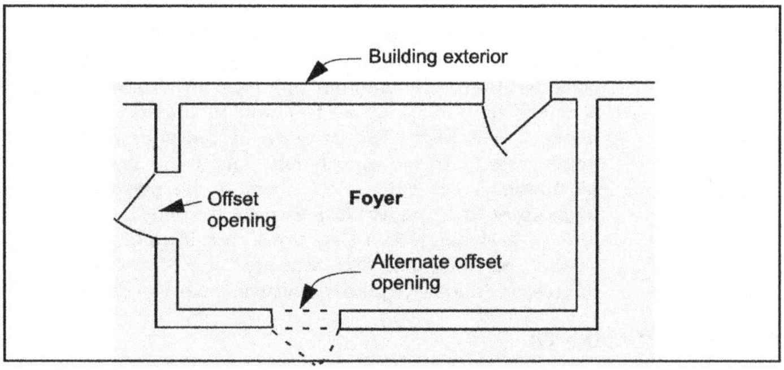

Figure 3-7. Door in Foyer

into the perimeter through an entry-control point. The recommended levels of searches depend on the required level of protection (see TM 5-853-1). Guards can be stationed at entry-control points continuously, or they can be summoned to an entry-control point when access is needed. The latter is commonly the case for the inner perimeter of exclusive standoff zones where only delivery and maintenance vehicles need access.

EXTERIOR ATTACK

3-23. An exterior attack is a physical attack using weapons such as rocks, clubs, improvised incendiary devices (IIDs) such as Molotov cocktails, explosives such as improvised explosive devices (IEDs), and hand grenades. The explosives can be thrown at or placed near a facility's exterior. Examples of IEDs for this tactic range from pipe bombs and hand grenades to briefcase-sized explosives.

GENERAL-DESIGN STRATEGY

3-24. Because the exterior attack is directed at a facility's exterior surfaces, the general-design strategy is to keep aggressors away from the facility (at a standoff distance) and, if necessary, to harden the facility's exterior components to resist the effects of weapons and explosives. A standoff distance from the facility reduces the degree of hardening required to resist weapons effects. When briefcase-sized bombs are a threat, an obstacle-free zone should be established around the facility and the explosives placed within should be detected and disarmed.

LEVELS OF PROTECTION

3-25. The levels of protection for exterior attacks are similar to those for vehicle bombs. Levels of protection vary based on the level of building damage and asset injury or damage allowed. However, due to the limited sizes of explosives involved in this tactic, the damage to the building will be much more localized and injuries or damage to assets will be confined to smaller areas.

SITE-WORK ELEMENTS

3-26. Site-work elements for exterior attacks are relatively limited because the explosive weights are more limited. Large standoff distances are not a consideration. The common approach to site-work elements is to lay out a standoff zone of about 50 feet and to provide a fence or perimeter barrier about 7 feet high. The purpose of the standoff is to make it harder for aggressors to throw pipe bombs and hand grenades at targets inside the perimeter. Trees can be left around the perimeter to make it harder for aggressors to throw explosives over the fence. The remaining component of site-work elements is a clear zone around the facility. A clear zone is applied so that anything placed in that area can be detected visually. This limits the aggressor's ability to place explosives near the target facility.

BUILDING ELEMENTS

3-27. Building elements for exterior attacks are similar to those for vehicle bombs. For small IEDs and IIDs, the building-element requirements do not increase the cost of the building significantly. For larger, briefcase-sized bombs, the measures are more significant than for incendiary devices but less than for vehicle bombs.

Walls and Roofs

3-28. Walls and roofs are not a problem with small explosives. Conventional construction normally provides adequate protection. Walls with 6-inch reinforced concrete or 8-inch, grout-filled, reinforced masonry will withstand the effects of typical pipe bombs or hand grenades. The corresponding roof construction is 6-inch reinforced concrete. In the case of briefcase-sized bombs, considerations similar to those discussed for vehicle bombs need to be employed.

Windows

3-29. A significant goal when constructing windows is to make them difficult to throw an explosive or incendiary device through, especially when considering smaller explosives. This is accomplished by constructing smaller windows or making narrow windows (see Figure 3-8, page 3-12). For existing windows, parts of the windows can be covered to achieve a narrow effect. These windows still may be susceptible to breakage due to explosive effects, even from the smaller explosives. This problem is solved by installing 3/4-inch-thick plastic (polycarbonate) glazing or by raising the windows over 6 feet high to develop a small standoff distance (as shown in Figure 3-9, page 3-12). A 3/4-inch glazing will also stop grenade fragments. Fragment-retention film, a blast curtain, or a heavy drape as described in vehicle-bomb tactics are also good applications for small bombs.

Conventionally Constructed Doors

3-30. Doors are not a significant problem with small bombs and incendiary devices. Generally, metal doors are adequate for incendiary devices, and doors placed in foyers (as shown in Figure 3-7) are adequate for pipe bombs and hand grenades. A similar application for briefcase-sized bombs would provide

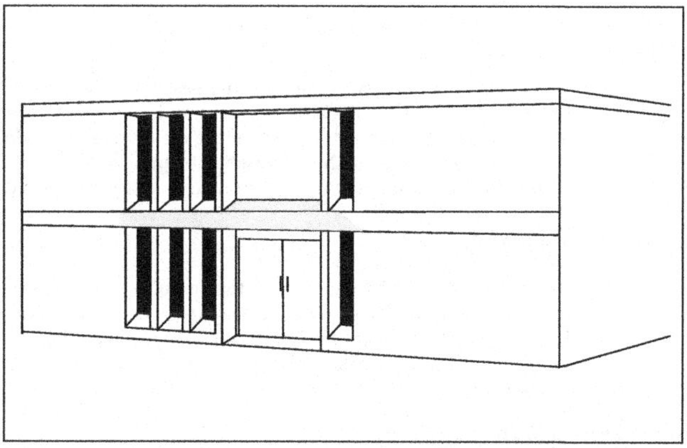

Figure 3-8. Narrow Recessed Windows

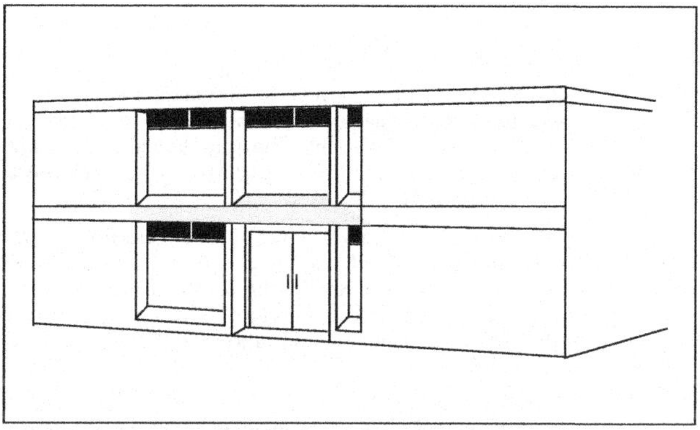

Figure 3-9. Narrow Raised Windows

only a low level of protection. To achieve higher levels of protection for briefcase-sized bombs, blast-resistant doors must be installed.

3-31. The requirements to meet the levels of protection for larger explosives are similar to those described for vehicle bombs, but they will not stop grenade fragments. Fragment-retention film and drapes or curtains can provide a low level of protection, but blast-resistant glazing is required to achieve a higher level of protection.

DETECTION ELEMENTS

3-32. Other than awareness of aggressor activity on or outside the site, detection is only a specific design goal where briefcase-sized bombs are anticipated. When that is the case, the clear zone around the building must be visually monitored so that any objects placed in it are detected. At higher levels of protection, visual surveillance is augmented by IDSs.

STANDOFF WEAPONS

3-33. The standoff-weapons tactic includes the use of AT weapons and mortars. In both of these tactics, the aggressor fires weapons at assets located in the protected facility from a distance. An AT-weapon attack requires a clear LOS to the target, while mortars can fire over obstacles and only need a clear line of flight.

GENERAL-DESIGN STRATEGY

3-34. Standoff-weapons attacks cannot be detected reliably before they occur. Protective design to resist these tactics relies on blocking LOSs to protected areas of a facility or hardening the facility to resist the particular weapon's effects. The approaches to protection against mortars and AT weapons differ from each other and will be discussed separately. Detection measures are not applicable for these tactics.

LEVELS OF PROTECTION

3-35. There are two levels of protection against both mortars and AT weapons. For AT weapons, the low level of protection depends on detonating the AT round before it hits the target facility. The high level of protection avoids the risk associated with that and hardens the building to resist the direct impact of the AT round.

3-36. For mortars, the low level of protection involves allowing some areas of the facility to be sacrificed. Those spaces provide a buffer to the assets to be protected. The assets within the sacrificial areas and the areas themselves may be destroyed. At the high level of protection, the building's exterior fully resists the mortar rounds and there are no sacrificial areas.

SITE-WORK ELEMENTS

3-37. The primary site-work element for standoff weapons is to obstruct LOSs from vantage points outside of the site. With AT weapons, the aggressor cannot hit what he cannot see. This is not true with mortars, but blocking LOSs from mortar firing points helps to make targeting more difficult. The LOSs are blocked by using trees, other buildings, vehicle parking areas, or fences. Another site-work element, a predetonation screen, applies only to an AT weapon. When using a predetonation screen, the AT round is detonated on the screen and its effects are dissipated in the distance between the screen and the target (see Figure 3-10, page 3-14). Any screen material (such as a

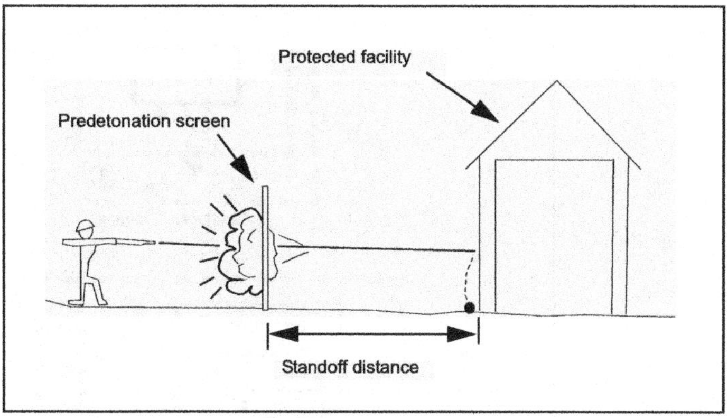

Figure 3-10. Predetonation Screen

wooden fence) will detonate the round unless it has spaces in it. The screen distances vary from less than 10 feet to almost 40 feet, depending on the building construction (see TM 5-853-1). This measure only applies to the low level of protection.

BUILDING ELEMENTS

3-38. Building elements for AT weapons and mortars involve the building's layout. This includes the materials used in the construction.

Layout

3-39. A building's interior layout is only an issue for the low level of protection against a mortar round. The layout issue involves designating sacrificial areas in which unimportant assets are located. The assets to be protected are located in a hardened interior layer. Figure 3-11 includes a plan view (from above). The sacrificial area has to be both around and above the protected area in case a mortar round comes from above. If such a layout is not feasible, other options include going to a higher level of protection and either hardening the entire building or building the facility underground (which are both very expensive).

Walls and Roofs

3-40. Walls and roofs must offer protection against both AT weapons and mortar rounds. The design of walls that protect against AT weapons varies with the level of protection. For the low level of protection where the round is predetonated, the walls can be of conventional construction, varying with the standoff distance from the predetonation screen to the wall. For higher levels of protection, the walls must resist the full effect of the round, requiring the walls to be 24-inch-thick reinforced concrete. Roofs are not an issue in

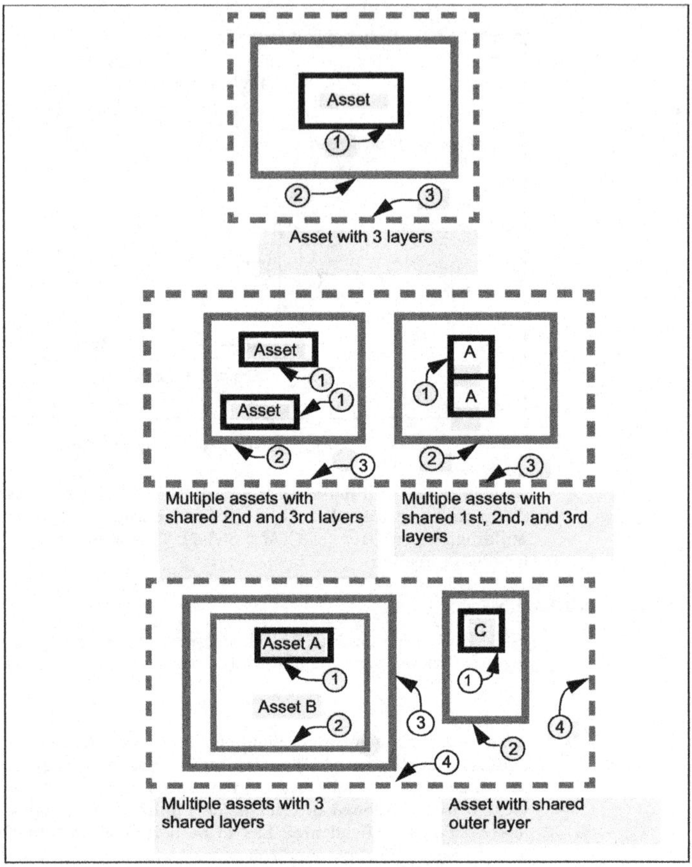

Figure 3-11. Assets Protected by Hardened Interior Layer

protecting against AT weapons because it is difficult to get direct LOSs to roofs. If such LOSs are possible, the roof should be designed like the walls.

3-41. To provide protection against mortar rounds, walls and roofs should be designed to resist the explosive effects in the rounds at the standoff distance that the sacrificial space provides. In the case of sacrificial areas, the walls can be of common construction. The interior protected-area walls are then designed of reinforced concrete or reinforced masonry for the standoff distance those sacrificial walls provide. When the walls must resist the full effect of the rounds (as in the higher level of protection), they are likely to be very thick (up to 30 inches of reinforced concrete for some improvised mortars). Similar considerations should be made for roofs. Roofs are designed to take the direct effects of the round or to take the round at the standoff distance provided by the sacrificial area.

Doors and Windows

3-42. It is impractical to provide doors and windows that are resistant to mortar rounds and AT weapons. Windows should only be used in sacrificial areas where there is a mortar threat. When there is an AT weapon threat, windows can only be used where the round is predetonated. The windows should be narrowed or raised to present a smaller target (see Figures 3-8 and 3-9, page 3-12). Doors should be placed in foyers (see Figure 3-7, page 3-10) for protection against AT rounds and to achieve a low level of protection against mortars. Blast-resistant doors are necessary to achieve a high level of protection against mortar rounds.

BALLISTICS

3-43. In a ballistics tactic, aggressors fire small arms at assets from vantage points outside of the target facility's control. Ballistic attacks cannot be detected reliably before they occur.

GENERAL-DESIGN STRATEGY

3-44. Protective measures to resist these tactics rely on blocking LOSs to protected areas of a facility or by hardening the facility to resist the ballistic effects. This strategy focuses on assets within buildings. Protecting people or property in the open is difficult and can only be addressed through operational measures. Detection measures are not applicable for this tactic.

LEVELS OF PROTECTION

3-45. There are only two levels of protection for this tactic. The low level of protection depends on blocking LOSs to assets. This strategy assumes that the aggressor cannot hit what he cannot see. The risk of an aggressor firing into a building randomly and hitting something is what makes this the low level of protection. The high level of protection involves hardening building components to resist the ballistic effects. These strategies can be thought of as either hardening or hiding.

SITE-WORK ELEMENTS

3-46. Site-work elements are of limited use for the ballistics tactic. When they are applied, they are used to obstruct LOSs from vantage points outside of the site, which is consistent with the low level of protection. The LOSs can be blocked using trees, other buildings, motor pools, or fences.

BUILDING ELEMENTS

3-47. Building elements are the principal means of protecting assets against a ballistics attack. They can be applied to achieve either the low or high level of protection.

Walls and Roofs

3-48. Walls and roofs are inherently opaque, so it is easy to achieve the low level of protection (hiding) with them. Achieving the high level of protection (hardening) for walls and roofs can be done within conventional construction using reinforced concrete, concrete-masonry units (CMUs), or clay brick. The material's required thickness is shown in Table 3-1. The thicknesses of CMUs and clay brick are nominal, meaning they do not represent the actual thickness of the material; they represent the thicknesses at which those materials are commercially available. Steel plates (mild steel and armor steel) and bullet-resistant fiberglass can be used to retrofit existing building components that would not provide the needed bullet resistance.

Table 3-1. Required Thicknesses, in Inches

Ballistics Type	Reinforced Concrete	Grouted CMU*	Clay Brick*	Steel Plate		Bullet-Resistant Fiberglass
				Mild	Armor	
.38 special	2	4	4	1/4	3/16	5/16
9 mm	2 1/2	4	4	5/16	1/4	7/16
7.62 and 5.56 mm	4	6	6	9/16	7/16	1 1/8
7.62-mm AP	6 1/2	8	8	13/16	11/16	N/A
*Nominal thicknesses						

Windows

3-49. Windows can include openings in walls and skylights, although skylights are only an issue where there are LOSs to them. When skylights require protection, treat them like windows. Achieving the low level of protection (hiding) for windows requires making it difficult to see through them, such as installing reflective film on the glass. An aggressor cannot see through the windows during daylight while it is lighter outside than inside, but he may see through them at night when the opposite might be true. Drapes or blinds that can be closed at night address that vulnerability. To achieve the high level of protection requires bullet-resistant window assemblies. These are commercially available for a wide range of ballistics types. They are purchased as manufactured-and-tested assemblies (including glazing and frames, both of which are equally bullet-resistant). The glazing materials and thicknesses and the framing details are proprietary to their manufacturers. The manufacturers make them according to industry test standards to ensure an effective product.

Doors

3-50. Doors without glass easily meet the requirements for the low level of protection. Meeting the high level of protection requires the installation of bullet-resistant door assemblies. Doors can be installed in foyers so that there is no direct LOS into assets within the building (see Figure 3-7, page 3-10).

FORCED ENTRY

3-51. In the forced-entry tactic, an aggressor tries to forcibly gain access to assets. He may use tools or explosives to breach building components or other barriers.

GENERAL-DESIGN STRATEGY

3-52. The general-design strategy for forced entry is to detect the aggressor early in the forced-entry attempt and delay him long enough for a response force to intercept him. The combination of detection and defensive measures must provide sufficient time for a response force to intercept the aggressor before he reaches the asset or before he escapes with it, depending on the protective goals for the asset. The first goal would apply where the asset is likely to be destroyed or where access to it is not acceptable. The second goal would be applied when the idea is to prevent it from being stolen.

LEVELS OF PROTECTION

3-53. Several levels of protection apply to forced entry. These levels vary in terms of system design, delay time, and response-force arrival time.

SITE-WORK ELEMENTS

3-54. Site-work elements do not normally play a major role in protecting against a forced entry. However, the site should be laid out and maintained so that an aggressor does not have a hiding place nearby that will conceal his attempts to break into the building. Another site-work element is the application of perimeter barriers, most commonly fences. Fences are effective at delineating a boundary and at keeping honest people honest, but they are ineffective for preventing a forced entry. The design strategy for forced entry is based on delaying the aggressor, and any serious aggressor could climb a fence in less than 4 seconds or can cut through a fence in less than 10 seconds. Therefore, fences are not used as delay elements, but they are used to establish boundaries and as platforms on which to hang sensors. The final site-work consideration is securing utility-access ports such as manholes. If there are utility tunnels through which aggressors can enter a building, those accesses should be locked using padlocks or locking bolts.

BUILDING ELEMENTS

3-55. Building elements are the principal construction elements of a system for protecting against a forced entry. The building elements are used to provide delay. The process for designing to resist forced entry involves laying out concentric "rings" of delay (called defensive layers). These defensive layers can include the facility's exterior, interior rooms within that layer, and containers within the interior rooms. The individual building components for each of the layers (walls, doors, windows, floors, ceilings, and roofs) provide the delay time (see TM 5-853-1).

DETECTION ELEMENTS

3-56. For a protective system to be effective against a forced entry, the aggressors must be detected at a point of adequate delay. Detection at that point can be achieved by using an IDS. Once a sensor detects an aggressor, the alarm annunciator communicates that event to security personnel, who then dispatch a response force. The alarm can be assessed through a guard response or via CCTV. Chapter 6 and TM 5-853-4 provide detailed discussion of IDSs, CCTV systems, and other elements of ESSs.

COVERT ENTRY AND INSIDER COMPROMISE

3-57. In the covert-entry tactic, an aggressor who is not authorized to be in the facility attempts to enter using false credentials. In the insider-compromise tactic, personnel with legitimate access to a facility try to compromise an asset. The insider may or may not have legitimate access to the asset itself. The purpose of the entry in either case can be to steal or otherwise compromise the asset or to destroy it. In the latter case, the aggressor may bring IEDs or IIDs.

GENERAL-DESIGN STRATEGY

3-58. The general-design strategy for both the insider-compromise and covert-entry tactics is to keep people from entering areas they are not authorized to enter. For covert entry, aggressors are denied access to controlled areas. For insider compromise, aggressors are denied access to assets within controlled areas based on their need to have access to them. The general-design strategy also includes detecting aggressors removing assets from protected areas and detecting aggressors carrying tools, weapons, and explosives into protected areas.

LEVELS OF PROTECTION

3-59. The levels of protection for these tactics address different issues, depending on whether the aggressor's goal is to steal or otherwise compromise an asset or to destroy it. When the goal is to steal or compromise an asset, the levels of protection vary with the number and sophistication of the access controls required to verify personnel access into a controlled area. When the goal is to destroy the assets, the levels of protection vary with the amount of damage the building (and the assets inside) are allowed to sustain and the sophistication of detecting weapons or explosives at entry points.

BUILDING ELEMENTS

3-60. Building elements vary with an aggressor's goal. To protect against theft or compromise of assets, building elements are used to establish and maintain controlled areas into which only authorized personnel can enter. For insider compromise, there may be an additional requirement that access be further limited among personnel otherwise authorized access to the controlled area. That access is based on the need to have access to a specific asset. The result is that the controlled area may be compartmentalized, and each compartmentalized area may have separate access requirements. There are no special construction requirements for these tactics if the goal is theft of compromise. The only requirement is that the building elements of controlled areas should provide enough resistance to require aggressors to force their way through them to gain entry and to provide evidence of the forced entry if it is attempted. Forcing entry would be contrary to the aggressor's assumed goal to be covert. In addition, a common design goal would be to limit the number of entrances into controlled areas because there will need to be access control at each entry.

3-61. To protect against the destruction of assets, building elements are used to shield assets from the effects of explosives going off at access-control points.

The basic approach is to lay out areas at access points in which guards can search for carried-in weapons, explosives, or incendiary devices. The construction of that area is designed to limit damage to the rest of the building if an explosive is detonated in that area. Those levels of damage are similar to those discussed in relation to vehicle bombs. The walls and doors between the access point and the protected area will be hardened, and the walls and doors to the outside will be of lightweight construction so that they may fail and vent the blast pressure away from the building. At the higher level of protection, the access-control area is located in a separate facility and the target building is hardened to resist an explosion in that separate facility.

DETECTION ELEMENTS

3-62. Detection elements for these tactics also vary based on the aggressor's goal. For theft, the detection elements are mainly related to access control. For destruction, the detection elements are used to detect weapons, explosives, or incendiary devices.

3-63. The main detection elements for theft or compromise are access-control devices. These can include procedural systems (such as guards checking ID), mechanical systems (such as keyed or combination locks), or electronic entry-control elements (such as electronic card readers, keypads, and biometric devices). Chapter 6 provides detailed discussion of electronic devices. The sophistication of these elements and the number used varies with the level of protection. For example, achieving the higher levels of protection requires the application of multiple forms of access-control elements such as a card reader and an electronic keypad for electronic-entry control or a badge check and badge exchange for a procedural system.

3-64. When destruction of the assets is the goal, detection is oriented toward detecting weapons, explosives, or incendiary devices. At the lower levels of protection, it is sufficient for guards to search for carried-in items. Achieving higher levels of protection requires the application of such equipment as metal detectors, X-ray machines, and explosive detectors.

SURVEILLANCE AND EAVESDROPPING

3-65. Surveillance and eavesdropping tactics include visual surveillance, acoustic eavesdropping, and electronic-emanations eavesdropping. In these tactics, aggressors remain outside of controlled areas and try to gather information from within those areas. The tools used for these tactics include ocular devices for the visual-surveillance tactic and listening devices and electronic-emanations-eavesdropping equipment for the eavesdropping tactic.

GENERAL-DESIGN STRATEGY

3-66. The general-design strategy for these tactics is to deny aggressors access to information assets. The kind of information (objects, operations, or files; secure conversations; or electronically processed data) and how it can be compromised differs for each tactic as do the specific protective strategies. Therefore, each tactic is addressed separately.

LEVELS OF PROTECTION

3-67. Each of these tactics has only one level of protection. Either one protects or fails to protect against these tactics.

SITE-WORK ELEMENTS

3-68. Site-work elements play a minor role in protecting assets from all surveillance or eavesdropping tactics. The main issue is to eliminate or control vantage points from which aggressors can surveil or eavesdrop on assets or operations. In addition, for the visual-surveillance tactic, a design goal can be to block LOSs from vantage points. Items used to block LOSs include trees, bushes, fences, and other buildings (see Figure 3-12).

BUILDING ELEMENTS

3-69. Building elements are the principal components of the protective strategies for surveillance and eavesdropping tactics. For visual surveillance,

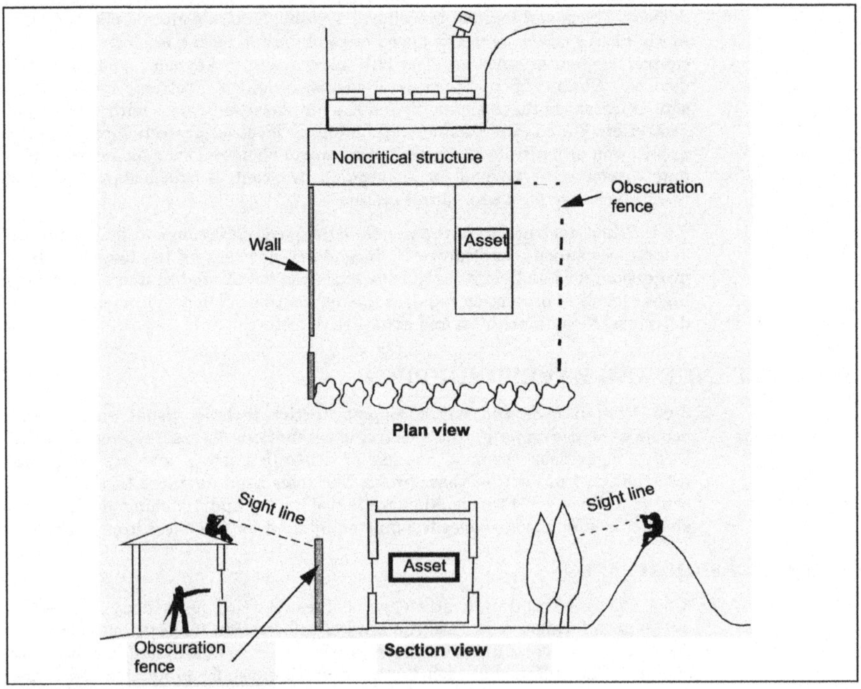

Figure 3-12. LOS Blocked From Potential Vantage Points

the building elements must block LOSs from outside the building. Walls and roofs perform this function effectively. Doors are only a problem when they have windows in them or are made of transparent materials. When this is the case, they can be treated like windows or they can be placed in foyers so that there are no direct LOSs through them. Windows can be treated with reflective film and drapes or blinds as described in the ballistics tactics. When there are LOSs through skylights, they should be treated like windows.

3-70. Building elements for acoustic eavesdropping relate to the construction of areas (preferably separated from the building exterior) that minimize the sound that can be transmitted through them. This requires specialized construction that has a sound-transmission-coefficient (STC) rating. Walls, floors, and ceilings can be constructed to achieve specific STC ratings using conventional construction materials as described in TM 5-853-1. Doors and windows that are STC rated are commonly manufactured and tested as assemblies. This type of design and construction can be expensive.

3-71. Protection against electronic-emanations eavesdropping involves the application of Terminal Electromagnetic-Pulse Emanation Standard (TEMPEST) guidance, most of which is classified. The protection is based on a TEMPEST assessment done for the Army by the US Army Intelligence and Security Command (INSCOM) and on guidance in AR 380-19. The results of a TEMPEST assessment will commonly lead to countermeasures from one or more of the following categories:

- Follow information security policies and procedures recommended during the assessments.
- Provide controlled space both inside and outside the facility.
- Provide TEMPEST-shielded equipment.
- Provide separation between electronic circuits that handle classified information and those that do not. This is commonly called red/black separation.
- Provide TEMPEST-shielded enclosures. This is specialized, metal-shielded construction that is very expensive.

MAIL AND SUPPLY BOMBS

3-72. In mail- and supply-bomb tactics, aggressors place bombs in materials delivered to a facility. Explosives used in supply bombs are significantly larger (briefcase size) than those in mail bombs (pipe bombs or smaller). Mail bombs are usually directed at individuals, while supply bombs may be used to target larger numbers of people. These tactics assume that the facility containing the asset has a mail-handling area or a supplies-handling and -receiving area. These tactics do not apply if mail or supplies are handled and screened in a different facility.

GENERAL-DESIGN STRATEGY

3-73. A bomb exploding within a building has more severe effects than the same size bomb exploding outside of the facility because the blast pressures cannot dissipate inside. Also, there is no standoff distance between the explosive and the facility to mitigate blast effects. The general-design strategy for mail and supply bombs is to detect delivered bombs before they explode and to harden the area where the explosion takes place. This minimizes the damage to the remainder of the facility. Occupants and contents within the mail room or supplies-handling area are likely to be killed or destroyed if an undetected bomb explodes.

LEVELS OF PROTECTION

3-74. The levels of protection for mail and supply bombs are based on the amount of damage allowed to the building and, therefore, the occupants of the building. They also vary based on the sophistication of the detection measures used.

BUILDING ELEMENTS

3-75. The purpose of building elements in relation to these bomb tactics is to shield assets from the effects of explosives going off at supply areas, receiving points, or mail rooms. The basic approach is to lay out either a mail room or a supplies-receiving area in which people can search suspicious packages for explosives or incendiary devices. Constructing this type of area will limit the damage to the rest of the building if an explosive is detonated there. Those levels of damage are similar to those discussed in relation to vehicle bombs.

Mail Rooms

3-76. Mail rooms should be located on the facility's exterior, away from any critical assets. The walls and ceiling between the mail room and the remainder of the building are hardened to keep the blast effects out of the facility. The exterior walls and doors should be of lightweight construction so that they may fail and vent the blast pressure away from the building. There may be an explosives container in the mail room where suspicious packages can be placed. If the package explodes, the container will keep its effects from causing damage or injury. The hardened construction will protect assets outside of the mail room if the explosion occurs outside of the container. Check with EOD personnel to determine the local policy for using explosive containers. At higher levels of protection, the mail room is constructed to completely contain the effects of an explosion either through hardened construction or by using a specialized construction called vented suppressive shielding. Mail rooms should not have windows into protected areas. Doors between the mail room and the rest of the building should be avoided, placed in foyers, or replaced with blast-resistant doors, depending on the desired level of protection.

Supplies-Handling Areas

3-77. Supplies-handling areas should also be on the building's exterior, away from critical areas of the facility. Walls and doors between the handling area and the protected area should be hardened, and the exterior walls and doors should be of lightweight construction so that they may fail and vent the blast pressure away from the building. There should be no windows between the handling area and the protected area. At the higher level of protection, the handling area is located in a separate facility and the target building is hardened to resist an explosion in that separate facility.

DETECTION ELEMENTS

3-78. Detection for these assets varies with the level of protection. At the lower levels of protection, bombs are detected by inspection. As the level of protection goes up, the sophistication of the detection increases. At the higher levels of protection, equipment such as X-ray examining devices, metal detectors, and explosives detectors can be used. Explosive-detection dogs are an alternative to explosive detectors.

CHEMICAL AND BIOLOGICAL CONTAMINATION

3-79. When using chemical- and biological-contamination tactics, aggressors introduce contaminants into the air or water supply to a facility or a group of facilities. Both airborne and waterborne contaminants include chemical, biological, and radiological agents. Aggressors may also forcibly enter a facility to contaminate water or air using the forced-entry tactic.

GENERAL-DESIGN STRATEGY

3-80. Both chemical and biological agents are difficult to detect in water and air supplies. Radiological agents are relatively easy to detect in water, but they are not commonly included in water-quality examinations. It is unlikely that all agents will be detected, so the general-design strategy for these tactics is to filter out suspected airborne contaminants or to shut off suspected waterborne contaminants. Also, because contaminants can easily be entered into the environment from inside a facility, the strategy includes limiting access to the facility (especially mechanical rooms, water intakes, and so forth).

LEVELS OF PROTECTION

3-81. The levels of protection for each of these tactics differ only in the frequency with which some protective measures are exercised. For the low level of protection, they are exercised only in response to a known threat. In the high level of protection, they are exercised continuously.

SITE-WORK ELEMENTS

3-82. Site-work elements are only significant for waterborne contamination. They include protecting water-treatment plants and water-storage structures. This protection may include constructing perimeter barriers (such as chain-link fences) and controlling access to the plant site. These measures are used because most contaminants require quantities on the order of truckloads to contaminate a water supply, so the focus of security is to keep such large vehicles under control. The perimeter barriers do not need to stop the vehicles because the assumption is that the aggressor wants to be covert. An overt act would alert people to avoid the water supply.

BUILDING ELEMENTS

3-83. Building elements for both tactics include controlling access so that aggressors cannot sneak in and plant devices in the building. Protection against airborne contamination at a facility involves making elements of the air-handling system (including air intakes) inaccessible and laying out toxin-free areas for people to be protected. A toxin-free area is an area in which the internal air pressure is higher than the external air pressure. Therefore, if a chemical, biological, or radiological device is set off outside, its contaminant will not be able to penetrate the protected area. Achieving that "net positive pressure" requires a significant air-handling system with air filters to filter contaminants out of the air. It also requires an air-lock entrance into the area so contaminants cannot enter through the door. At the low level of protection, the filters and the air-handling system are only used in response to a credible threat. At the high level of protection, that risk is not acceptable and the filters are run continuously.

3-84. The building-element issues for waterborne contamination are limited to providing protection against forced and covert entries into water-treatment plants and water-storage areas. These methods have been previously described. The only additional issue is the provision for alternative water sources. If it is suspected or detected that the water is contaminated, a backup water source should be in place (such as bottled water). For the high level of protection, bottled water should always be used for drinking.

Chapter 4

Protective Barriers

Protective barriers are used to define the physical limits of an installation, activity, or area. Barriers restrict, channel, or impede access and are fully integrated to form a continuous obstacle around the installation. They are designed to deter the worst-case threat. The barriers should be focused on providing assets with an acceptable level of protection against a threat.

OVERVIEW

4-1. Protective barriers form the perimeter of controlled, limited, and exclusion areas. Utility areas (such as water sources, transformer banks, commercial power and fuel connections, heating and power plants, or air-conditioning units) may require these barriers for safety standards. Protective barriers consist of two major categories—natural and structural.

- Natural protective barriers are mountains and deserts, cliffs and ditches, water obstacles, or other terrain features that are difficult to traverse.
- Structural protective barriers are man-made devices (such as fences, walls, floors, roofs, grills, bars, roadblocks, signs, or other construction) used to restrict, channel, or impede access.

4-2. Barriers offer important benefits to a physical-security posture. They create a psychological deterrent for anyone thinking of unauthorized entry. They may delay or even prevent passage through them. This is especially true of barriers against forced entry and vehicles. Barriers have a direct impact on the number of security posts needed and on the frequency of use for each post.

4-3. Barriers cannot be designed for all situations. Considerations for protective structural barriers include the following:

- Weighing the cost of completely enclosing large tracts of land with significant structural barriers against the threat and the cost of alternate security precautions (such as patrols, MWD teams, ground sensors, electronic surveillance, and airborne sensors).
- Sizing a restricted area based on the degree of compartmentalization required and the area's complexity. As a rule, size should be kept to a minimum consistent with operational efficiency. A restricted area's size may be driven by the likelihood of an aggressor's use of certain tactics. For example, protecting assets from a vehicle bomb often calls for a substantial explosives standoff distance. In these cases, mitigating the vehicle bomb would often be more important than minimizing the restricted area to the extent necessary for operational efficiency. Protective barriers should be established for—
  - Controlling vehicular and pedestrian traffic flow.

- Providing entry-control points where ID can be checked.
- Defining a buffer zone for more highly classified areas.
- Precluding visual compromise by unauthorized individuals.
- Delaying forced entry.
- Protecting individual assets.

4-4. If a secured area requires a limited or exclusion area on a temporary or infrequent basis, it may not be possible to use physical structural barriers. A temporary limited or exclusion area may be established where the lack of proper physical barriers is compensated for by additional security posts, patrols, and other security measures during the period of restriction. Temporary barriers (including temporary fences, coiled concertina wire, and vehicles) may be used. Barriers are not the only restrictive element, and they may not always be necessary. They may not be ideal when working with limited or exclusion areas or when integrated with other controls.

4-5. Because barriers can be compromised through breaching (cutting a hole through a fence) or by nature (berms eroded by the wind and rain), they should be inspected and maintained at least weekly. Guard-force personnel should look for deliberate breaches, holes in and under barriers, sand dunes building up against barriers, and the proper functioning of locks.

FENCING

4-6. Three types of fencing are authorized for use in protecting restricted areas—chain link, barbed wire, and barbed tape or concertina. The type used for construction depends primarily on the threat and the degree of permanence. It may also depend on the availability of materials and the time available for construction. Fencing may be erected for other uses besides impeding personnel access. It can impede observation, can serve as a means to defeat standoff-weapon systems (such as rocket-propelled grenades [RPGs]), and can serve as a barrier to hand-thrown weapons (such as grenades and firebombs).

4-7. Generally, chain-link fencing will be used for protecting permanent limited and exclusion areas. All three types of fencing may be used to augment or increase the security of existing fences that protect restricted areas. Examples would be to create an additional barrier line, to increase existing fence height, or to provide other methods that effectively add to physical security. It is important to recognize that fencing provides very little delay when it comes to motivated aggressors, but it can act as a psychological deterrent.

CHAIN LINK

4-8. Chain-link fence (including gates) must be constructed of 6-foot material, excluding the top guard. Fence heights for conventional arms and ammunition security must be 6 feet for standard chain-link, wire-mesh fencing. Chain-link fences must be constructed with 9-gauge or heavier wire. They must be galvanized with mesh openings not larger than 2 inches per side and have twisted and barbed selvages at the top and the bottom. The wire must be taut and securely fastened to rigid metal or reinforced-concrete posts set in concrete. It must reach within 2 inches of hard ground or pavement. On soft ground, it must reach below the surface deep enough to compensate for shifting soil or sand. Materials and construction must meet with the US Army Corps of Engineers (USACE) guide specifications shown in the USACE Standard (STD) 872-90 series. Weaknesses in the chain-link fence occur as a result of weather (rusting) or failure to keep it fastened to the post that affects the desired tightness. Damage to the fence and fence fabric may be the result of allowing vegetation and trees to grow on or near the fence. The interaction between the fence and the overgrowth often leads to fence damage and reduces the integrity and continuity of the fence as a perimeter boundary and barrier. The perimeter fence is the most obvious protective measure. A well-maintained fence indicates that the asset owner is dedicated to physical security.

BARBED WIRE

4-9. Standard barbed wire is twisted, double-strand, 13.5-gauge wire, with four-point barbs spaced an equal distance apart. Barbed-wire fencing (including gates) intended to prevent human trespassing should not be less than 6 feet high and must be affixed firmly to posts not more than 6 feet apart. The distance between strands should not exceed 6 inches, and at least one wire should be interlaced vertically and midway between posts. The ends must be staggered or fastened together, and the base wire must be picketed to the ground.

BARBED TAPE OR CONCERTINA

4-10. A barbed-taped obstacle (BTO) is fabricated from 0.025-inch stainless steel and is available in 24-, 30-, 40-, and 60-inch-diameter coils. The barbs shall have a minimum length of 1.2 inches, and the barb cluster's width shall be 1.21 inches. A BTO deploys tangle-free for fast installation. It may be recovered and used again. Fifty feet (plus or minus 2 inches) can be covered by 101 coil loops. Handling barbed tape requires the use of heavy barbed-tape gauntlets instead of standard barbed-wire gauntlets.

Barbed-Tape Concertina

4-11. Barbed-tape concertina (standard concertina barbed tape) is a commercially manufactured wire coil of high-strength-steel barbed wire that is clipped together at intervals to form a cylinder. When opened, it is 50 feet long and 3 feet in diameter. When used as the perimeter barrier for a restricted area, the concertina must be laid between poles with one roll on top of another or in a pyramid arrangement (with a minimum of three rolls).

4-12. Reinforced barbed-tape concertina consists of a single strand of spring-steel wire and a single strand of barbed tape. The sections between barbs of the barbed tape are securely clinched around the wire. Each coil is about 37 1/2 inches in diameter and consists of 55 spiral turns connected by steel clips to form a cylindrical diamond pattern when extended to a coil length of 50 feet. One end turn is fitted with four bundling wires for securing the coil when closed and each end turn is fitted with two steel carrying loops. The concertina extends to 50 feet without permanent distortion. When released, it can be retracted into a closed coil.

4-13. When possible, a top guard should be constructed on all perimeter fences and may be added on interior enclosures for additional protection. A top guard is an overhang of barbed wire or tape along the top of a fence, facing outward and upward at about a 45-degree angle. Placing barbed wire or tape above it can further enhance the top guard. Top-guard supporting arms will be permanently affixed to the top of fence posts to increase the overall height of the fence by at least 1 foot. (Due to liability issues in some locations, the top guards will not be allowed to face outward where the fence is adjacent to public areas.) Three strands of barbed wire spaced 6 inches apart must be installed on the supporting arms. The number of strands of wire or tape may be increased when required. The top guard of fencing adjoining gates may range from a vertical height of 18 inches to the normal 45-degree outward protection but only for sufficient distance along the fence to open the gates adequately. Bottom and top tension wires should be used in lieu of fence rails. A concrete sill may be cast at the bottom of the fence to protect against soil erosion. A bottom rail is used on high-security fences to prevent intruders from lifting the fence.

Gates and Entrances

4-14. The number of gates and perimeter entrances must be the minimum required for safe and efficient operation of the facility. Active perimeter entrances must be designed so that the guard force maintains full control. Semiactive entrances, such as infrequently used vehicular gates, must be locked on the inside when not in use. When closed, gates and entrances must provide a barrier structurally comparable to their associated barriers. Care must be afforded against the ability to crawl under gates. Top guards, which may be vertical, are required for all gates.

Triple-Standard Concertina (TSC) Wire

4-15. This type of fence uses three rolls of stacked concertina. One roll will be stacked on top of two rolls that run parallel to each other while resting on the ground, forming a pyramid. In many situations, this fence has been used effectively in place of a chain-link fence. (If perimeter fencing consists of TSC, a top guard is not feasible.)

Tangle-Foot Wire

4-16. Barbed wire or tape may be used in appropriate situations to construct a tangle-foot obstruction either outside a single perimeter fence or in the area between double fences to provide an additional deterrent to intruders. The wire or tape should be supported on short metal or wooden pickets spaced at irregular intervals of 3 to 10 feet and at heights between 6 and 12 inches. The wire or tape should be crisscrossed to provide a more effective obstacle. The space and materials available govern the depth of the field.

AIRCRAFT CABLE

4-17. Although not used very often, aircraft cable can be used as a temporary barrier. Refer to FM 5-34 for information required for determining the barrier's strength. The barrier is created using wire rope. Clips are spaced six times the diameter of the wire rope. Aircraft cable (deployed as described above or attached to a chain-link fence) can also be made to act as a barrier to moving vehicles. To do so, the cable must be anchored into the ground at both ends at about 200-foot intervals (see TM 5-853-1).

UTILITY OPENINGS

4-18. Sewers, air and water intakes and exhausts, and other utility openings of 10 inches or more in diameter that pass through perimeter barriers must have security measures equivalent to that of the perimeter (see TM 5-820-4). Specific requirements of various openings are discussed below:

- Manhole covers 10 inches or more in diameter must be secured to prevent unauthorized opening. They may be secured with locks and hasps, by welding them shut, or by bolting them to their frame. Ensure that hasps, locks, and bolts are made of materials that resist corrosion. Keyed bolts (which make removal by unauthorized personnel more difficult) are also available.
- Drainage ditches, culverts, vents, ducts, and other openings that pass through a perimeter and that have a cross-sectional area greater than 96 square inches and whose smallest dimension is greater than 6 inches will be protected by securely fastened welded bar grilles (refer to TM 5-853-3, Figure 8-1). As an alternative, drainage structures may be constructed of multiple pipes, with each pipe having a diameter of 10 inches or less. Multiple pipes of this diameter may also be placed and secured in the inflow end of a drainage culvert to prevent intrusion into the area. Ensure that any addition of grilles or pipes to culverts or other drainage structures is coordinated with the engineers so that they can compensate for the diminished flow capacity and additional maintenance that will result from the installation.

OTHER PERIMETER BARRIERS

4-19. Buildings less than two stories high that form part of a perimeter must have a top guard along the outside edge to deny access to the roof. When using masonry walls as part of a perimeter barrier, they must be at least 7 feet high and have a barbed-wire top guard. The top guard should be sloped outward at a 45-degree angle and carry at least three strands of barbed wire. This will increase the vertical height of the barrier by at least 1 foot.

4-20. Protect windows, active doors, and other designated openings by securely fastening bars, grilles, or chain-link screens. Fasten window barriers from the inside. If hinged, the hinges and locks must be on the inside. Building elements that provide delay against forced entry have stringent requirements. These elements should be designed according to TM 5-853-1.

SECURITY TOWERS

4-21. It is not acceptable to observe a perimeter from towers only. However, all towers should be located to provide maximum observation and should be constructed for protection from small-arms fire.

4-22. Mobile towers are useful in some temporary situations such as a large, open storage area where receiving and storing activities take place. All facilities using towers must have a support force available for emergencies. Tower personnel should be rotated at frequent intervals.

4-23. The height of a tower increases the range of observation during daylight hours and at night with artificial illumination. However, during inclement weather and during a blackout, towers lose this advantage and must be supplemented by on-ground observation.

4-24. The following considerations should be made when planning for the use of towers:

- Hardening the tower against small-arms effects by using sandbags, salvaged armor, or commercially fabricated bullet-resistant construction. This may require strengthening the tower supports, which should be performed only under the supervision of an engineer. The level of protection required must equate to the threat level identified during the IPB or the military decision-making process (MDMP). The best approach is to design for the worst identified threat rather than to try and modify the tower at a later date on short notice.
- Installing communications and alarm systems, both audible and visual (primary and alternate).
- Using appropriate surveillance, target-acquisition, and night-observation (STANO) equipment with the tower and perimeter barriers being surveilled. Infrared (IR) items may be especially valuable. Considerations for the selection and use of STANO equipment must be made while evaluating the effects of perimeter protective lighting.
- Providing security lighting for route protection to the tower. Security lighting also allows for support of the guard force entering or exiting the perimeter.
- Ensuring that the tower's height is determined according to the area of observation.
- Ensuring that towers have overlapping, mutually supporting fields of observation and fire.
- Providing towers with a backup fortified defensive fighting position, as appropriate.

INSTALLATION ENTRANCES

4-25. The number of installation or activity gates and perimeter entrances in active use should be limited to the minimum number required for safe and efficient operations. When necessary, install vehicle barriers in front of vehicle gates. Security lighting should be considered at entry points (see Chapter 5). Refer to TM 5-853-1 for the application and selection of these barriers.

4-26. Plans to use guards for controlling entry to an installation or activity must be predetermined based on the threat conditions (THREATCON). The construction of the guard post must be included in the security plan.

PERIMETER ENTRANCES

4-27. Active perimeter entrances should be designated so that security forces maintain full control without an unnecessary delay in traffic. This is accomplished by having sufficient entrances to accommodate the peak flow of pedestrian and vehicular traffic and having adequate lighting for rapid and efficient inspection. When gates are not operational during nonduty hours, they should be securely locked, illuminated during hours of darkness, and inspected periodically by a roving patrol. Additionally, warning signs should be used to warn drivers when gates are closed. Doors and windows on buildings that form a part of the perimeter should be locked, lighted, and inspected.

ENTRY-CONTROL STATIONS

4-28. Entry-control stations should be provided at main perimeter entrances where security personnel are present. Considerations for construction and use should be based on the information outlined in USACE STD 872-50-01.

4-29. Entry-control stations should be located as close as practical to the perimeter entrance to permit personnel inside the station to maintain constant surveillance over the entrance and its approaches. Additional considerations at entry-control stations include—

- Establishing a holding area for unauthorized vehicles or those to be inspected further. A turnaround area should be provided to keep from impeding other traffic.
- Establishing control measures such as displaying a decal on the window or having a specially marked vehicle.

4-30. Entry-control stations that are manned 24 hours each day should have interior and exterior lighting, interior heating (where appropriate), and a sufficient glassed area to afford adequate observation for personnel inside. Where appropriate, entry-control stations should be designed for optimum personnel ID and movement control. Each station should also include a telephone, a radio, and badge racks (if required).

4-31. Signs should be erected to assist in controlling authorized entry, to deter unauthorized entry, and to preclude accidental entry. Signs should be plainly displayed and be legible from any approach to the perimeter from a reasonable distance. The size and coloring of a sign, its letters, and the interval of posting must be appropriate to each situation.

4-32. Entry-control stations should be hardened against attacks according to the type of threat. The methods of hardening may include—

- Reinforced concrete or masonry.
- Steel plating.
- Bullet-resistant glass.
- Sandbags, two layers in depth.
- Commercially fabricated, bullet-resistant building components or assemblies.

WARNING SIGNS

4-33. A significant amount of warning signs should be erected to ensure that possible intruders are aware of entry into restricted areas. Warning signs augment control signs. They warn intruders that the area is restricted and that trespassing may result in the use of deadly force.

4-34. Warning signs should be installed along the limited area's physical barriers and at each entry point where they can be seen readily and understood by anyone approaching the perimeter. In areas where English is one of two or more languages commonly spoken, warning signs must contain the local language in addition to English. The wording on the signs will denote warning of a restricted area. The signs should be posted at intervals of no more than 100 feet. They must not be mounted on fences equipped with intrusion-detection equipment. Additionally, the warning signs prescribed in AR 190-13 should be posted at all entrances to limited, controlled, and exclusion areas. See Chapter 7 for more details.

OTHER SIGNS

4-35. Signs setting forth the conditions of entry to an installation or area should be plainly posted at all principal entrances. The signs should be legible under normal conditions at a distance not less than 50 feet from the point of entry. Such signs should inform the entrant of the provisions (search of the person, the vehicle, packages, and so forth) or prohibitions (such as against cameras, matches, and lighters and entry for reasons other than official business) that may be prescribed by the installation commander.

4-36. Signs or notices legibly setting forth the designation of restricted areas and provisions of entry should be plainly posted at all entrances and at other points along the perimeter line as necessary. The wording of these signs or notices is prescribed in AR 190-13.

INSTALLATION PERIMETER ROADS AND CLEAR ZONES

4-37. When the perimeter barrier encloses a large area, an interior all-weather perimeter road should be provided for security-patrol vehicles. Clear zones should be maintained on both sides of the perimeter barrier to provide an unobstructed view of the barrier and the ground adjacent to it. Roads within the clear zone should be as close to the perimeter barrier as possible without interfering with it. The roads should be constructed to allow effective road barriers to deter motor movement of unauthorized personnel during mobilization periods.

4-38. Clear zones should be kept clear of weeds, rubbish, or other material capable of offering concealment or assistance to an intruder attempting to breach the barrier. A clear zone of 20 feet or more should exist between the perimeter barrier and exterior structures, parking areas, and natural or man-made features. When possible, a clear zone of 50 feet or more should exist between the perimeter barrier and structures within the protected area, except when a building's wall constitutes part of the perimeter barrier. Ammunition supply points (ASPs) will have clear zones 12 feet outside of the ASP and 30 feet inside, and the vegetation will not exceed 8 inches (4 inches for high-threat and highly controlled areas). Refer to AR 190-11 and DOD 0-2000.12-H, Appendix EE, for further information.

4-39. When it is impossible to have adequate clear zones because of property lines or natural or man-made features, it may be necessary to increase the height of the perimeter barrier, increase security-patrol coverage, add more security lighting, or install an intrusion-detection device along that portion of the perimeter.

4-40. When considering the construction of a new site or perimeter, ensure that the plans include a fence located well inside the property line, thus permitting control of enough space outside the fence to maintain at least a minimal clear zone. The following considerations apply:

- On a large installation (such as a proving ground), it is unreasonable to construct an expensive perimeter fence and keep it under constant observation. Such an installation is usually established in a sparsely inhabited area. Its comparative isolation and the depth of the installation give reasonable perimeter protection. Under these circumstances, it is usually sufficient to post warning signs or notices, reduce access roads to a minimum, and periodically patrol the area between the outer perimeter and the conventionally protected vital area of the installation.
- An alternative to erecting new or replacing old chain-link fence involving an entire installation perimeter is to relocate or isolate the sensitive area or item by—
  - Relocating the item within a safe perimeter.
  - Consolidating the item with other items.
  - Erecting a chain-link fence (regulations permitting) around individual assets rather than the installation's perimeter.

ARMS-FACILITY STRUCTURAL STANDARDS

4-41. It is next to impossible to build a protective barrier that cannot be penetrated by a human or heavy armor. Therefore, as opposed to protecting a facility using only one barrier, enhance security by using a combination of barriers to increase delay. Multiple barriers also cause aggressors to expend more energy trying to breach all of the barriers. They also provide the appearance of additional security and may further deter some aggressors.

4-42. The interest of security must be kept in mind when constructing walls, ceilings, floors, and roofs. Facilities that house arms and ammunition are constructed as security barriers in the interest of deterring and delaying penetration. Construction guidelines for arms facilities are outlined in AR 190-11. AR 190-11 requires coordination with the engineer office, the safety office, the provost marshal office (PMO), or the security-force office when definitive drawings and specifications for new construction or upgrades or modifications of AA&E storage structures are proposed. This coordinated effort ensures that safety and physical-security requirements are met. AR 190-11 also addresses waivers and exceptions for AA&E storage structures, as well as the requirements for a tactical (training or operational) or shipboard environment. Waivers and exceptions are not discussed in this manual. The following guidelines are provided for securing AA&E in tactical and shipboard environments:

- The criteria and standards for protecting AA&E will be developed by the major Army command (MACOM) according to AR 190-11.
- The deploying commander will establish and enforce procedures for securing deployed AA&E based on the assessment of the threat, the objectives, the location, and the duration of the deployment.
- The AA&E in the tactical environment will be secured at all times.
- The AA&E will be under continuous positive control.
- Persons charged with the custody of AA&E will have the capability to sound the alarm if a forceful theft is attempted.
- A response force will be available to protect the AA&E.
- A system of supervisory checks will be established to ensure that all personnel comply with security measures. Supervisory checks of the AA&E holding area will be made to ensure that the AA&E being guarded have not been tampered with.
- All officers, noncommissioned officers (NCOs), or civilian equivalents will closely monitor the control of ammunition and explosives during field training or range firing.
- Selection of personnel to perform guard duties at AA&E holding areas will be closely monitored by commanders to ensure that only responsible individuals are assigned duties.

Chapter 5

Physical-Security Lighting

Security lighting allows security personnel to maintain visual-assessment capability during darkness. When security-lighting provisions are impractical, additional security posts, patrols, MWD patrols, NVDs, or other security means are necessary.

OVERVIEW

5-1. Security lighting should not be used as a psychological deterrent only. It should also be used along perimeter fences when the situation dictates that the fence be under continuous or periodic observation.

5-2. Lighting is relatively inexpensive to maintain and, when properly used, may reduce the need for security forces. It may also enhance personal protection for forces by reducing the advantages of concealment and surprise for a determined intruder.

5-3. Security lighting is desirable for those sensitive areas or structures within the perimeter that are under observation. Such areas or structures include pier and dock areas, vital buildings, storage areas, motor pools, and vulnerable control points in communication and power- and water-distribution systems. In interior areas where night operations are conducted, adequate lighting facilitates the detection of unauthorized persons approaching or attempting malicious acts within the area. Security lighting has considerable value as a deterrent to thieves and vandals and may make the job of the saboteur more difficult. It is an essential element of an integrated physical-security program.

5-4. A secure auxiliary power source and power-distribution system for the facility should be installed to provide redundancy to critical security lighting and other security equipment. During deployed operations, primary power may not exist or may be subject to constraints or interruptions due to poor infrastructure or hostile activity. Auxiliary power sources must be available for critical electrical loads and must be secured against direct and indirect fires as well as sabotage. If automatic-transfer switches are not installed, security procedures must designate the responsibility for the manual start of the source.

COMMANDER'S RESPONSIBILITY

5-5. Commanders determine perimeter lighting needs based on the threat, site conditions along the perimeter, surveillance capabilities, and available guard forces. Commanders ensure that security lighting is designed and used to discourage unauthorized entry and to facilitate the detection of intruders approaching or attempting to gain entry into protected areas.

PLANNING CONSIDERATIONS

5-6. Security lighting usually requires less intensity than working lights, except for ID and inspection at entry-control points. Each area of a facility presents its own unique set of considerations based on physical layout, terrain, atmospheric and climatic conditions, and security requirements. Information is available from the manufacturers of lighting equipment and from the installation's director of public works, who will assist in designing a lighting system. This information includes—

- Descriptions, characteristics, and specifications of various lighting fixtures, arc, and gaseous-discharge lamps.
- Lighting patterns of various fixtures.
- Typical layouts showing the most efficient height and spacing of equipment.
- Minimum levels of illumination and lighting uniformity required for various applications.

5-7. In planning a security-lighting system, the physical-security manager considers the—

- Cost of replacing lamps and cleaning fixtures, as well as the cost of providing the required equipment (such as ladders and mechanical buckets) to perform this maintenance.
- Provision of manual-override capability during a blackout, including photoelectric controls. These controls may be desirable in a peacetime situation but undesirable when a blackout is a possibility.
- Effects of local weather conditions on lighting systems.
- Fluctuating or erratic voltages in the primary power source.
- Grounding requirements.
- Provisions for rapid lamp replacement.
- Use of lighting to support a CCTV system.
- Limited and exclusion areas. Specific lighting requirements are referenced in AR 190-59 and TM 5-853-2. TM 5-853-4 provides guidance for facility applications that include CCTV cameras.
  - Lighting in these areas must be under the control of the guard force.
  - For critical areas (such as weapons storage areas), instantaneous lighting with a backup source is required. Any period without lighting in a critical area is unacceptable. Therefore, these areas generally have a requirement for backup power (such as diesel-engine generators, uninterrupted power supplies, and batteries) in case of power loss.
  - Security-lighting systems are operated continuously during hours of darkness.
  - Protective lights should be used so that the failure of one or more lights will not affect the operation of the remaining lights.
- Lighting requirements for adjoining properties and activities.
- Restrike time (the time required before the light will function properly after a brief power interruption).

- Color accuracy.
- Other facilities requiring lighting, such as parking areas.

PRINCIPLES OF SECURITY LIGHTING

5-8. Security lighting enables guard-force personnel to observe activities around or inside an installation while minimizing their presence. An adequate level of illumination for all approaches to an installation will not discourage unauthorized entry; however, adequate lighting improves the ability of security personnel to assess visually and intervene on attempts at unauthorized entry. Lighting is used with other security measures (such as fixed security posts or patrols, fences, and ESSs) and should never be used alone. Other principles of security lighting include the following:

- Optimum security lighting is achieved by adequate, even light on bordering areas; glaring lights in the eyes of an intruder; and little light on security-patrol routes. In addition to seeing long distances, security forces must be able to see low contrasts (such as indistinct outlines of silhouettes) and must be able to detect an intruder who may be exposed to view for only a few seconds. Higher levels of illumination improve these abilities.
- High brightness contrast between an intruder and the background should be the first consideration when planning for security lighting. With predominantly dark, dirty surfaces or camouflage-type painted surfaces, more light is needed to produce the same brightness around installations and buildings than when clean concrete, light brick, and grass predominate. When the same amount of light falls on an object and its background, the observer must depend on contrasts in the amount of light reflected. His ability to distinguish poor contrasts is significantly improved by increasing the illumination level.
- The observer primarily sees an outline or a silhouette when the intruder is darker than his background. Using light finishes on the lower parts of buildings and structures may expose an intruder who depends on dark clothing and darkened face and hands. Stripes on walls have also been used effectively, as they provide recognizable breaks in outlines or silhouettes. Providing broad-lighted areas around and within the installation against which intruders can be seen can also create good observation conditions.

5-9. To be effective, two basic systems or a combination of both may be used to provide practical and effective security lighting. The first method is to light the boundaries and approaches; the second is to light the area and structures within the property's general boundaries. Protective lighting should—

- Discourage or deter attempts at entry by intruders. Proper illumination may lead a potential intruder to believe detection is inevitable.
- Make detection likely if entry is attempted.
- Prevent glare that may temporarily blind the guards.

TYPES OF LIGHTING

5-10. The type of lighting system used depends on the installation's overall security requirements. Four types of lighting units are used for security-lighting systems—continuous, standby, movable (portable), and emergency.

5-11. Continuous lighting is the most common security-lighting system. It consists of a series of fixed lights arranged to flood a given area continuously during darkness with overlapping cones of light. Two primary methods of using continuous lighting are glare projection and controlled lighting.

- The glare security-lighting method is used when the glare of lights directed across the surrounding territory will not be annoying nor interfere with adjacent operations. It is a strong deterrent to a potential intruder because it makes it difficult to see inside of the area. Guards are protected by being kept in comparative darkness and being able to observe intruders at a considerable distance beyond the perimeter.
- Controlled lighting is best when it limits the width of the lighted strip outside the perimeter, such as along highways. In controlled lighting, the width of the lighted strip is controlled and adjusted to fit the particular need. This method of lighting may illuminate or silhouette security personnel.

5-12. Standby lighting has a layout similar to continuous lighting. However, the luminaries are not continuously lit but are either automatically or manually turned on when suspicious activity is detected or suspected by the security force or alarm systems.

5-13. Movable lighting consists of manually operated, movable searchlights that may be lit during hours of darkness or only as needed. The system normally is used to supplement continuous or standby lighting.

5-14. Emergency lighting is a system of lighting that may duplicate any or all of the above systems. Its use is limited to times of power failure or other emergencies that render the normal system inoperative. It depends on an alternative power source such as installed or portable generators or batteries.

FENCED PERIMETERS

5-15. Fenced perimeters require the lighting specifications indicated in TM 5-853-2. Specific lighting requirements are based on whether the perimeter is isolated, semi-isolated, or nonisolated.

- Isolated fenced perimeters are fence lines around areas where the fence is 100 feet or more from buildings or operating areas. The approach area is clear of obstruction for 100 or more feet outside of the fence. Other personnel do not use the area. Use glare projection for these perimeters and keep patrol routes unlit.
- Semi-isolated fenced perimeters are fence lines where approach areas are clear of obstruction for 60 to 100 feet outside of the fence. The general public or installation personnel seldom have reason to be in the area. Use controlled lighting for these perimeters and keep patrol routes in relative darkness.

- Nonisolated fenced perimeters are fence lines immediately adjacent to operating areas. These areas may be within an installation or public thoroughfares. Outsiders or installation personnel may move about freely in this approach area. The width of the lighted strip depends on the clear zones inside and outside the fence. Use controlled lighting for these perimeters. It may not be practical to keep the patrol area dark.

ENTRANCES

5-16. Entrances for pedestrians will have two or more lighting units providing adequate illumination for recognition of persons and examination of credentials. Vehicle entrances will have two lighting units located to facilitate the complete inspection of passenger cars, trucks, and freight cars as well as their contents and passengers. Semiactive and inactive entrances will have the same degree of continuous lighting as the remainder of the perimeter, with standby lighting to be used when the entrance becomes active. Gatehouses at entrances should have a low level of interior illumination, enabling guards to see approaching pedestrians and vehicles.

OTHER

5-17. Areas and structures within the installation's property line consist of yards; storage spaces; large, open working areas; piers; docks; and other sensitive areas and structures.

- Open yards (unoccupied land only) and outdoor storage spaces (material storage areas, railroad sidings, motor pools, and parking areas) should be illuminated. An open yard adjacent to a perimeter (between guards and fences) will be illuminated according to the perimeter's illumination requirements. Where lighting is necessary in other open yards, illumination will not be less than 0.2 foot-candle at any point.
- Lighting units are placed in outdoor storage spaces to provide an adequate distribution of light in aisles, passageways, and recesses to eliminate shadowed areas where unauthorized persons may hide.
- Illuminating both water approaches and the pier area safeguards piers and docks located on an installation. Decks on open piers will be illuminated to at least 1 foot-candle and the water approaches (extending to a distance of 100 feet from the pier) to at least 0.5 foot-candle. The area beneath the pier floor will be lit with small wattage floodlights arranged on the piling. Movable lighting is recommended as a part of the protective lighting system for piers and docks. The lighting must not in any way violate marine rules and regulations (it must not be glaring to pilots). Consult the US Coast Guard (USCG) for approval of protective lighting adjacent to navigable waters.

WIRING SYSTEMS

5-18. The wiring circuit should be arranged so that failure of any one lamp will not leave a large portion of the perimeter line or a major segment of a critical or vulnerable position in darkness. Feeder lines will be placed underground (or sufficiently inside the perimeter in the case of overhead wiring) to minimize the possibility of sabotage or vandalism from outside the perimeter. Another advantage to underground wiring is reduced effects from adverse weather conditions.

MAINTENANCE

5-19. Periodic inspections will be made of all electrical circuits to replace or repair worn parts, tighten connections, and check insulation. Keep fixtures clean and properly aimed.

POWER SOURCES

5-20. Primary and alternate power sources must be identified. The following is a partial list of considerations:

- The primary source is usually a local public utility.
- An alternate source (standby batteries or diesel-fuel-driven generators may be used) is provided where required and should—
  - Start automatically upon failure of primary power.
  - Be adequate to power the entire lighting system.
  - Be equipped with adequate fuel storage and supply.
  - Be tested under load to ensure efficiency and effectiveness.
  - Be located within a controlled area for additional security.

CCTV-CAMERA LIGHTING REQUIREMENTS

5-21. TM 5-853-4 provides a detailed discussion of CCTV-camera lighting requirements and guidelines for minimum lighting levels and lighting uniformity. The following considerations apply when lighting systems are intended to support CCTV assessment or surveillance:

- The camera's field of view.
- Lighting intensity levels.
- Maximum light-to-dark ratio.
- Scene reflectance.
- Daylight-to-darkness transitions.
- Camera mounting systems relative to lighting.
- The camera's spectral response.
- The cold-start time.
- The restrike time.

Chapter 6

Electronic Security Systems

An overall site-security system is comprised of three major subelements—detection, delay, and response. The detection subelement includes intrusion detection, assessment, and entry control. An ESS is an integrated system that encompasses interior and exterior sensors; CCTV systems for assessing alarm conditions; electronic entry-control systems (EECSs); data-transmission media (DTM); and alarm reporting systems for monitoring, controlling, and displaying various alarm and system information. Interior and exterior sensors and their associated communication and display subsystems are collectively called IDSs.

OVERVIEW

6-1. Many Army and DOD regulations specify protective measures, policies, and operations related to security. Although the regulations specify minimum requirements, it is possible that more stringent requirements will be necessary at specific sites. A designer will use a previously performed site survey to determine which regulations apply and to determine whether circumstances require more stringent measures. Refer to TM 5-853-4 for additional detailed information.

6-2. AR 190-13 requires the use of a standardized ESS, if practical and available. The receiving element must determine whether a standardized system can meet the requirements and whether it is available. After coordinating with the product manager for physical-security equipment to verify that a standardized system is available, the associated MACOM can issue approval to procure a commercial system in lieu of a standardized system.

USE OF ESS

6-3. An ESS is used to provide early warning of an intruder. This system consists of hardware and software elements operated by trained security personnel.

6-4. A system is configured to provide one or more layers of detection around an asset. Each layer is made up of a series of contiguous detection zones designed to isolate the asset and to control the entry and exit of authorized personnel and materials.

GENERAL ESS DESCRIPTION

6-5. An ESS consists of sensors interfaced with electronic entry-control devices, CCTV, alarm reporting displays (both visual and audible), and security lighting. The situation is assessed by sending guards to the alarm point or by using CCTV. Alarm reporting devices and video monitors are located in the security center. The asset's importance will determine whether multiple or redundant security centers are required and, ultimately, the required sophistication of all elements in the ESS. Digital and analog data are transmitted from local (field) interior and exterior locations to the security center for processing. Reliability and accuracy are important functional requirements of the data-transmission system.

ESS IMPLEMENTATION PROCESS

6-6. The ESS implementation process is shown in Figure 6-1. Implementing an ESS is based on general requirements tailored to a site-specific mission and physical profile. The process begins with a site survey that includes a top-down view of basic needs and classic configurations that are tailored to such site-specific characteristics as terrain, site geography, climatic conditions, the type of asset, and priorities. This data is used to determine the hardware and software requirements, taking into account the additional capacity that should be factored into the design system for future expansion. Once the requirements for an ESS have been identified, the user must determine whether an existing standardized system is suitable for the application. (AR 190-13 outlines the process for gaining approval to use nonstandard equipment.) The user must also secure funding for the equipment (refer to Appendix J). Depending on the current funding regulations, operation-and-maintenance, procurement, or other funds may be required. For example, operations and procurement, Army (OPA) funds may be required for IDS devices; and operations and maintenance, Army (OMA) funds may be required for installation items. A contract is normally awarded to procure and install the equipment. The procurement or installation must be overseen. This may be accomplished by reviewing submittals, inspecting the contractor's work, or responding to the contractor's requests for information. Once the equipment is installed, the acceptance-testing activities must be witnessed and verified. Site conditions during acceptance testing affect the demonstrated detection capability of an exterior IDS. As feasible, acceptance testing should be designed to determine a sensor system's probability of detection (PD) under a range of conditions. For some types of sensor systems, this may be as straightforward as conducting both daytime and nighttime trials to experience differences in temperature and solar heating. After the ESS has been accepted, it must be operated and maintained throughout the remainder of its life cycle. Planning for manpower to operate the system and forecasting the funding and personnel to properly maintain the system is critical for success.

ESS DESIGN CONSIDERATIONS

6-7. A facility may require interior and exterior ESS elements, depending on the level of protection required. The applicable regulations, threat, and design criteria will define the ESS's general requirements. For an existing ESS, hardware and software may need to be supplemented, upgraded, or completely replaced. A site layout (in which all assets are identified and located) is required. It is a useful design tool for such tasks as configuring the DTM.

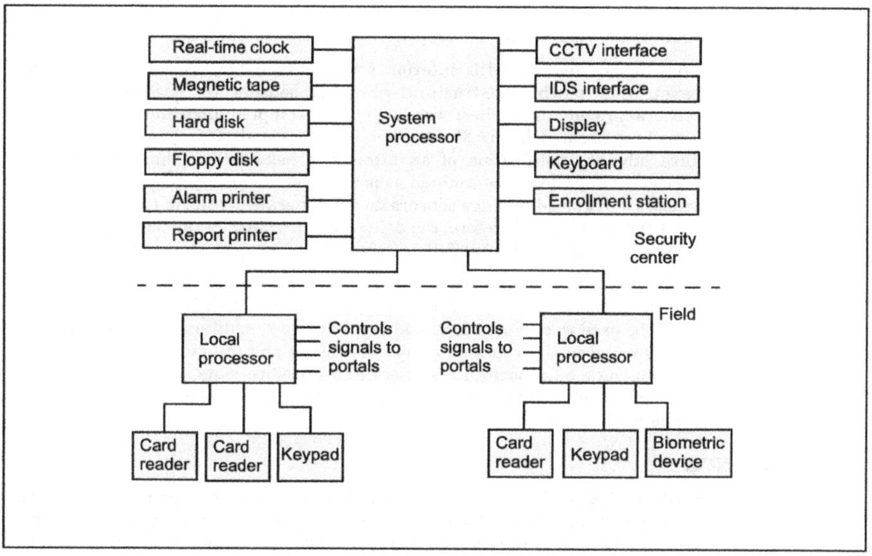

Figure 6-1. Entry-Control System Configured With Distributed Control

6-8. The exterior and interior IDSs should be configured as layers of unbroken rings concentrically surrounding the asset. These rings should correspond to defensive layers that constitute the delay system. The first detection layer is located at the outermost defensive layer necessary to provide the required delay. Detection layers can be on a defensive layer, in the area between two defensive layers, or on the asset itself, depending on the delay required. For example, if a wall of an interior room provides sufficient delay for effective response to aggression, detection layers could be between the facility exterior and interior-room wall or on the interior-room wall. These would detect the intruder before penetration of the interior wall is possible.

RESPONSE AND DELAY

6-9. When dealing with an ESS, the response time is defined as the time it takes the security force to arrive at the scene after an initial alarm is received at the security center. The total delay time is defined as the sum of all of the barriers' delay times, the time required to cross the areas between barriers after an intrusion alarm has been reported, and the time required to accomplish the mission and leave the protected area.

6-10. An ESS's basic function is to notify security personnel that an intruder is attempting to penetrate, or has penetrated, a protected area in sufficient time to allow the response force to intercept and apprehend him. To accomplish this, there must be sufficient physical delay between the point where the intruder is first detected and his objective. This provides delay time equal to or greater than the response time (refer to TM 5-853-1).

6-11. When dealing with interior sensors, boundary sensors that detect penetration (such as structural-vibration sensors or passive ultrasonic sensors) provide the earliest warning of an attempted intrusion. This alarm is usually generated before the barrier is penetrated. This gives the security force advance notification of an attempted penetration, thus allowing the barrier's delay time to be counted as part of the total delay time. Door-position sensors and glass-breakage sensors do not generate an alarm until the barrier has been breached; therefore, the delay time provided by the barrier cannot be counted as part of the total delay time.

6-12. Volumetric motion sensors do not generate an alarm until the intruder is already inside the area covered by the sensors. Therefore, if these sensors are to be used to provide additional response time, additional barriers must be placed between the volumetric motion sensors and the protected asset. Point sensors, such as capacitance sensors and pressure mats, provide warning of attempted penetration only if they detect the intruder before access is gained to the protected area.

BASIC GUIDANCE

6-13. An IDS is deployed in and around barriers (as detailed in TM 5-853-1). Voice communication links (radio, intercom, and telephone) with the response force are located in the security center. Security personnel will man the center and will alert and dispatch response forces in case of an alarm.

6-14. The barrier should always be deployed behind the IDS to ensure that integrity is maintained against intruders. An intruder will then activate the alarm sensor before penetrating or bypassing the barriers, thus providing delay for alarm assessment and response. The delay time is the determining factor in whether an assessment is conducted by dispatching a guard or by observing the CCTV. Normally, an intruder can climb a fence before a guard can be dispatched; therefore, a CCTV is usually required with an exterior IDS. Barriers can be located ahead of an alarm sensor as a boundary demarcation and can serve to keep people and animals from causing nuisance alarms by inadvertently straying into a controlled area. These barriers provide no additional response time because the barrier could be breached before the IDS sensors could be activated.

6-15. Data for monitoring and controlling an ESS are gathered and processed in the security center where the operator interacts with information from the ESS components located at remote facilities. The ESS's alarm-annunciation computer and its DTM line-termination equipment should be located in a controlled area and provided with tamper protection. Supervisory personnel should permit changes to software only, and these changes should be documented. If redundant DTM links connect the central computer to the local processor, diverse paths should be used to route these links.

6-16. The preferred medium for transmitting data in an ESS is a dedicated fiber-optics system. It provides for communications not susceptible to voltage transients, lightning, electromagnetic interference, and noise. Additionally, the fiber optics will provide a measure of communication-line security and wide bandwidth for video signals and increased data-transmission rates.

ESS EFFECTIVENESS

6-17. An ESS has a degree-of-protection effectiveness that is based on its probability of detecting intruders attempting to go over, under, around, or through the physical-security system. The intruder may use forced-entry, covert-entry, or insider-compromise tactics. A well-designed system will minimize the possibility of a successful penetration through covert entry or insider compromise. Interior and exterior alarm sensors have a PD based on the capability to detect an intruder passing through a sensing field. An intruder disturbs the steady-state quiescent condition of a sensor for a finite period. Sensors are designed to detect a person of minimum stature moving within a specific range of speeds and distances from the sensor, and any target outside of those parameters will probably not be detected. The PD for a specific sensor is usually specified at 0.9 or greater, but the designer must be aware that the PD is based on certain constraints and environmental conditions.

6-18. Manufacturer specifications usually do not discuss environmental or nuisance alarms that can be caused by climatic conditions (such as wind or rain) or by the intrusion of animals (including birds). The alarm annunciation is valid because the sensor's thresholds have been exceeded; however, the alarm does not represent a valid penetration attempt. If the assessment system is slow, the operator may not be able to determine the cause of the alarm and must, therefore, treat an environmental or nuisance alarm as real.

6-19. Another type of false alarm is caused by electronic-circuit tolerances being exceeded, resulting in the sensor's actuation. False alarms may also result from improper installation of the sensor or from effects of other equipment in the immediate area.

6-20. After an alarm is sensed and information is displayed in the security center, the console operator must determine the cause of the alarm (intrusion, nuisance, environmental, or false). Timely assessment is required when determining its cause. For example, if an intruder scales a fence in 10 seconds and runs 20 feet per second, the intruder will have overcome the barrier and be 2,200 feet from the point of penetration in 2 minutes. To conduct an accurate assessment of the alarm after 2 minutes, guards will have to search an area of about 200 acres. A fixed-television camera properly located and integrated with the alarm processor can assess the situation while the intruder is still in the controlled area.

6-21. For a CCTV camera to be effective, the area it views must be adequately lighted. To correlate the alarms and cameras in a large system (more than 10 cameras) in a timely manner, a computer-based processing system must be used to select and display alarms and camera scenes for the operator. A complex ESS has the following basic components:

- Intrusion-detection sensors.
- Electronic entry-control devices.
- CCTV.

- Alarm-annunciation system.
- DTM.

6-22. The intrusion-detection sensors are normally deployed in a series of concentric layers. The overall PD improves with each added layer of sensors. The layers (interior and exterior) should be functionally uniform; however, their overall effectiveness and cost are different. The exterior zones significantly differ from the interior zones due to the following considerations:

- The consistency of the PD.
- The PD.
- The cost per detection zone.
- The number of zones.
- The overall sensor coverage.

6-23. Exterior IDSs usually have PDs equal to those of interior IDSs. However, exterior sensors are more likely to experience weather-related situations that cause the system's PD to vary. Sensor phenomenology (passive infrared [PIR], microwave radar, and so forth) determines which environmental factors may alter the system's PD. The frequency of occurrence, severity, and duration of a weather event jointly determine whether it represents security vulnerability with the IDS in use. Typically, sophisticated intruders will attempt their penetration and challenge an ESS under conditions most favorable to themselves. Inclement weather (fog, snow, and rain) affects the usefulness of CCTVs and security lighting such that the capability for remote assessment of alarm events may be lost. Exterior IDSs are not necessarily less likely to detect a penetration attempt during fog, rain and snow; the effect of such site conditions on the IDS depends on sensor phenomenology. For example, fence motion caused by rain impact may drive the response of a fence-mounted sensor closer to satisfying the system's alarm criteria, with the result that the margin of disturbance available to the intruder is less. Also, certain buried sensors are more likely to detect an intruder when the ground is wet because of rain or melting snow. Since interior sensor systems are less influenced by environmental conditions, their PD is typically more consistent than that of some types of exterior sensor systems. Other considerations in comparing an interior and exterior ESS are the cost, the number and size of detection zones required, and the detection height.

- Because of environmental conditions, the exterior electronics must be designed and packaged for extremes of temperature, moisture, and wind. The result is that exterior electronic packages are more costly than equivalent packages for interior applications.
- State-of-the-art exterior sensors do not detect penetration attempts above the height of a fence (typically 8 feet). Fence-mounted sensors are usually limited to this height because the fence fabric or poles are used to support the sensor. For aboveground sensors in the controlled area between the fences, the sensor's mounting brackets and posts limit the detection height. In some applications of field sensors (especially buried sensors), the detection height is no more than 3 feet. For a facility, interior sensors can be deployed on walls, floors, or ceilings, thus permitting complete protection of the asset.

6-24. An interior ESS may be far less costly than that of a comparable exterior ESS. This comparison indicates to the designer the value of selecting and deploying a well-planned, well-designed, layered system. The basic rule in overall design of an ESS is to design from the inside out; that is, layered from the asset to the site boundary.

INTERIOR ESS CONSIDERATIONS

6-25. An interior ESS is typically deployed within a boundary in the immediate vicinity of the asset being protected. If the interior ESS operates in a controlled environment, its PD will be independent of any weather-induced variation in exterior conditions. Also, the physical-security system's effectiveness is enhanced by the interior barriers (walls, ceiling, and floor) that inherently impose a longer delay than exterior barriers (fences and gates).

6-26. Functionally, an interior asset should be viewed as being contained within a cube with sensors protecting all six faces. Interior sensors can be deployed at the cube's perimeter, in its interior space, or in the space immediately outside of the cube.

6-27. If an increased level of protection is dictated by the threat, and if the building is large enough, multiple layers of interior sensors may be deployed for a given asset. A multilayered interior IDS will improve the overall PD. Tamper protection and access-/secure-mode capabilities must be considered when planning and laying out interior sensors.

TAMPER PROTECTION

6-28. To minimize the possibility of someone tampering with circuitry and associated wiring, all sensor-related enclosures must be equipped with tamper switches. These switches must be positioned so that an alarm is generated before the cover has been moved enough to permit access to the circuitry of adjustment controls. In addition, several types of sensors should be equipped with tamper switches to protect against being repositioned or removed. Security screens containing grid-wire sensors and vibration sensors that can be easily removed from a wall are examples of sensors that require tamper switches.

ACCESS/SECURE MODE

6-29. During regular working hours, many of the interior sensors must be deactivated by placing the area in the access mode. For example, door-position sensors and volumetric sensors in occupied areas must be deactivated to prevent multiple nuisance alarms caused by the normal movement of people. This can be done locally or remotely. With local control, a switch is used to bypass or shunt alarm contacts when the sensor is placed in the access mode. When done remotely, the security-center operator usually enters a command that causes the processor software to ignore incoming alarms from those sensors placed in the access mode. However, when a sensor is placed in the access mode, its tamper-protection circuitry must remain in the activated or secure mode. During nonworking hours when the facility is unoccupied, all sensors must be placed in the secure mode. Certain devices (such as duress-alarm switches, tamper switches, grid-wire sensors covering vent openings, and glass-breakage sensors) should never be placed in the access mode. The designer must ensure that selected sensors can be placed in an access mode (if required) and that certain types of sensors (such as duress and tamper switches) are configured so that they cannot be put in the access mode under any condition.

EXTERIOR ESS CONSIDERATIONS

6-30. An exterior ESS is typically deployed at a site's boundary or some other significant boundary such as the demarcation fence for a group of bunkers. An exterior ESS has the advantage that it remains in the secure mode at all times.

6-31. The ideal configuration for an exterior ESS is a rectangle or a polygon, with all sides being straight. The ESS is located in and around barriers that typically include a dual fence. The outside fence is used for demarcation, and the interior fence is used to aid in detection and provide some delay. If dual fences are not used, the sensors should be deployed on the fence or inside it.

DESIGN GUIDELINES

6-32. The general-design criteria of a perimeter IDS involves primarily the selection and layout of exterior sensors that are compatible with the physical and operational characteristics of a specific site. Important factors to consider during the selection process include physical and environmental conditions at the site, the sensor's performance, and the overall cost of the system. Refer to TMs 5-853-1 and 5-853-2 for additional guidance on the requirements for and placement of exterior sensor systems. Since exterior barriers provide very little delay, exterior sensor systems generally do not provide a significant increase in the available response time.

Physical and Environmental Considerations

6-33. Physical and environmental considerations are often the determining factors for selecting exterior sensors. The site's characteristics can significantly affect a sensor's operational performance, both in terms of PD and the susceptibility to nuisance alarms. Exterior sensor systems should be selected on the basis of the frequency and duration of weather-related periods of poor detection capability. An exterior IDS may have an unacceptably low PD during a particular weather event or site condition, yet otherwise be superior to other IDSs in terms of good detection capability and a low nuisance-alarm rate. It may be appropriate to select that IDS in spite of its known vulnerability, precisely because the circumstances of its vulnerability are known and precautionary measures can be taken at those times. The overall performance of that IDS, together with its cost, may justify its selection.

6-34. Weather and climatic conditions at a specific site can significantly influence sensor selection. For example, IR detectors are not very effective in heavy rain, fog, dust, or snow. Deep snow can affect detection patterns and performance of both IR and microwave sensors. High winds can cause numerous false alarms in fence-mounted sensors. Electrical storms can cause alarms in many types of sensors and may also damage the equipment.

6-35. Vegetation can be a significant cause of nuisance alarms. Tall grass or weeds can disturb the energy pattern of microwave and both thermal IR and near-IR beam-break sensors. Vegetation growing near electric-field sensors and capacitance sensors can cause nuisance alarms. Large weeds or bushes rubbing against a fence can produce nuisance alarms from fence-mounted sensors. Large trees and bushes moving within the field of view of video motion sensors can cause nuisance or environmental alarms. A clear area must be established for exterior sensors. This area must be void of vegetation or contain vegetation of carefully controlled growth.

6-36. Topographic features are extremely important. Ideally, perimeter terrain should be flat, although gently sloping terrain is acceptable. Irregular terrain with steep slopes may preclude the use of LOS sensors and make CCTV assessment difficult. Gullies and ditches crossing the perimeter represent a vulnerability to LOS sensors and may be a source of false alarms (from flowing water) for buried line sensors. Large culverts can provide an intruder with an entry or exit route across the perimeter without causing an alarm. Likewise, overhead power and communication lines may permit an intruder to bridge the perimeter without causing an alarm.

6-37. Large animals (such as cows, horses, and deer) can cause nuisance alarms in both aboveground and buried sensors. Sensors sensitive enough to detect a crawling or rolling intruder are susceptible to nuisance alarms from small animals such as rabbits, squirrels, cats, and dogs. To minimize the interference from animals, a dual chain-link-fence configuration may be established around the site perimeter with the sensors installed between the fences.

Sensor Performance

6-38. Exterior sensors must have a high PD for all types of intrusion and have a low unwanted-alarm rate for all expected environmental and site conditions. Unfortunately, no single exterior sensor that is presently available meets both these criteria. All are limited in their detection capability, and all have high susceptibility to nuisance and environmental conditions. Table 6-1, page 6-10, provides estimates of PDs for various types of intrusions. Table 6-2, page 6-10, lists the relative susceptibility of various types of sensors to nuisance and environmental alarms.

Economic Considerations

6-39. Exterior sensor costs are usually given in cost per linear foot per detection zone (typically 300 feet). These costs include both equipment and installation. Fence-mounted sensors (such as strain-sensitive cable, electromechanical, and mechanical) are generally less costly than stand-alone and buried line sensors. Installation costs can vary significantly, depending on the type of sensor. Table 6-3, page 6-11, provides a comparison of relative costs for procuring and installing various types of exterior sensor systems. It should be remembered that the sensor system's cost is only a portion of the total cost for employing a perimeter IDS. Additional costs include fencing, site preparation, CCTV assessment, and perimeter lighting.

Table 6-1. Estimate of PD by Exterior Sensors

	Intruder Technique											
Type of Sensor	Slow walk	Walking	Running	Crawling	Rolling	Jumping	Tunnelling	Trenching	Bridging	Cutting	Climbing	Lifting
Fence mounted	N/A	N/A	N/A	N/A	N/A	VH	VL	L	VL	M/H	H	M/H
Taut wire	N/A	N/A	N/A	N/A	N/A	VH	VL	VL	VL	H	H	H
Electric field	VH	VH	VH	H	VH	VH	VL	L	L	N/A	N/A	N/A
Capacitance	VH	VH	VH	H	H	VH	VL	L	L	N/A	N/A	N/A
Ported cable	H	VH	VH	VH	VH	H	M	VH	L	N/A	N/A	N/A
Seismic	H	VH	H	M	M	M	L	M	L	N/A	N/A	N/A
Seismic/magnetic	H	VH	H	M	M	M	L	M	L	N/A	N/A	N/A
Microwave	H	VH	H	M/H	M/H	M/H	VL	L/M	L	N/A	N/A	N/A
IR	VH	VH	VH	M/H	M/H	H	VL	L	VL	N/A	N/A	N/A
Video motion	H	VH	VH	H	H	H	VL	L/M	M	N/A	N/A	N/A
VL = very low, L = low, M = medium, H = high, VH = very high, N/A = not applicable												

Table 6-2. Relative Susceptibility of Exterior Sensors to False Alarms

	Intruder Technique											
Type of Sensor	Wind	Rain	Standing water/runoff	Snow	Fog	Small animals	Large animals	Small birds	Large birds	Lightning	Overhead power lines	Buried power lines
Fence mounted	H	M	L	L	VL	L	M	L	L	L	VL	VL
Taut wire	VL	VL	VL	VL	VL	VL	L	VL	VL	VL	VL	VL
Electric field	M	L/H	VL	M	VL	M	VH	L	M	M	L	VL
Capacitance	M	M	VL	M	VL	M	VH	L	M	M	L	VL
Ported cable	VL	M	H	L	VL	VL	M	VL	VL	M	VL	L
Seismic	M	L	L	L	VL	L	VH	VL	VL	L	L	M
Seismic/magnetic	M	L	L	L	VL	L	VH	VL	VL	H	M	H
Microwave	L	L	M/H	L/M	L	M/H	VH	VL	M	L/M	L	VL
IR	L	L	L	M	M	M	VH	L	M	L	VL	VL
Video motion	M	L	L	L	M/H	L	VH	VL	M	L	L	VL
VL = very low, L = low, M = medium, H = high, VH = very high												

Table 6-3. Exterior IDS Sensor Cost Comparison

Type of Sensor	Equipment	Installation	Maintenance
Fence mounted	L	L	L
Taut wire	H	H	M
Electric field	H	M	M
Capacitance	M	L	M
Ported cable	H	M	M
Seismic	M	M	L
Seismic/magnetic	H	M	L
Microwave	M	M	L
IR	M	L	M
Video motion	M	L	M
L = low, M = medium, H = high			

PERIMETER LAYOUT AND ZONING

6-40. A protected area's perimeter is usually defined by an enclosing wall or fence or a natural barrier such as water. For exterior sensors to be effective, the perimeter around which they are to be deployed must be precisely defined. In most applications, a dual chain-link-fence configuration will be established around the perimeter. Typically, fences should be between 30 and 50 feet apart; as the distance increases, it is harder for an intruder to bridge the fences. If fence separation is less than 30 feet, some microwave and ported-coax sensors cannot be used. The area between fences (called the controlled area or isolation zone) may need to be cleared of vegetation and graded, depending on the type of sensor used. Proper drainage is required to preclude standing water and to prevent the formation of gullies caused by running water after a heavy rain or melting snow. Cleared areas are required inside and outside of the controlled area. These areas enhance routine observation, as well as sensor-alarm assessment, and minimize the protective cover available to a would-be intruder.

6-41. After the perimeter has been defined, the next step is to divide it into specific detection zones. The length of each detection zone is determined by evaluating the contour, the existing terrain, and the operational activities along the perimeter. Detection zones should be long and straight to minimize the number of sensors or cameras necessary and to aid guard assessment if cameras are not used. It may be more economical to straighten an existing fence line than to create numerous detection zones in accommodating a crooked fence line. If the perimeter is hilly and LOS sensors or CCTV assessment are used, the length of individual detection zones will be commensurate with sensor limitations. Entry points for personnel and vehicles must be configured as independent zones. This enables deactivation of the sensors in these zones; that is, placing them in the access mode during customary working hours (assuming the entry points are manned) without having to deactivate adjacent areas.

6-42. The specific length of individual zones can vary around the perimeter. Although specific manufacturers may advertise maximum zone lengths exceeding 1,000 feet, it is not practical to exceed a zone length of 300 feet. If the zone is longer, it will be difficult for an operator using CCTV assessment or for the response force to identify the location of an intrusion or the cause of a false alarm.

6-43. When establishing zones using multiple sensors, the designer should establish coincident zones where the length and location of each individual sensor will be identical for all sensors within a given zone. If an alarm occurs in a specific zone, the operator can readily determine its approximate location by referring to a map of the perimeter. This also minimizes the number of CCTV cameras required for assessment and simplifies the interface between the alarm-annunciation system and the CCTV switching system.

ESS ALARM-ANNUNCIATION SYSTEM

6-44. Status information from the various intrusion-detection sensors and entry-control terminal devices must be collected from the field and transmitted to the alarm-annunciation system in the security center, where it is processed, annunciated, and acted on by security personnel. The alarm-annunciation system may also interface with a CCTV system. There are typically two types of alarm-annunciation configurations available. The simplest configuration, which is suitable for small installations, is the point-to-point configuration. With this configuration, a separate transmission line is routed from the protected area to the security center (see Figure 6-2). The Joint-Service Interior Intrusion-Detection System (J-SIIDS) is typical of this type of configuration but will not be further discussed in this manual. The second, and more popular type, is a digital multiplexed configuration that allows multiple protected areas to communicate with the security center over a common data line. A block diagram of a typical multiplexed alarm-annunciation system is shown in Figure 6-3.

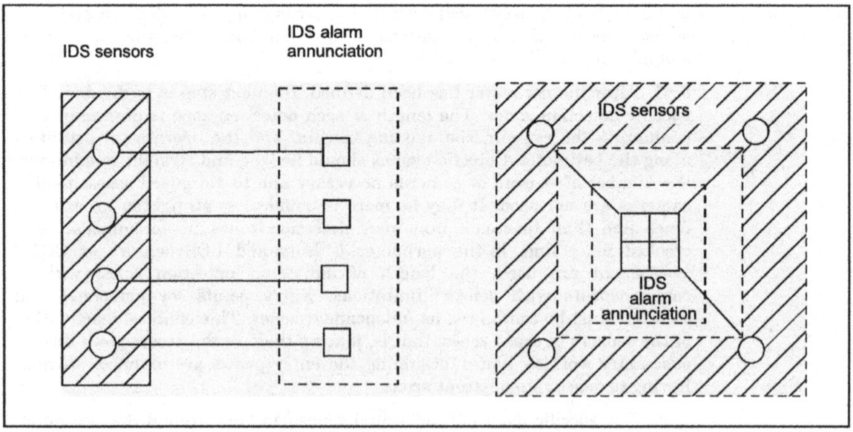

Figure 6-2. Typical Point-to-Point IDS

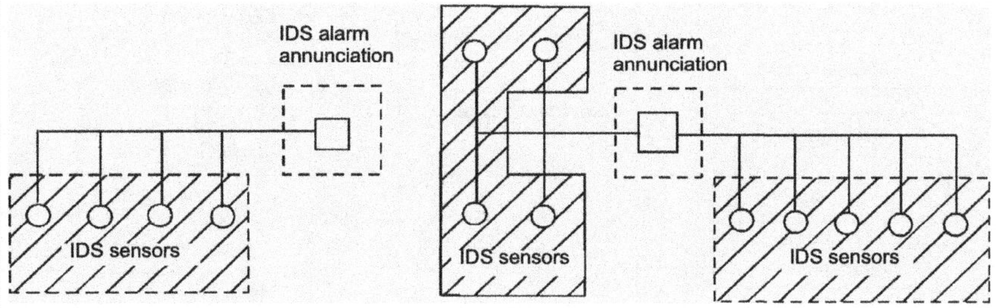

Figure 6-3. Typical Multiplexed IDS

ALARM-ANNUNCIATION CONFIGURATION

6-45. A block diagram of a typical alarm-annunciation system is shown in Figure 6-4, page 6-14. As shown in the figure, the central computer is the hub of the information flow. The central computer receives and displays alarm and device status information and sends operator-control commands to the ESS's local processors. It also interfaces with the CCTV system. For larger facilities, the management of the DTM communications tasks may be delegated to a separate communication processor so that the central computer can turn its full attention to interpreting the incoming information and updating the control and display devices located at the security console (display, logging, control, and storage devices).

6-46. The central computer may consist of one or more digital computers. The real-time clock is usually integral to the central computer and provides a time stamp for alarms and other events. It allows for time synchronization with the CCTV and other systems, if included. The console operator must be able to set the clock, which should include a battery backup. All system events must be properly time-correlated. For example, there will be an exact time correlation for an ESS alarm event reported on the alarm printer and the corresponding video scene recorded by the CCTV's video processor.

DATA STORAGE

6-47. Computer-based systems are required to store large amounts of information such as system software, application programs, data structures, and system events (alarm transactions and status changes). Therefore, a large amount of nonvolatile memory is required. The semiconductor memory provided with a central computer is designed for rapid storage and retrieval

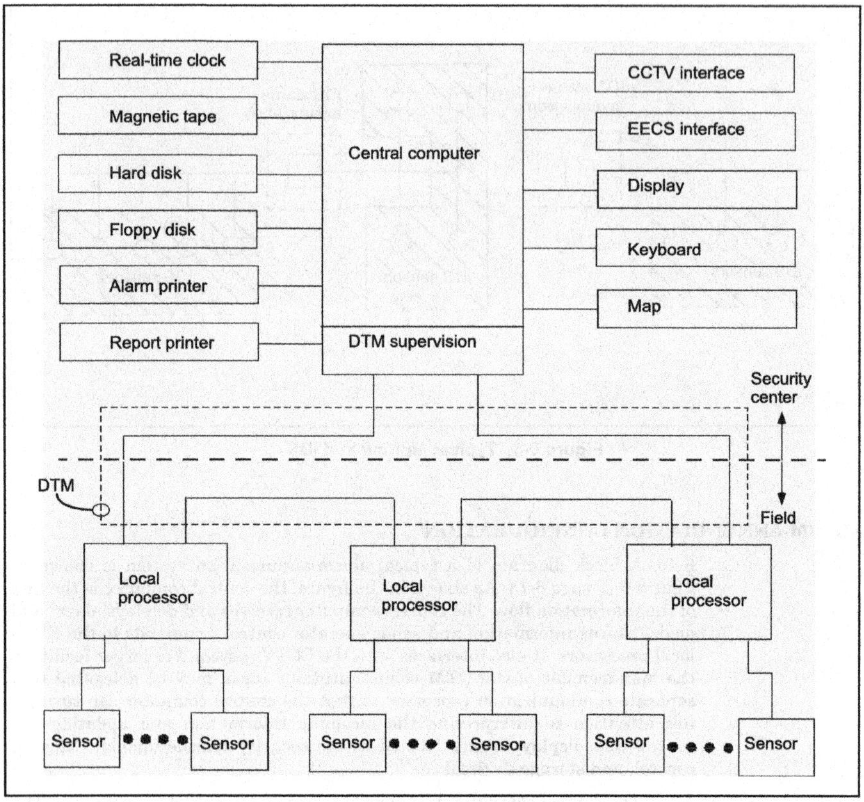

Figure 6-4. Typical IDS Alarm-Annunciation System

and possesses extremely fast access times. The most commonly used media for archival storage are magnetic tape; compact-disk, read-only memory (CD-ROM); and magnetic disk. These media are capable of economically storing large amounts of data.

OPERATOR INTERFACE

6-48. The operator interacts with the alarm-annunciation system through devices that can be seen, heard, or touched and manipulated. Visual displays and printers can be used to inform the operator of an alarm or the equipment's status. Audible devices are used to alert an operator to an alarm or the equipment's failure. Devices such as push buttons and keyboards permit an operator to acknowledge and reset alarms, as well as change operational parameters.

- **Visual displays.** The type of display used to inform the operator visually of the ESS's status is determined primarily by the system's complexity. Status information is usually displayed on monitors. Alphanumeric displays and map displays are seldom used. Monitors provide great flexibility in the type and format of alarm information that may be displayed. Both text and graphic information can be displayed in a variety of colors. Multiple alarms may also be displayed. If alarms are prioritized, higher-priority alarms may be highlighted by blinking, by using bold print or reverse video, or by changing colors. To assist the operator in determining the correct response, alarm-specific instructions may be displayed adjacent to the alarm information.
- **Audible alarm devices.** In conjunction with the visual display of an alarm, the alarm-annunciation system must also generate an audible alarm. The audible alarm may be produced by the ringing of a bell or by the generation of a steady or pulsating tone from an electronic device. In any case, the audible alarm serves to attract the operator's attention to the visual-alarm display. A silence switch is usually provided to allow the operator to silence the bell or tone before actually resetting the alarm.
- **Logging devices.** All alarm-system activity (such as a change of access/secure status, an alarm event, an entry-control transaction, or a trouble event) should be logged and recorded. Logged information is important not only for security personnel investigating an event, but also for maintenance personnel checking equipment performance for such causes as false and nuisance alarms. Most alarm-annunciation systems are equipped with logging and alarm printers.
- **Alarm printers.** Alarm printers are typically of the high-speed, continuous-feed variety. The printer provides a hard-copy record of all alarm events and system activity, as well as limited backup in case the visual display fails.
- **Report printers.** Most ESSs include a separate printer (report printer) for generating reports using information stored by the central computer. This printer will usually be typical of those found in modern office environments.
- **Operator control.** A means is required to transmit information from the operator to the system. The type of controls provided usually depends on the type of display provided. The following are consistent with the controls:
  - Keypads consist of a numeric display system that will generally be provided with a 12-digit keypad and several function keys such as access, secure, acknowledge, and reset. The keypad enables an operator to key in numeric requests for the status of specific zones.
  - Monitor-based systems are usually provided with a typewriter-type keyboard that enables an operator to enter more information using a combination of alphanumeric characters and function keys.
  - An ESS may be equipped with enhancement hardware/devices to help the operator enter information or execute commands quickly. A mouse or a trackball are typical examples.

FIELD-DATA COLLECTION

6-49. Sensor and terminal device data must be transmitted to the central alarm monitor located in the security center using a selected DTM. The following are DTM methods that may be used:

Local Processors

6-50. Multiplexing techniques can be used to minimize the number of data links needed to communicate field-device status to the security center. This is done through devices called local processors. The following is descriptive of a local processor's capabilities:

- A local processor may have very few device inputs, or it may have many (depending on the manufacturer). Rather than having a fixed number of inputs, many local processors are expandable. For example, a basic local processor may be provided with eight device inputs with additional blocks of eight inputs available by using plug-in modules.
- The local processor must provide line supervision for all communication links to sensors, terminal devices, and so forth. Usually, direct-current (DC) line supervision is supplied as the standard with more secure techniques available as options. The data communication links between the local processor and the central alarm monitor must also be supervised.
- Local processors can also provide output signals that can be used for such functions as activating sensor remote test features, light control, or portal control or activating a deterrent (such as a loud horn).
- The local processor contains a microprocessor, solid-state memory, and appropriate software. It has the capability to perform a number of functions locally (such as access-/secure-mode selection, alarm reset, card or keypad electronic-entry control, portal control, and device testing). If the communication link to the security center is temporarily lost, local processors can continue to operate in a stand-alone mode, storing data for transmission after the link is restored.
- The number of local processors required for a specific site depends on the number of protected areas and their proximity to each other and the number of sensors within a protected area. For example, a small building may require one local processor, whereas a large building may require one or more for each floor. An exterior IDS perimeter with two or three different sensors may require one local processor for every two perimeter zones. All local processors may be linked to the central computer using one common DTM link, or the DTM may consist of several links. The designer should note that the temporary loss of a DTM link would render all local processors on that link inactive for the duration of the loss.

Central Computer and Local-Processor Data Exchange

6-51. When the ESS is powered up or reset at the security center, the central computer will download all necessary operational information over the DTM to all local processors. After the download is complete, the central computer will automatically begin polling the local processors for ESS device status. In addition to alarm status, tamper indications, and local-processor status, the DTM may be required to convey security-center console-operator commands to field devices. Examples include security-area access-/secure-mode changes and initiation of the intrusion-sensor self test.

CCTV Interface

6-52. If a CCTV assessment system is deployed with the ESS, an interface between the two is required. This interface allows CCTV system alarms (such as loss of video) to be displayed by the ESS's alarm-annunciation system. The interface also provides IDS alarm signals to the CCTV's video switcher so that the correct CCTV camera will be displayed on the CCTV monitors to allow real-time alarm assessment and video recording as required.

ESS SOFTWARE

6-53. The software provided with computer-based ESS alarm-annunciation systems consists of three types—a standard operating system (such as the Microsoft®-disk operating system [MS-DOS]); vendor-developed application programs; and user-filled, site-specific data structures.

- **System software.** The designer will ensure that system software provided by the vendor conforms to accepted industry standards so that standard, follow-on maintenance and service contracts can be negotiated to maintain the central computer system.
- **Application software.** The vendor-developed application programs are typically proprietary and include ESS monitoring, display, and entry-control capabilities.
- **User-filled data structures.** These data structures are used to populate the site-specific database. Specific electronic address information, personnel access schedules, and normal duty hours are typically included in the site-specific database. The information may include preferred route descriptions for the response force, the phone number of the person responsible for the alarmed area, and any hazardous material that may be located in the alarmed area.

6-54. ESS software functions typically include the following:

- **Alarm monitoring and logging.** The software should provide for monitoring all sensors, local processors, and data communication links and notifying the operator of an alarm condition. All alarm messages should be printed on the alarm printer, archived, and displayed at the console. As a minimum, printed alarm data should include the date and time (to the nearest second) of the alarm and the location and type of alarm.
- **Alarm display.** The software should be structured to permit several alarms to be annunciated simultaneously. A buffer or alarm queue should be available to store additional alarms until they are annunciated and, subsequently, acted upon and reset by the console operator.
- **Alarm priority.** A minimum of five alarm-priority levels should be available. Higher-priority alarms should always be displayed before

lower-priority alarms. This feature permits an operator to respond quickly to the more important alarms before those of lesser importance. For example, the priority of alarm devices may be as follows:

- Duress.
- Intrusion detection.
- Electronic-entry control.
- Tamper.
- CCTV alarms and equipment-malfunction alarms.

- **Reports.** The application software should provide for generating, displaying, printing, and storing reports.

PASSWORDS

6-55. Software security will be provided by limiting access to personnel with authorized passwords assigned by a system manager. A minimum of three password levels shall be provided. Additional security can be provided by programmed restrictions that limit the keyboard actions of logged-in passwords to the user ranks of system managers, supervisors, and console operators, as appropriate.

OPERATOR INTERFACE

6-56. The software should enable an operator with the proper password to enter commands and to obtain displays of system information. As a minimum, an operator should be able to perform the following functions through the keyboard or the keypad:

- Log on by password to activate the keyboard.
- Log off to deactivate the keyboard.
- Request display of all keyboard commands that are authorized for the logged-in password.
- Request display of detailed instructions for any authorized keyboard command.
- Acknowledge and clear alarm messages.
- Display the current status of any device in the system.
- Command a status change for any controlled device in the system.
- Command a mode change for any access/secure device in the system.
- Command printouts of alarm summaries, status summaries, or system activity on a designated printer.
- Add or delete ESS devices or modify parameters associated with a device.

INTERIOR INTRUSION-DETECTION SENSORS

6-57. Interior intrusion-detection sensors are devices used to detect unauthorized entry into specific areas or volumetric spaces within a building. These sensors are usually not designed to be weatherproof or rugged enough to survive an outdoor environment. Therefore, this type of sensor should not be used outdoors unless described by the manufacturer as suitable for outdoor use.

6-58. Interior intrusion-detection sensors generally perform one of three detection functions—detection of an intruder penetrating the boundary of a protected area, detection of intruder motion within a protected area, and detection of an intruder touching or lifting an asset within a protected area. Therefore, interior sensors are commonly classified as boundary-penetration sensors, volumetric motion sensors, and point sensors. Although duress switches are not intrusion-detection sensors, they are included in this discussion because they are usually wired to the same equipment that monitors the interior intrusion-detection sensors.

BOUNDARY-PENETRATION SENSORS

6-59. Boundary-penetration sensors are designed to detect penetration or attempted penetration through perimeter barriers. These barriers include walls, ceilings, duct openings, doors, and windows.

Structural-Vibration Sensors

6-60. Structural-vibration sensors detect low-frequency energy generated in an attempted penetration of a physical barrier (such as a wall or a ceiling) by hammering, drilling, cutting, detonating explosives, or employing other forcible methods of entry. A piezoelectric transducer senses mechanical energy and converts it into electrical signals proportional in magnitude to the vibrations. To reduce false alarms from single accidental impacts on the barrier, most vibration sensors use a signal processor that has an adjustable pulse-counting accumulator in conjunction with a manual sensitivity adjustment. The count circuit can be set to count a specific number of pulses of specific magnitude within a predefined time interval before an alarm is generated. However, the circuitry is usually designed to respond immediately to large pulses, such as those caused by an explosion. The sensitivity adjustment is used to compensate for the type of barrier and the distance between transducers. Typically, several transducers can be connected together and monitored by one signal processor. Figure 6-5, page 6-20, shows an example of wall-mounted, structural-vibration sensors.

Glass-Breakage Sensors

6-61. Glass-breakage sensors detect the breaking of glass. The noise from breaking glass consists of frequencies in both the audible and ultrasonic range. Glass-breakage sensors use microphone transducers to detect the glass breakage. The sensors are designed to respond to specific frequencies only, thus minimizing such false alarms as may be caused by banging on the glass.

Passive Ultrasonic Sensors

6-62. Passive ultrasonic sensors detect acoustical energy in the ultrasonic frequency range, typically between 20 and 30 kilohertz (kHz). They are used to detect an attempted penetration through rigid barriers (such as metal or masonry walls, ceilings, and floors). They also detect penetration through windows and vents covered by metal grilles, shutters, or bars if these openings are properly sealed against outside sounds.

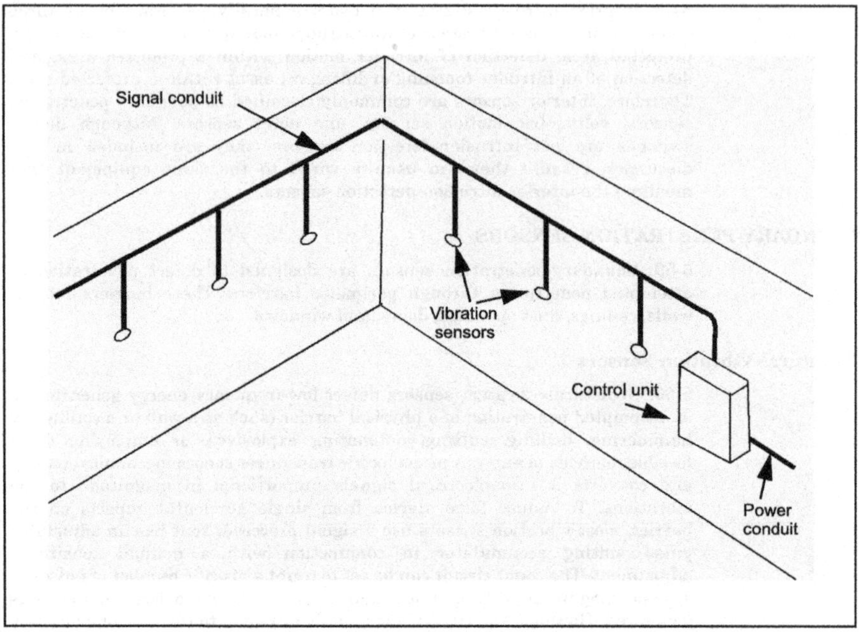

Figure 6-5. Wall-Mounted, Structural-Vibration Sensors

6-63. **Detection Transducer.** The detection transducer is a piezoelectric crystal that produces electrical signals proportional to the magnitude of the vibrations. A single transducer provides coverage of an area about 15 by 20 feet in a room with an 8- to 12-foot ceiling. A typical detection pattern is shown in Figure 6-6. Ten or more transducers can be connected to a signal processor. As with vibration sensors, the signal processor for a passive ultrasonic sensor has manual sensitivity adjustment and an adjustable pulse-counting accumulator.

6-64. **Sensors.** Passive ultrasonic sensors detect ultrasonic energy that results from the breaking of glass, the snipping of bolt cutters on metal barriers, the hissing of an acetylene torch, and the shattering of brittle materials (such as concrete or cinderblock). However, the sensors will not reliably detect drilling through most material nor attacks against soft material such as wallboard. Their effective detection range depends largely on the barrier material, the method of attempted penetration, and the sensitivity adjustment of the sensor. Examples of maximum detection distances for a typical sensor for different types of attempted penetration are shown in Table 6-4.

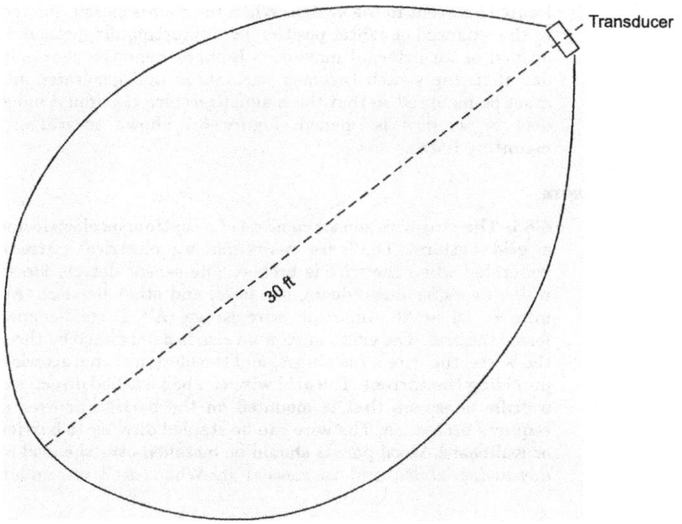

Figure 6-6. Typical Passive-Ultrasonic-Sensor Detection Pattern

Table 6-4. Detection Range for Passive Ultrasonic Sensors

Penetration	Distance (in Feet)
Cut 1/4-inch-thick expanded metal with bolt cutters	55
Cut 5/8-inch reinforcing bar with bolt cutters	45
Use acetylene cutting torch	39
Cut wood with circular saw	30
Cut 5/8-inch reinforcing bar with hacksaw	19
Drill through brick	15
Drill through 1/8-inch steel plate	6
Cut 1/8-inch steel plate with hacksaw	4
Drill through cinderblock	3

6-65. **Balanced Magnetic Switches.** Balanced magnetic switches (BMSs) are typically used to detect the opening of a door. These sensors can also be used on windows, hatches, gates, or other structural devices that can be opened to gain entry. When using a BMS, mount the switch mechanism on the door frame and the actuating magnet on the door. Typically, the BMS has a three-position reed switch and an additional magnet (called the bias magnet) located adjacent to the switch. When the door is closed, the reed switch is held in the balanced or center position by interacting magnetic fields. If the door is opened or an external magnet is brought near the sensor in an attempt to defeat it, the switch becomes unbalanced and generates an alarm. A BMS must be mounted so that the magnet receives maximum movement when the door or window is opened. Figure 6-7 shows several configurations for mounting BMSs.

Grid-Wire Sensors

6-66. The grid-wire sensor consists of a continuous electrical wire arranged in a grid pattern. The wire maintains an electrical current. An alarm is generated when the wire is broken. The sensor detects forced entry through walls, floors, ceilings, doors, windows, and other barriers. An enamel-coated number 24 or 26 American wire gauge (AWG) solid-copper wire typically forms the grid. The grid's maximum size is determined by the spacing between the wires, the wire's resistance, and the electrical characteristics of the source providing the current. The grid wire can be installed directly on the barrier, in a grille or screen that is mounted on the barrier, or over an opening that requires protection. The wire can be stapled directly to barriers made of wood or wallboard. Wood panels should be installed over the grid to protect it from day-to-day abuse and to conceal it. When used on cinder, concrete, and

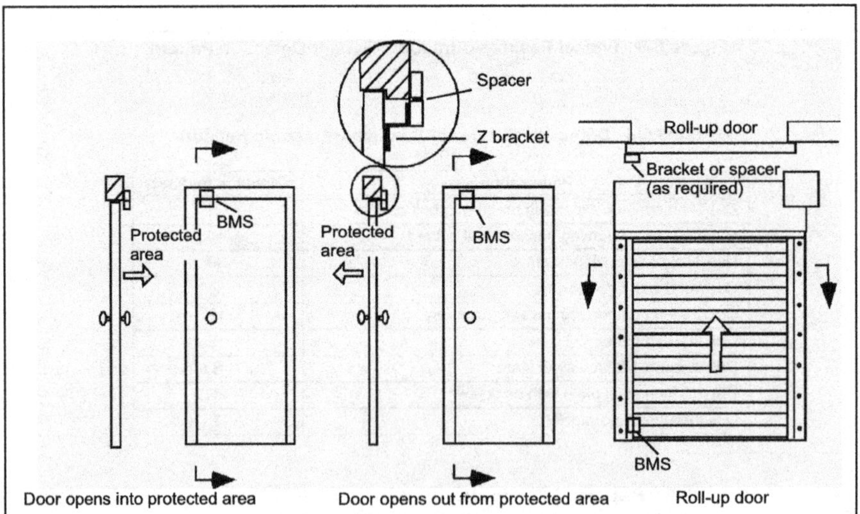

Figure 6-7. BMS Mounting Configurations

masonry surfaces, these surfaces must first be covered with plywood or other material to which the wire can be stapled. An alternative method is to staple the wire grid to the back side of a panel and install the panel over the surface.

VOLUMETRIC MOTION SENSORS

6-67. Volumetric motion sensors are designed to detect intruder motion within the interior of a protected volume. Volumetric sensors may be active or passive. Active sensors (such as microwave) fill the volume to be protected with an energy pattern and recognize a disturbance in the pattern when anything moves within the detection zone. Whereas active sensors generate their own energy pattern to detect an intruder, passive sensors (such as IR) detect energy generated by an intruder. Some sensors, known as dual-technology sensors, use a combination of two different technologies, usually one active and one passive, within the same unit. If CCTV assessment or surveillance cameras are installed, video motion sensors can be used to detect intruder movement within the area. Since ultrasonic motion sensors are seldom used, they will not be discussed here.

Microwave Motion Sensors

6-68. With microwave motion sensors, high-frequency electromagnetic energy is used to detect an intruder's motion within the protected area. Interior or sophisticated microwave motion sensors are normally used.

6-69. **Interior Microwave Motion Sensors.** Interior microwave motion sensors are typically monostatic; the transmitter and the receiver are housed in the same enclosure (transceiver). They may each be provided with a separate antenna or they may share a common antenna. The high-frequency signals produced by the transmitter are usually generated by a solid-state device, such as a gallium arsenide field-effect transistor. The power generated is usually less than 10 milliwatts, but it is sufficient to transmit the signal for distances up to about 100 feet. The shape of the transmitted beam is a function of the antenna configuration. The range of the transmitted beam can be controlled with a range adjustment. A variety of detection patterns can be generated (see Figure 6-8, page 6-24). The frequency of the transmitted signal is compared with the frequency of the signal reflected back from objects in the protected area. If there is no movement within the area, the transmitted and received frequencies will be equal and no alarm will be generated. Movement in the area will generate a Doppler frequency shift in the reflected signal and will produce an alarm if the signal satisfies the sensor's alarm criteria. The Doppler shift for a human intruder is typically between 20 and 120 hertz (Hz). Microwave energy can pass through glass doors and windows as well as lightweight walls or partitions constructed of plywood, plastic, or fiberboard. As a result, false alarms are possible because of the reflection of the microwave signals from the movement of people or vehicles outside of the protected area. The designer can sometimes take advantage of this when the protected area is large and contains a number of partitions, but this is not normally done.

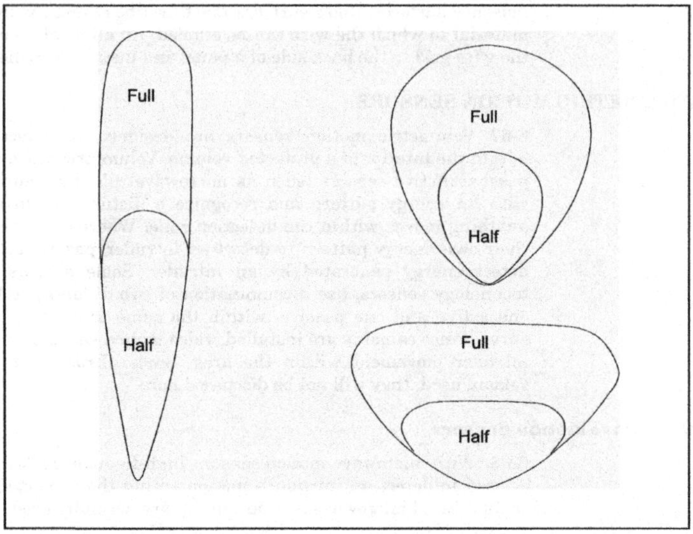

Figure 6-8. Typical Detection Patterns for Microwave Motion Sensors

6-70. **Sophisticated Microwave Motion Sensors.** Sophisticated microwave motion sensors may be equipped with electronic range gating. This feature allows the sensor to ignore the signals reflected beyond the settable detection range. Range gating may be used to effectively minimize unwanted alarms from activity outside the protected area.

PIR Motion Sensors

6-71. PIR motion sensors detect a change in the thermal energy pattern caused by a moving intruder and initiate an alarm when the change in energy satisfies the detector's alarm criteria. These sensors are passive devices because they do not transmit energy; they monitor the energy radiated by the surrounding environment.

6-72. All objects with temperatures above absolute zero radiate thermal energy. The wavelengths of the IR energy spectrum lie between 1 and 1,000 microns. Because the human body radiates thermal energy of between 7 and 14 microns, PIR motion sensors are typically designed to operate in the far IR wavelength range of 4 to 20 microns.

6-73. The IR energy must be focused onto a sensing element, somewhat as a camera lens focuses light onto a film. Two techniques are commonly used. One technique uses reflective focusing; parabolic mirrors focus the energy. The other uses an optical lens. Of the various types of optical lenses, Fresnel lenses are preferred because they can achieve short focal lengths with minimal thickness. Because IR energy is severely attenuated by glass, lenses are usually made of plastic.

6-74. The sensor's detection pattern is determined by the arrangement of lenses or reflectors. The pattern is not continuous but consists of a number of rays or fingers, one for each mirror or lens segment. Numerous detection patterns are available, several of which are shown in Figure 6-9. The PIR is not provided with a range adjustment, but the range can be adjusted

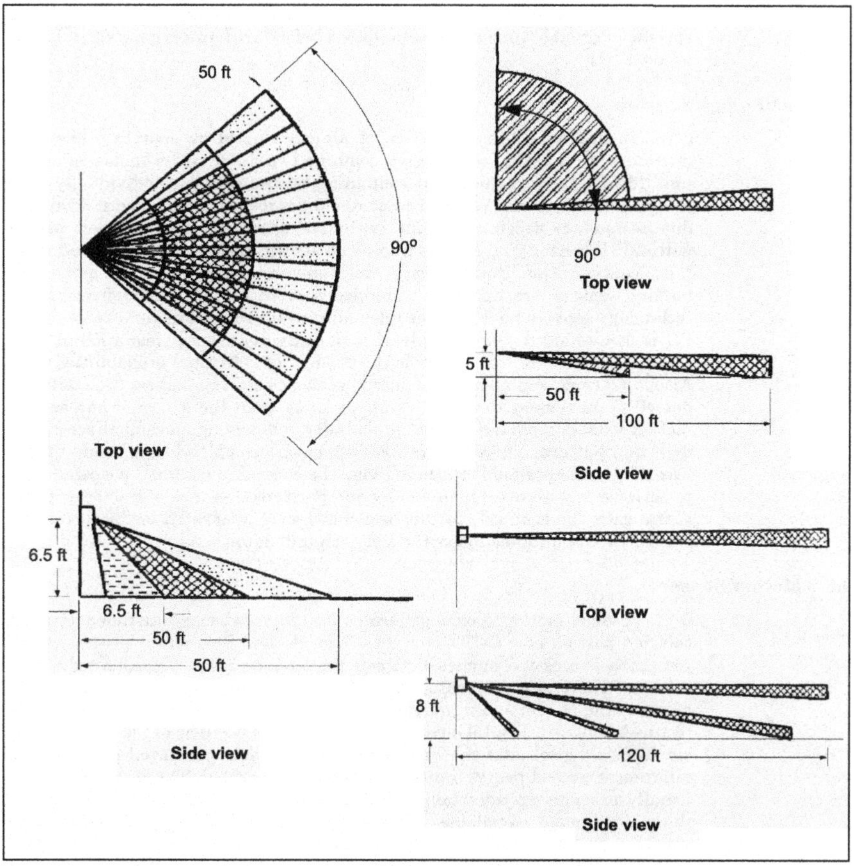

Figure 6-9. Typical Detection Patterns for a PIR Motion Detector

somewhat by manipulating the sensor's position; therefore, careful selection of the appropriate detection pattern is critical to proper sensor performance.

6-75. Most manufacturers use a pyroelectric material as the thermal sensing element. This material produces a change in electric charge when exposed to changes in temperature. To minimize false alarms caused by changes in ambient temperature, most manufacturers use a dual-element sensor. The sensing element is split into halves, one that produces a positive voltage pulse and the other a negative pulse when a change in temperature changes. An intruder entering one of the detection fingers produces an imbalance between the two halves, resulting in an alarm condition. Quadelement sensors that combine and compare two dual-element sensors are also in use. Pulse-count activation, a technique in which a predefined number of pulses within a specific interval of time must be produced before an alarm is generated, is also used.

Dual-Technology Sensors

6-76. To minimize the generation of alarms caused by sources other than intruders, dual-technology sensors combine two different technologies in one unit. Ideally, this is achieved by combining two sensors that individually have a high PD and do not respond to common sources of false alarms. Available dual-technology sensors combine an active ultrasonic or microwave sensor with a PIR sensor. The alarms from each sensor are logically combined in an "and" configuration; that is, nearly simultaneous alarms from both active and passive sensors are needed to produce a valid alarm. Although combined technology sensors have a lower false-alarm rate than individual sensors, the PD is also reduced. For example, if each individual sensor has a PD of 0.95, the PD of the combined sensors is the product of individual probabilities (0.9). Also, ultrasonic and microwave motion sensors have the highest probability of detecting movement directly toward or away from the sensor, whereas PIR motion sensors have the highest probability of detecting movement across the detection pattern. Therefore, the PD of sensors combined in a single unit is less than that obtainable if the individual sensors are mounted perpendicular to each other with overlapping detection patterns. Because of the lower false-alarm rate, the reduced PD can be somewhat compensated for by increasing the sensitivity or detection criteria of each individual sensor.

Video Motion Sensors

6-77. A video motion sensor generates an alarm when an intruder enters a selected portion of a CCTV camera's field of view. The sensor processes and compares successive images between the images against predefined alarm criteria. There are two categories of video motion detectors—analog and digital. Analog detectors generate an alarm in response to changes in a picture's contrast. Digital devices convert selected portions of the analog video signal into digital data that are compared with data converted previously; if differences exceed preset limits, an alarm is generated. The signal processor usually provides an adjustable window that can be positioned anywhere on the video image. Available adjustments permit changing horizontal and vertical window size, window position, and window sensitivity. More sophisticated units provide several adjustable windows that can be individually sized and positioned. Multiple windows permit concentrating on several specific areas of an image while ignoring others. For example, in a scene containing six doorways leading into a long hallway, the sensor can be set to monitor only two critical doorways.

POINT SENSORS

6-78. Point sensors are used to protect specific objects within a facility. These sensors (sometimes referred to as proximity sensors) detect an intruder coming in close proximity to, touching, or lifting an object. Several different types are available, including capacitance sensors, pressure mats, and pressure switches. Other types of sensors can also be used for object protection.

Capacitance Sensors

6-79. Capacitance sensors detect an intruder approaching or touching a metal object by sensing a change in capacitance between the object and the ground. A capacitor consists of two metallic plates separated by a dielectric medium. A change in the dielectric medium or electrical charge results in a change in capacitance. In practice, the metal object to be protected forms one plate of the capacitor and the ground plane surrounding the object forms the second plate. The sensor processor measures the capacitance between the metal object and the ground plane. An approaching intruder alters the dielectric value, thus changing the capacitance. If the net capacitance change satisfies the alarm criteria, an alarm is generated.

6-80. The maximum capacitance that can be monitored by this type of sensor is usually between 10,000 and 50,000 picofarads. The minimum detectable change in capacitance can be as low as 20 picofarads. The signal processor usually has a sensitivity adjustment that can be set to detect an approaching intruder several feet away or to require that the intruder touch the object before an alarm is generated.

6-81. Because air forms most of the dielectric of the capacitor, changes in relative humidity will affect the sensor's sensitivity. An increase in humidity causes the conductivity of the air to increase, lowering the capacitance. Moving a metal object (such as a file cabinet) closer to or away from the protected object can also affect the sensitivity of a capacitance sensor. Figure 6-10, page 6-28, illustrates a typical application using a capacitance sensor.

Pressure Mats

6-82. Pressure mats generate an alarm when pressure is applied to any part of the mat's surface, as when someone steps on the mat. One type of construction uses two layers of copper screening separated by soft-sponge rubber insulation with large holes in it. Another type uses parallel strips of ribbon switches made from two strips of metal separated by an insulating material and spaced several inches apart. When enough pressure is applied to the mat, either the screening or the metal strips make contact, generating an alarm. Pressure mats can be used to detect an intruder approaching a protected object, or they can be placed by doors or windows to detect entry. Because pressure mats are easy to bridge, they should be well concealed, such as placing them under a carpet.

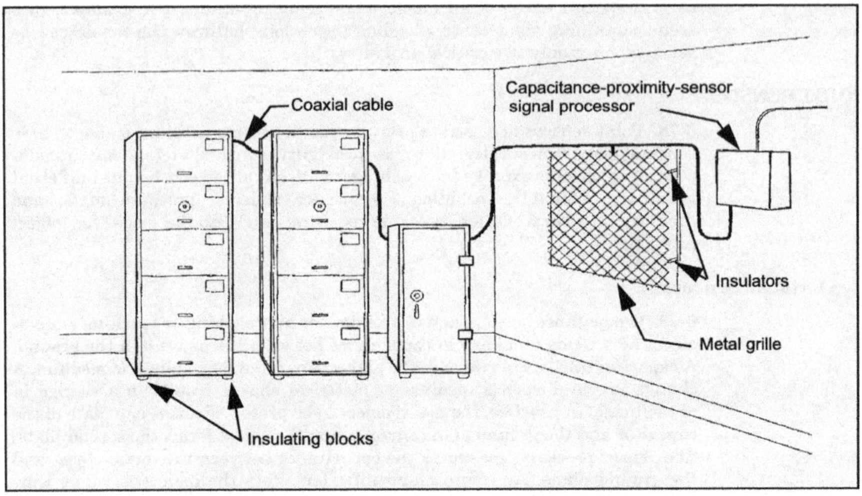

Figure 6-10. Capacitance-Proximity-Sensor Application

Pressure Switches

6-83. Mechanically activated contact switches or single ribbon switches can be used as pressure switches. Objects that require protection can be placed on top of the switch. When the object is moved, the switch actuates and generates an alarm. In this usage, the switch must be well concealed. The interface between the switch and the protected object should be designed so that an adversary cannot slide a thin piece of material under the object to override the switch while the object is removed.

DURESS-ALARM DEVICES

6-84. Duress-alarm devices may be fixed or portable. Operations and security personnel use them to signal a life-threatening emergency. Activation of a duress device will generate an alarm at the alarm-monitoring station. Because of the nature of the alarm, duress devices should never annunciate at the point of threat. These devices are customarily manually operated.

6-85. Fixed duress devices are mechanical switches permanently mounted in an inconspicuous location, such as under a counter or desk. They can be simple push-button switches activated by the touch of a finger or hand or foot-operated switches attached to the floor.

6-86. Portable duress devices are wireless units consisting of a transmitter and a receiver. The transmitter is portable and small enough to be conveniently carried by a person. The receiver is mounted in a fixed location within the facility. Either ultrasonic or RF energy can be used as the communication medium. When activated, the transmitter generates an alarm that is detected (within range) by the receiver. The receiver then activates a relay that is hardwired to the alarm-monitoring system.

EXTERIOR INTRUSION-DETECTION SENSORS

6-87. Exterior intrusion-detection sensors are customarily used to detect an intruder crossing the boundary of a protected area. They can also be used in clear zones between fences or around buildings, for protecting materials and equipment stored outdoors within a protected boundary, or in estimating the PD for buildings and other facilities.

6-88. Exterior sensors are designed to operate in outdoor environmental conditions. The detection function must be performed with a minimum of unwanted alarms such as those caused by wind, rain, ice, standing water, blowing debris, animals, and other sources. Important criteria for selecting an exterior sensor are the PD, the sensor's susceptibility to unwanted alarms, and the sensor's vulnerability to defeat.

6-89. The PD of an exterior sensor is much more vulnerable to the physical and environmental conditions of a site than that of an interior sensor. Many uncontrollable forces (such as wind, rain, ice, frozen soil, standing or running water, falling and accumulated snow, and blowing dust and debris) may affect an exterior sensor's performance. Although attention generally is directed to circumstances that cause a dramatic drop in the PD, environmental factors can also cause short-term increases in the PD. If controlled intrusions (intrusions by security personnel to verify the current detection capability of an IDS) are done while an IDS temporarily has a higher than usual PD as the result of current site conditions, the results may give a false indication of the general effectiveness of that IDS.

6-90. Because of the nature of the outdoor environment, exterior sensors are also more susceptible to nuisance and environmental alarms than interior sensors. Inclement weather conditions (heavy rain, hail, and high wind), vegetation, the natural variation of the temperature of objects in the detection zone, blowing debris, and animals are major sources of unwanted alarms.

6-91. As with interior sensors, tamper protection, signal-line supervision, self-test capability, and proper installation make exterior sensors less vulnerable to defeat. Because signal-processing circuitry for exterior sensors is generally more vulnerable to tampering and defeat than that for interior sensors, it is extremely important that enclosures are located and installed properly and that adequate physical protection is provided. Several different types of exterior intrusion-detection sensors are available. They can be categorized as—

- Fence sensors.
- Buried line sensors.
- LOS sensors.
- Video motion sensors.

FENCE SENSORS

6-92. Fence sensors detect attempts to penetrate a fence around a protected area. Penetration attempts (such as climbing, cutting, or lifting) generate mechanical vibrations and stresses in fence fabric and posts that are usually different than those caused by natural phenomena like wind and rain. The basic types of sensors used to detect these vibrations and stresses are strain-sensitive cable, taut wire, and fiber optics. Other types of fence sensors detect penetration attempts by sensing changes in an electric field or in capacitance. Mechanical and electromechanical fence sensors are seldom used and will not be discussed here.

Strain-Sensitive Cable

6-93. Strain-sensitive cables are transducers that are uniformly sensitive along their entire length. They generate an analog voltage when subject to mechanical distortions or stress resulting from fence motion. Strain-sensitive cables are sensitive to both low and high frequencies. The signal processor usually has a band-pass filter that passes only those signals characteristic of fence-penetration actions. An alarm is initiated when the signal's frequency, amplitude, and duration characteristics satisfy the processor's criteria. Because the cable acts like a microphone, some manufacturers offer an option that allows the operator to listen to fence noises causing the alarm. Operators can then determine whether the noises are naturally occurring sounds from wind or rain or are from an actual intrusion attempt. This feature is relatively costly to implement because it requires additional cable from each signal processor to the security center and, if CCTV is being used, it may be of limited benefit. Strain-sensitive cable is attached to a chain-link fence about halfway between the bottom and top of the fence fabric with plastic ties. One end of the cable is terminated at the signal processor and the other end with a resistive load. The DC through the cable provides line supervision against cutting or electrically shorting the cable or disconnecting it from the processor. A typical installation is shown in Figure 6-11.

Taut-Wire Sensor

6-94. A taut-wire sensor combines a physically taut-wire barrier with an intrusion-detection sensor network. The taut-wire sensor consists of a column of uniformly spaced horizontal wires up to several hundred feet in length and securely anchored at each end. Typically, the wires are spaced 4 to 8 inches apart. Each is individually tensioned and attached to a detector located in a sensor post. Two types of detectors are commonly used—mechanical switches and strain gauges.

- The mechanical switch consists of a specially designed switch mechanism that is normally open. The tensioned wires are mechanically attached to the switch, and movement of the wire beyond a preset limit causes the switch to close. To counteract small gradual movements of a wire (such as that caused by settling of the fence or by freezing or thawing of soil) switches are usually supported in their housing by a soft plastic material. This material allows the switch to self-adjust when acted upon by gradual external forces and wire effects

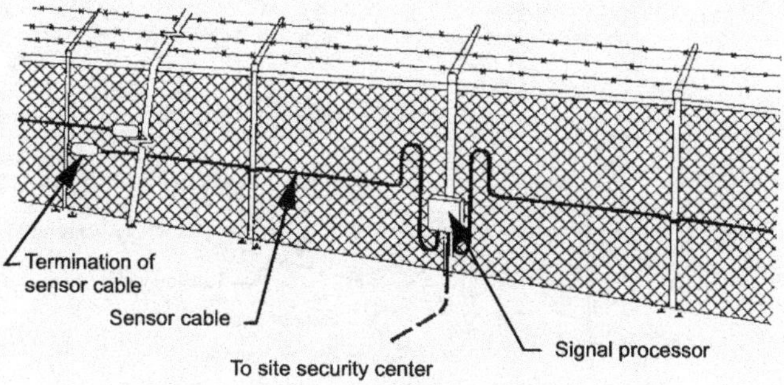

Figure 6-11. Typical Strain-Sensor Cable Installation

such as the relaxation of the wire with time and its thermal expansion or contraction.

- Strain-gauge detectors are attached to the taut wire with a nut on a threaded stud. When a force is applied to the taut wire, the resulting deflection is converted by the strain gauge into a change in electrical output that is monitored by a signal processor.

6-95. With sensors that use mechanical switches as detectors, the switches in a single sensor-post assembly are wired in parallel and are connected directly to the alarm-annunciation system. Pulse-count circuitry is not used because a single switch closure, such as that caused by an intruder moving or cutting one wire, is indicative of an intrusion attempt. Strain-gauge detectors in a sensor post are monitored by a signal processor. When the signal from one or more strain gauges satisfies the processor's criteria, an alarm is initiated.

6-96. The taut-wire sensor can be installed as a freestanding fence or can be mounted on an existing fence or wall. Figure 6-12, page 6-32, shows a freestanding configuration.

Fiber-Optic Cable Sensors

6-97. Fiber-optic cable sensors are functionally equivalent to the strain-sensitive cable sensors previously discussed. However, rather than electrical signals, modulated light is transmitted down the cable and the resulting received signals are processed to determine whether an alarm should be initiated. Since the cable contains no metal and no electrical signal is present, fiber-optic sensors are generally less susceptible to electrical interference from lightning or other sources.

Electric-Field Sensors

6-98. Electric-field sensors consist of an alternating-current (AC) field generator, one or more field wires, one or more sense wires, and a signal

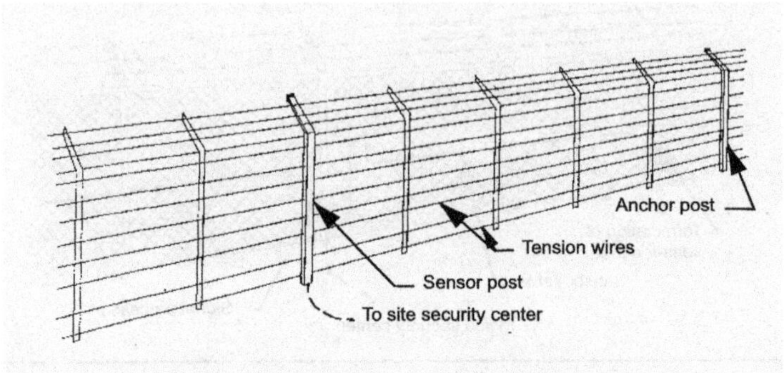

Figure 6-12. Typical Taut-Wire Installation

processor. The generator excites the field wires around which an electrostatic-field pattern is created. The electrostatic field induces electrical signals in the sense wires, which are monitored by the signal processor. Under normal operating conditions, the induced signals are constant. However, when an intruder approaches the sensor, the induced electrical signals are altered, causing the signal processor to generate an alarm.

6-99. Several different field- and sense-wire configurations are available. They range from one field wire and one sense wire to as many as four field wires and one sense wire or four field wires and four sense wires. Figure 6-13 shows the detection pattern produced by vertical three-wire (one field and two sense wires) configurations. The three-wire system has a wider detection envelope and is less costly (one less field wire and associated hardware). However, because of the tighter coupling between wires, the four-wire system is less susceptible to nuisance alarms caused by extraneous noise along the length of the zone.

6-100. A signal processor monitors the signals produced by the sense wires. The processor usually contains a band-pass filter that rejects high-frequency signals such as those caused by objects striking the wires. Additional criteria that must be satisfied before the processor initiates an alarm include signal amplitude and signal duration. By requiring the signal to be present for a preset amount of time, false alarms (such as those caused by birds flying through the detection pattern) can be minimized.

6-101. As with taut-wire sensors, electric-field sensors can be freestanding (mounted on their own posts) or attached by standoffs to an existing fence. They can also be configured to follow contours of the ground. The area under the sensor must be clear of vegetation, since vegetation near or touching sense wires can cause false alarms. These sensors can also be installed on the walls and roof of a building.

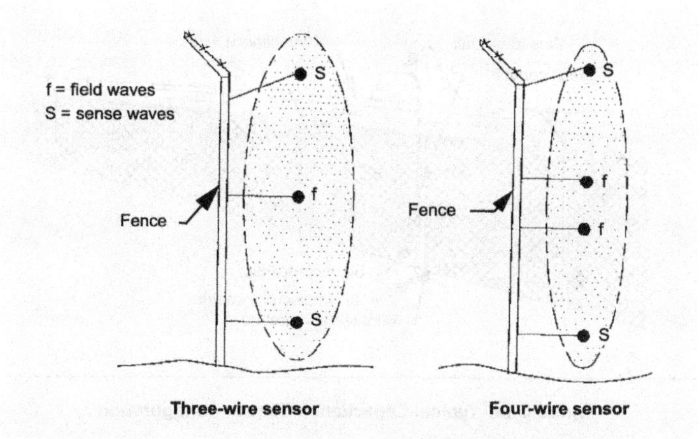

Figure 6-13. Typical Electric-Field-Sensor Detection Patterns

Capacitance Proximity Sensors

6-102. Capacitance proximity sensors measure the electrical capacitance between the ground and an array of sense wires. Any variations in capacitance, such as that caused by an intruder approaching or touching one of the sense wires, initiates an alarm. These sensors usually consist of two or three wires attached to outriggers along the top of an existing fence, wall, or roof edge. Figure 6-14, page 6-34, shows a typical capacitance sensor consisting of three sensor wires attached to the outrigger of a fence. To minimize environmental alarms, the capacitance sensor is divided into two arrays of equal length. The signal processor monitors the capacitance of each array. Changes in capacitance common to both arrays (such as produced by wind, rain, ice, fog, and lightning) are canceled within the processor. However, when changes occur in one array and not the other because of an intruder, the processor initiates an alarm.

BURIED-LINE SENSORS

6-103. A buried-line sensor system consists of detection probes or cable buried in the ground, typically between two fences that form an isolation zone. These devices are wired to an electronic processing unit. The processing unit generates an alarm if an intruder passes through the detection field. Buried-line sensors have several significant features:

- They are hidden, making them difficult to detect and circumvent.
- They follow the terrain's natural contour.
- They do not physically interfere with human activity, such as grass mowing or snow removal.

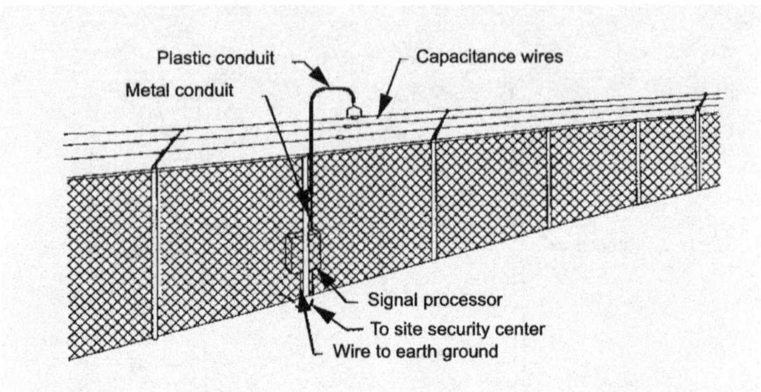

Figure 6-14. Typical Capacitance-Sensor Configuration

- They are affected by certain environmental conditions, such as running water and ground freeze/thaw cycles. (Seismic, seismic/magnetic, magnetic, and balanced pressure sensors are seldom used and will not be discussed here.)

6-104. The ported-coax cable sensor consists of two coax cables buried in the ground parallel to each other. An RF transmitter is connected to one cable and a receiver to the other. The outer conductor of each cable is ported (fabricated with small holes or gaps in the shield). The transmitter cable radiates RF energy into the medium surrounding the cables. A portion of this energy is coupled into the receiver cable through its ported shield. (Because of the ported shields, these cables are frequently referred to as leaky cables.) When an intruder enters the RF field, the coupling is disturbed, resulting in a change of signal monitored by the receiver, which then generates an alarm. Two basic types of ported-coax sensors are available—pulse and continuous wave.

- Pulse-type sensors transmit a pulse of RF energy down one cable and monitors the received signal on the other. The cables can be up to 10,000 feet long. The signal processor initiates an alarm when the electromagnetic field created by the pulse is disturbed and identifies the disturbance's approximate location.
- Continuous-wave sensors apply continuous RF energy to one cable. The signal received on the other cable is monitored for electromagnetic-field disturbances that indicate an intruder's presence. Cable lengths are limited to 300 to 500 feet. Additionally, the sensor is available in a single-cable configuration as well as two separate cables. The pattern typically extends 2 to 4 feet above the ground and can be 5 to 13 feet wide, depending on cable spacing and soil composition. Figure 6-15 represents a typical cross-section of a detection pattern created by a ported-cable sensor.

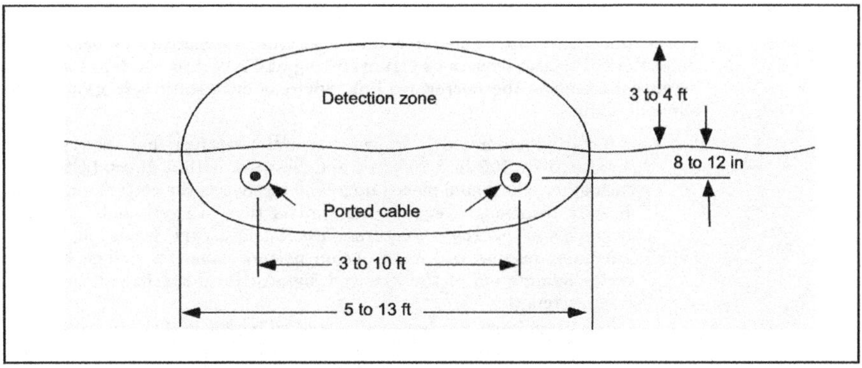

Figure 6-15. Typical Ported-Cable Detection Pattern

6-105. Sensor performance depends on properties of the medium surrounding the cables. Velocity and attenuation of the RF wave that propagates along the cables and the coupling between the cables are functions of the dielectric constant of the soil and its conductivity which, in turn, depends on its moisture content. For example, the velocity is greater and the attenuation is less for cables buried in dry, low-loss soil than in wet, conductive soil. Freeze/thaw cycles in the soil also affect the sensor's performance. When wet soil freezes, the wave velocity and the cable coupling increase and the attenuation decreases, resulting in greater detection sensitivity. Seasonal sensitivity adjustments may be necessary to compensate for changing ground conditions.

6-106. Although usually buried in soil, ported cables can also be used with asphalt and concrete. If the asphalt or concrete pavement area is relatively small and only a few inches thick (such as a pedestrian pavement crossing the perimeter), the ported cables can be routed under the pavement. However, for the large and deep pavements, slots must be cut into the asphalt or concrete to accept the cable.

6-107. A portable ported-coax sensor is available that can be rapidly deployed and removed. The cables are placed on the surface of the ground rather than buried. This sensor is useful for temporary perimeter detection coverage for small areas or objects (such as vehicles or aircraft).

LOS SENSORS

6-108. The LOS sensors, which are mounted above ground, can be either active or passive. Active sensors generate a beam of energy and detect changes in the received energy that an intruder causes by penetrating the beam. Each sensor consists of a transmitter and a receiver and can be in a monostatic or bistatic configuration. Passive sensors generate no beam of energy; they simply look for changes in the thermal characteristics of their field of view. For effective detection, the terrain within the detection zone must be flat and free of obstacles and vegetation.

Microwave Sensors

6-109. Microwave intrusion-detection sensors are categorized as bistatic or monostatic. Bistatic sensors use transmitting and receiving antennas located at opposite ends of the microwave link, whereas monostatic sensors use the same antenna.

- A bistatic system uses a transmitter and a receiver that are typically separated by 100 to 1,200 feet and that are within direct LOS with each other. The signal picked up by the receiver is the vector sum of the directly transmitted signal and signals that are reflected from the ground and nearby structures. Detection occurs when an object (intruder) moving within the beam pattern causes a change in net-vector summation of the received signals, resulting in variations of signal strength.
  - The same frequency bands allocated by the Federal Communications Commission (FCC) for interior microwave sensors are also used for exterior sensors. Because high-frequency microwave beams are more directive than low-frequency beams and the beam pattern is less affected by blowing grass in the area between the transmitter and the receiver, most exterior sensors operate at the next to highest allowable frequency, 10.525 gigahertz (GHz).
  - The shape of the microwave beam and the maximum separation between the transmitter and the receiver are functions of antenna size and configuration. Various antenna configurations are available, including parabolic-dish arrays, strip-line arrays, and slotted arrays. The parabolic antenna uses a microwave-feed assembly located at the focal point of a metallic parabolic reflector. A conical beam pattern is produced (see Figure 6-16). A strip-line antenna configuration produces a nonsymmetrical beam that is higher than its height. Larger antenna configurations generally produce narrower beam patterns.

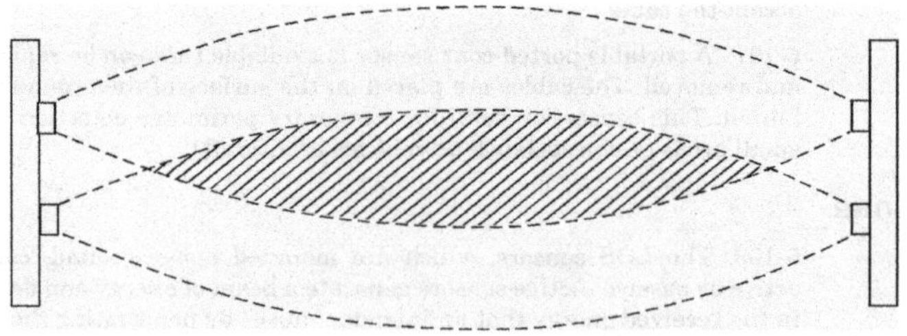

Figure 6-16. Stacked Microwave Configuration

- Monostatic microwave sensors use the same antenna or virtually coincident antenna arrays for the transmitter and receiver, which are usually combined into a single package. Two types of monostatic sensors are available. Amplitude-modulated (AM) sensors detect changes in the net-vector summation of reflected signals similar to bistatic sensors. Frequency-modulated (FM) sensors operate on the Doppler principle similar to interior microwave sensors. The detection pattern is typically shaped like a teardrop (see Figure 6-17). Monostatic sensors can provide volumetric coverage of localized areas, such as in corners or around the base of critical equipment.

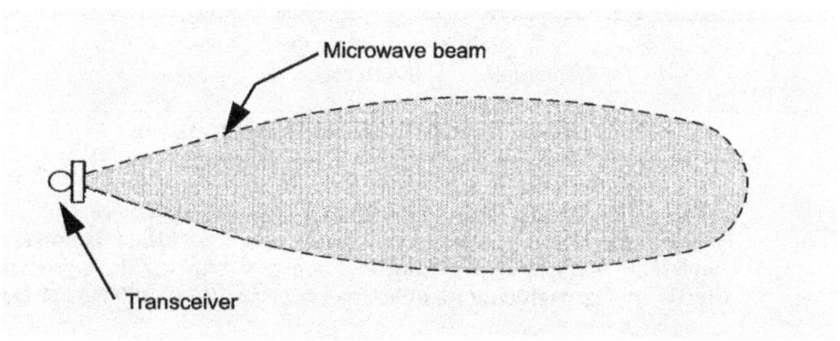

Figure 6-17. Typical Monostatic-Microwave-Sensor Detection Pattern

IR Sensors

6-110. The IR sensors are available in both active and passive models. An active sensor generates one or more near-IR beams that generate an alarm when interrupted. A passive sensor detects changes in thermal IR radiation from objects located within its field of view.

6-111. Active sensors consist of transmitter/receiver pairs. The transmitter contains an IR light source (such as a gallium arsenide light-emitting diode [LED]) that generates an IR beam. The light source is usually modulated to reduce the sensor's susceptibility to unwanted alarms resulting from sunlight or other IR light sources. The receiver detects changes in the signal power of the received beam. To minimize nuisance alarms from birds or blowing debris, the alarm criteria usually require that a high percentage of the beam be blocked for a specific interval of time.

6-112. Active sensors can be single- or multiple-beam systems. Because single-beam sensors can be easily bypassed, multiple-beam systems are generally used in perimeter applications. There are two basic types of multiple-beam configurations—one type uses all transmitters on one post and all receivers on the other post; the second type uses one transmitter and several receivers on each post. Both types are illustrated in Figure 6-18, page 6-38.

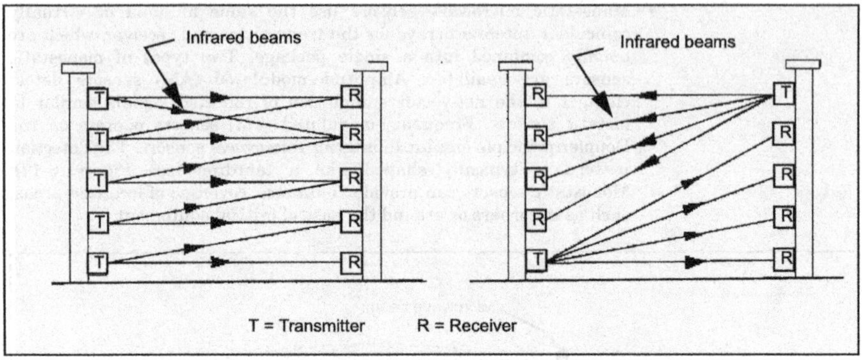

Figure 6-18. Typical IR-Sensor Beam Patterns

6-113. The spacing between transmitters and receivers can be as great as 1,000 feet when operation is under good weather conditions. However, conditions such as heavy rain, fog, snow, or blowing dust particles attenuate the IR energy, reducing its effective range to 100 to 200 feet or less.

VIDEO MOTION SENSORS

6-114. A video motion sensor generates an alarm whenever an intruder enters a selected portion of a CCTV camera's field of view. The sensor processes and compares successive images from the camera and generates an alarm if differences between the images satisfy predefined criteria. Digital devices convert selected portions of the analog video signal into digital data that are compared with data converted previously; if differences exceed preset limits, an alarm is generated.

6-115. The signal processor usually provides an adjustable window that can be positioned anywhere on the video image. Available adjustments permit changing the window's horizontal and vertical sizes, its position, and its sensitivity. More sophisticated units provide several adjustable windows that can be individually sized and positioned. Multiple windows permit concentrating on several specific areas of an image while ignoring others. For example, in a scene that contains several critical assets and multiple sources of nuisance alarms (such as large bushes or trees), the sensor can be adjusted to monitor only the assets and ignore the areas that contain the nuisance-alarm sources.

6-116. The use of video motion-detection systems for exterior applications has been limited, primarily because of difficulties with uncontrolled exterior environments. Lighting variations caused by cloud movement and shadows of slow-moving objects, birds and animals moving within the camera's field of view, camera motion and moving vegetation during windy conditions, and severe weather conditions have traditionally caused a multitude of unwanted alarms in this type of system. Systems using more advanced signal-processing algorithms have improved motion-detection capability and nuisance-alarm rejection; however, they are still subject to high unwanted-alarm rates under certain conditions and should be used with due caution and extreme care.

ELECTRONIC ENTRY CONTROL

6-117. The function of an entry-control system is to ensure that only authorized personnel are permitted into or out of a controlled area. Entry can be controlled by locked fence gates, locked doors to a building or rooms within a building, or specially designed portals.

6-118. These means of entry control can be applied manually by guards or automatically by using entry-control devices. In a manual system, guards verify that a person is authorized to enter an area, usually by comparing the photograph and personal characteristics of the individual requesting entry. In an automated system, the entry-control device verifies that a person is authorized to enter or exit. The automated system usually interfaces with locking mechanisms on doors or gates that open momentarily to permit passage. Mechanical hardware (such as locking mechanisms, electric door strikes, and specially designed portal hardware) and equipment used to detect contraband material (such as metal detectors, X-ray baggage-search systems, explosives detectors, and special nuclear-material monitors) are described in other documentation. Refer to TM 5-853-1 for additional information on determining entry-control requirements and integrating manual electronic-entry control into a cohesive system.

6-119. All entry-control systems control passage by using one or more of three basic techniques—something a person knows, something a person has, or something a person is or does. Automated entry-control devices based on these techniques are grouped into three categories—coded, credential, and biometric devices.

CODED DEVICES

6-120. Coded devices operate on the principle that a person has been issued a code to enter into an entry-control device. This code will match the code stored in the device and permit entry. Depending on the application, a single code can be used by all persons authorized to enter the controlled area or each authorized person can be assigned a unique code. Group codes are useful when the group is small and controls are primarily for keeping out the general public. Individual codes are usually required for control of entry to more critical areas. Coded devices verify the entered code's authenticity, and any person entering a correct code is authorized to enter the controlled area. Electronically coded devices include electronic and computer-controlled keypads.

Electronic Keypad Devices

6-121. The common telephone keypad (12 keys) is an example of an electronic keypad. This type of keypad consists of simple push-button switches that, when depressed, are decoded by digital logic circuits. When the correct sequence of buttons is pushed, an electric signal unlocks the door for a few seconds.

Computer-Controlled Keypad Devices

6-122. These devices are similar to electronic keypad devices, except they are equipped with a microprocessor in the keypad or in a separate enclosure at a different location. The microprocessor monitors the sequence in which the keys are depressed and may provide additional functions such as personal ID and digit scrambling. When the correct code is entered and all conditions are satisfied, an electric signal unlocks the door.

CREDENTIAL DEVICES

6-123. A credential device identifies a person having legitimate authority to enter a controlled area. A coded credential (plastic card or key) contains a prerecorded, machine-readable code. An electric signal unlocks the door if the prerecorded code matches the code stored in the system when the card is read. Like coded devices, credential devices only authenticate the credential; it assumes a user with an acceptable credential is authorized to enter. Various technologies are used to store the code upon or within a card. Hollerith, optically coded, magnetic-spot, capacitance, and electric-circuit cards are seldom used and will not be discussed here. The most commonly used types of cards are described as follows:

Magnetic-Stripe Card

6-124. A strip of magnetic material located along one edge of the card is encoded with data (sometimes encrypted). The data is read by moving the card past a magnetic read head.

Wiegand-Effect Card

6-125. The Wiegand-effect card contains a series of small-diameter, parallel wires about one-half inch long, embedded in the bottom half of the card. The wires are manufactured from ferromagnetic materials that produce a sharp change in magnetic flux when exposed to a slowly changing magnetic field. This type of card is impervious to accidental erasure. The card reader contains a small read head and a tiny magnet to supply the applied magnetic field. It usually does not require external power.

Proximity Card

6-126. A proximity card is not physically inserted into a reader; the coded pattern on the card is sensed when it is brought within several inches of the reader. Several techniques are used to code cards. One technique uses a number of electrically tuned circuits embedded in the card. Data are encoded by varying resonant frequencies of the tuned circuits. The reader contains a transmitter that continually sweeps through a specified range of frequencies and a receiver that senses the pattern of resonant frequencies contained in the card. Another technique uses an integrated circuit embedded in the card to generate a code that can be magnetically or electrostatically coupled to the reader. The power required to activate embedded circuitry can be provided by a small battery embedded in the card or by magnetically coupling power from the reader.

Laser Card

6-127. The optical memory card, commonly called the laser card, uses the same technology developed for recording video and audio disks for entertainment purposes. Data is recorded on the card by burning a microscopic hole (using a laser) in a thin film covering the card. Data is read by using a laser to sense the hole locations. The typical laser card can hold several megabytes of user data.

Smart Card

6-128. A smart card is embedded with a microprocessor, memory, communication circuitry, and a battery. The card contains edge contacts that enable a reader to communicate with the microprocessor. Entry-control information and other data may be stored in the microprocessor's memory.

Bar Code

6-129. A bar code consists of black bars printed on white paper or tape that can be easily read with an optical scanner. This type of coding is not widely used for entry-control applications because it can be easily duplicated. It is possible to conceal the code by applying an opaque mask over it. In this approach, an IR scanner is used to interpret the printed code. For low-level security areas, the use of bar codes can provide a cost-effective solution for entry control. Coded strips and opaque masks can be attached to existing ID badges, alleviating the need for complete badge replacement.

BIOMETRIC DEVICES

6-130. The third basic technique used to control entry is based on the measurement of one or more physical or personal characteristics of an individual. Because most entry-control devices based on this technique rely on measurements of biological characteristics, they have become commonly known as biometric devices. Characteristics such as fingerprints, hand geometry, voiceprints, handwriting, and retinal blood-vessel patterns have been used for controlling entry. Typically, in enrolling individuals, several reference measurements are made of the selected characteristic and then stored in the device's memory or on a card. From then on, when that person attempts entry, a scan of the characteristic is compared with the reference data template. If a match is found, entry is granted. Rather then verifying an artifact, such as a code or a credential, biometric devices verify a person's physical characteristic, thus providing a form of identity verification. Because of this, biometric devices are sometimes referred to as personnel identity-verification devices. The most common biometric devices are discussed below.

Fingerprints

6-131. Fingerprint-verification devices use one of two approaches. One is pattern recognition of the whorls, loops, and tilts of the referenced fingerprint, which is stored in a digitized representation of the image and compared with the fingerprint of the prospective entrant. The second approach is minutiae comparison, which means that the endings and branching points of ridges and valleys of the referenced fingerprint are compared with the fingerprint of the prospective entrant.

Hand Geometry

6-132. Several devices are available that use hand geometry for personnel verification. These devices use a variety of physical measurements of the hand, such as finger length, finger curvature, hand width, webbing between fingers, and light transmissivity through the skin to verify identity. Both two- and three-dimensional units are available.

Retinal Patterns

6-133. This type of technique is based on the premise that the pattern of blood vessels on the human eye's retina is unique to an individual. While the eye is focused on a visual target, a low-intensity IR light beam scans a circular area of the retina. The amount of light reflected from the eye is recorded as the beam progresses around the circular path. Reflected light is modulated by the difference in reflectivity between blood-vessel pattern and adjacent tissue. This information is processed and converted to a digital template that is stored as the eye's signature. Users are allowed to wear contact lenses; however, glasses should be removed.

Device Combinations

6-134. Frequently, an automated entry-control system uses combinations of the three types of entry-control devices. Combining two different devices can significantly enhance the system's security level. In some cases, combining devices results in reduced verification times.

APPLICATION GUIDELINES

6-135. The primary function of an automated entry-control system is to permit authorized personnel to enter or exit a controlled area. Features available to the designer are described below.

- **Enrollment.** All entry-control systems must provide a means of entering, updating, and deleting information about authorized individuals into the system's database files. This is usually accomplished with a dedicated enrollment station for enrolling and disenrolling purposes that is directly connected to the central-processing unit. When credential devices are used, all authorized users must be provided with an appropriate credential. A means should also be provided to disenroll a person quickly without having to retrieve a credential. When using biometric devices, additional enrollment equipment will be required.
- **Entry-control techniques.** Some entry-control functions require additional hardware, while others are accomplished with software. Those features accomplished with software require that the appropriate database be available for every portal affected by them. Typically, these techniques include—
  - Area zones.
  - Time zones.
  - Team zones.
  - Anti-pass back.

  - Antitailgate.
  - Guard tour.
  - Elevator control.
- **Alarms.** Several types of alarms can be used with an entry-control system. These alarms must annunciate audibly and visually in the security center.
- **Entry denial.** Most entry-control devices are configured to permit the user three entry attempts. If more than three unsuccessful entry attempts are made within a specified period, the device generates an alarm. An alarm is also generated if an invalid credential is used or attempted entries are detected that violate specified area, time, or team zoning requirements.
- **Communication failure.** This alarm is generated when a loss of communication between the central processor and the local equipment is detected.
- **Portal open.** If a portal door remains open longer than a predefined time, an alarm is generated.
- **Duress.** This alarm is generated when a special duress code is entered at a keypad.
- **Guard overdue.** This duress alarm is generated when a security guard is determined to be overdue at a checkpoint during a predefined guard tour.
- **Software tamper.** This type of alarm is generated when unauthorized persons are detected attempting to invoke certain system commands or modify database files.

PERFORMANCE CRITERIA

6-136. The overall performance of an entry-control system can be evaluated by examining the verification error rate and the throughput rate. An entry-control system can produce two types of errors—denial of admission of a person who should be admitted or admission of a person who should not be admitted. These are commonly referred to as false-reject errors (type I errors) and false-accept errors (type II errors). Although a false-reject error does not constitute a breach of security, it does create an operational problem that must be handled by an alternative method. False-accept errors constitute a breach of security. Ideally, both false-reject and false-accept error rates should be zero; in practice, however, they are not. In fact, they tend to act in opposition to each other. When the system is adjusted to minimize the false-accept error rate, the false-reject error rate usually increases. Verification error rates are typically measured in percent (number of errors/number of attempts x 100 percent). These error rates are typically very low for coded and credential devices, but many become significant if biometric devices are used.

6-137. The throughput rate is the number of persons that can pass through an entry point in a given unit of time and is usually expressed in persons per minute. It is the time required to approach the entry-control device and for the device to verify information (verification time) and the time required passing through the entry point. Typically, an individual can approach the device and pass through in 3 to 5 seconds. Verification time depends on the type of device and may vary from 3 to 15 seconds. Table 6-5 provides a list of typical verification times for different types of entry-control devices.

Table 6-5. Typical Verification Times of Entry-Control Devices

Device	Verification Time
Keypad	3 seconds
Card reader	3 seconds
Keypad/card reader	6 seconds
Biometric/keypad	6 to 15 seconds
Biometric/card reader	6 to 15 seconds
Biometric	2 minutes

DATA TRANSMISSION

6-138. A critical element in an integrated ESS is the DTM that transmits information from sensors, entry-control devices, and video components to display and assessment equipment. A DTM link is a path for transmitting data between two or more components (such as a sensor and alarm reporting system, a card reader and controller, a CCTV camera and monitor, or a transmitter and receiver). The DTM links connect remote ESS components to the security center. An effective DTM link ensures rapid and reliable transmission media, transmission technique, associated transmission hardware, and degree of security to be provided for the communication system.

6-139. A number of different media are used in transmitting data between elements of an IDS, an EECS, and a CCTV system. These include wire lines, coaxial cable, fiber-optic cable, and RF transmission.

- **Wire line.** Wire lines are twisted pairs that consist of two insulated conductors twisted together to minimize interference by unwanted signals.
- **Coaxial cable.** Coaxial cable consists of a center conductor surrounded by a shield. The center conductor is separated from the shield by a dielectric. The shield protects against electromagnetic interference.
- **Fiber optics.** Fiber optics uses the wide bandwidth properties of light traveling through transparent fibers. Fiber optics is a reliable communication medium best suited for point-to-point, high-speed data transmission. Fiber optics is immune to RF electromagnetic interference and does not produce electromagnetic radiation emission. The preferred DTM for an ESS is fiber-optic cables unless there are justifiable economic or technical reasons for using other types of media.
- **RF transmission.** Modulated RF can be used as a DTM with the installation of radio receivers and transmitters. An RF transmission system does not require a direct physical link between the points of communication, and it is useful for communicating over barriers such as bodies of water and heavily forested terrain. A disadvantage is that the signal power received depends on many factors (including transmission power, antenna pattern, path length, physical

obstructions, and climatic conditions). Also, RF transmission is susceptible to jamming and an adversary with an appropriately tuned receiver has access to it. The use of RF will be coordinated with the communications officer to avoid interference with other existing or planned facility RF systems.

6-140. There are two basic types of communication links—point-to-point and multiplex lines. A point-to-point link is characterized by a separate path for each pair of components. This approach is cost effective for several component pairs or when a number of scattered remote areas communicate with a single central location. The multiplex link, commonly referred to as a multidrop or multipoint link, is a path shared by a number of components. Depending on the number and location of components, this type of configuration can reduce the amount of cabling required. However, the cost reduction from reduced cabling is somewhat offset by costs of equipment required to multiplex and demultiplex data.

6-141. Data links used to communicate the status of ESS devices or other sensitive information to the security center must be protected from possible compromise. Attempts to defeat the security system may range from simple efforts to cut or short the transmission line to more sophisticated undertakings, such as tapping and substituting bogus signals. Data links can be made more secure by physical protection, tamper protection, line supervision, and encryption.

CCTV FOR ALARM ASSESSMENT AND SURVEILLANCE

6-142. A properly integrated CCTV assessment system provides a rapid and cost-effective method for determining the cause of intrusion alarms. For surveillance, a properly designed CCTV system provides a cost-effective supplement to guard patrols. For large facilities, the cost of a CCTV system is more easily justified. It is important to recognize that CCTV alarm-assessment systems and CCTV surveillance systems perform separate and distinct functions. The alarm-assessment system is designed to respond rapidly, automatically, and predictably to the receipt of ESS alarms at the security center. The surveillance system is designed to be used at the discretion of and under the control of the security center's console operator. When the primary function of the CCTV system is to provide real-time alarm assessment, the design should incorporate a video-processing system that can communicate with the alarm-processing system.

6-143. A candidate site for a CCTV assessment system will typically have the following characteristics:

- Assets requiring ESS protection.
- A need for real-time alarm assessment.
- Protected assets spaced some distance apart.

6-144. Figure 6-19, page 6-46, shows a typical CCTV system configuration. A typical site will locate CCTV cameras—

- Outdoors, along site-perimeter isolation zones.
- Outdoors, at controlled access points (sally ports).

- Outdoors, within the protected area, and at viewing approaches to selected assets.
- Indoors, at selected assets within the protected area.

6-145. The security console is centrally located in the security center. The CCTV monitors and the ancillary video equipment will be located at this console, as will the ESS alarm-processing and -annunciation equipment.

CCTV CAMERA COMPONENTS

6-146. An optical-lens system that captures and focuses reflected light from the scene being viewed onto an image target is common to all CCTV cameras. The image target converts reflected light energy into electrical impulses in a two-dimensional array of height and width. An electronic scanning system (reading these impulses in a predetermined order) creates a time-sensitive voltage signal that is a replica of optical information captured by the lens and focused on the target. This voltage signal is then transmitted to a location where it is viewed and possibly recorded. For components and technical information regarding CCTV cameras, see the appropriate TMs.

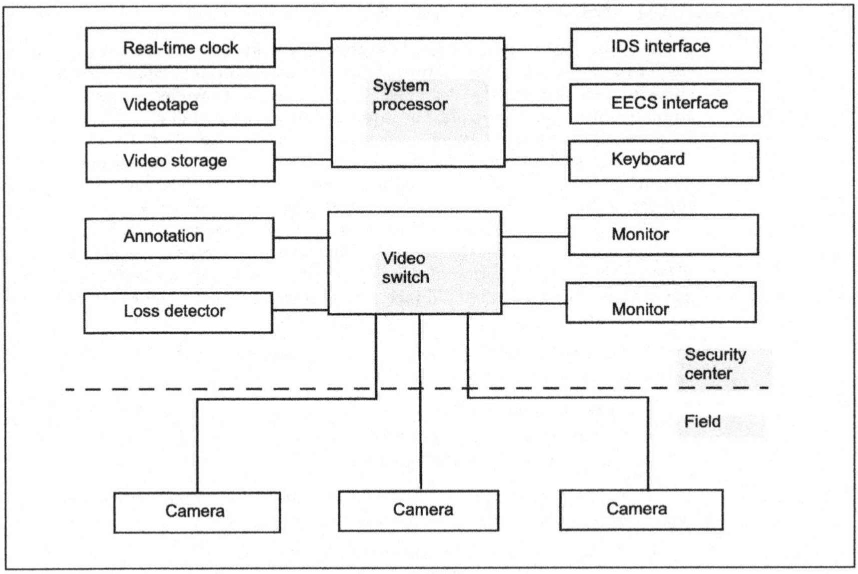

Figure 6-19. Typical CCTV System

VIDEO SIGNAL AND CONTROL LINKS

6-147. A CCTV transmission system is needed to convey video signals from various facility cameras to the security center and to carry commands from the security center to the cameras. Information may be sent via metallic cable, RF, or optical transmission.

Metallic Cable

6-148. Metallic video cables are electrical conductors manufactured specifically for the transmission of frequencies associated with video components. Coaxial cable is a primary example of this type of transmission media. Devices such as video-equalization amplifiers, ground loop correctors, and video-distribution amplifiers may be required.

RF Transmission

6-149. For a system that has widely separated nodes, RF transmission may be a good alternative to metallic cable and associated amplifiers. The information can be transmitted over a microwave link. A microwave link can be used for distances of about 50 miles, as long as the receiver and the transmitter are in the LOS.

Fiber-Optic Cable

6-150. In fiber-optic cable systems, electrical video signals are converted to optical light signals that are transmitted down the optical fiber. The signal is received and reconverted into electrical energy. An optic driver and a receiver are required per fiber. The fiber-optic transmission method provides a low-loss, high-resolution transmission system with usable length three to ten times that of traditional metallic in cable systems. Fiber-optic cable is the transmission media favored by DA.

CCTV-SYSTEM SYNCHRONIZATION

6-151. Timing signals are processed within the image-scan section of the CCTV camera. These signals may be generated internally from a crystal clock, derived from the camera's AC power source, or supplied by an external signal source. The camera should be capable of automatic switchover to its internal clock in case of external signal loss. When CCTV cameras are supplied by a common external (master) signal source or are all powered from the same AC power source, all cameras scan in synchronism. In this case, a console CCTV monitor will display a smooth transition when switched from one video source to another. Without this feature, the monitor display breaks up or rolls when switched between video sources. The rolling occurs for as long as it takes the monitor to synchronize its scan with that of the new video source, typically one second. The resynchronization delay will be experienced by all system components that receive video information, including recorders. To avoid this delay, the designer must specify that all cameras are powered from the AC power phase or must specify master synchronization for the design.

VIDEO PROCESSING AND DISPLAY COMPONENTS

6-152. As shown in Figure 6-19, page 6-46, CCTV camera signals propagate through the video transmission system and through coverage at the security center. In very simple configurations with only a few cameras and monitors, a hardwired connection between each camera and console monitor is adequate. As the number of cameras increases, the need to manage and add supplemental information to camera signals also increases. Psychological testing has demonstrated that the efficiency of console-operator assessment improves as the number of console monitors is reduced, with the optimum number being four to six monitors. Effectiveness is also enhanced by the use of alarm-correlated video. Major components of the video-processor system are the video switcher, the video-loss detector, the alarm-processor communication path, the master video-sync generator, video recorders, and monitors.

- **Video switchers.** Video switchers are required when the number of cameras exceeds the number of console monitors or when a monitor must be capable of selecting video from one of many sources. Video switchers are capable of presenting any of multiple video images to various monitors, recorders, and so forth.
- **Video-loss detector.** Video-loss detectors sense the continued integrity of incoming camera signals.
- **ESS interface and communication path.** There must be a means of rapid communication between the ESS alarm-annunciation and video-processor systems. The alarm processor must send commands that cause the video switcher to select the camera appropriate for the sensor reporting an alarm. The video-processor system must report system tampering or failures (such as loss of video) to the alarm processor. The path should also pass date-and-time synchronizing information between processors so that recorded video scenes and printed alarm logs are properly correlated.
- **Master video-sync generation and distribution.** Master video sync includes a crystal-controlled timing generator, distribution amplifiers, and a transmission link to each camera.
- **Video recorders.** Video recorders provide the means to record alarm-event scenes in real time for later analysis. A recorder typically receives its input through dedicated video-switcher outputs. To support recorder playback, the recorder output is connected to a dedicated switcher input and must be compatible with the switcher-signal format. In addition, the recorder receives start commands from the switcher, and compatibility must exist at this interface. Videocassette recorders should be used when alarm events are to be recorded for later playback and analysis. The cassettes can record in time lapse for up to 240 hours (depending on the user-selected speed) and will change to real-time recording on command. The cassettes can be erased and reused or archived if required.
- **Monitors.** Monitors are required to display the individual scenes transmitted from the cameras or from the video switcher. In alarm-assessment applications, the monitors are driven by dedicated outputs of the video switcher and the monitors display video sources selected by the switcher. For security-console operations, the 9-inch monitor is the

smallest screen that should be used for operator recognition of small objects in a camera's field of view. Two 9-inch monitors can be housed side by side in a standard 19-inch console. If the monitors are to be mounted in freestanding racks behind the security console, larger units will be used.

6-153. Video-processor equipment will be specified to append the following alphanumeric information so that it appears on both monitors and recordings. The equipment must allow the operator to program the annotated information and dictate its position on the screen. This information includes—

- Time and date information.
- Video-source or alarm-zone identification.
- Programmable titles.

CCTV APPLICATION GUIDELINES

6-154. Site-specific factors must be taken into consideration in selecting components that comprise a particular CCTV system. The first is the system's size in terms of the number of cameras fielded, which is the minimum number needed to view all ESS sensor-detection fields and surveillance cameras. Another factor is that some CCTV cameras may require artificial light sources. Finally, there are CCTV-system performance criteria and physical, environmental, and economic considerations. Each is discussed in detail in TM 5-853-4.

Scene Resolution

6-155. The level to which video details can be determined in a CCTV scene is referred to as resolving ability or resolution. It is generally accepted that for assessment purposes, three resolution requirements can be defined. In order of increasing resolution requirements, they are detection, recognition, and identification.

- Detection is the ability to detect the presence of an object in a CCTV scene.
- Recognition is the ability to determine the type of object in a CCTV scene (animal, blowing debris, or crawling human).
- Identification is the ability to determine object details (a particular person, a large rabbit, a small deer, or tumbleweed).

6-156. A CCTV assessment system should provide sufficient resolution to recognize human presence and to detect small animals or blowing debris. Given an alarmed intrusion sensor, it is crucial that the console operator be able to determine if the sensor detected an intruder or if it is simply responding to a nuisance condition. (Refer to TM 5-853-4 for detailed design applications.)

Illumination Levels

6-157. For interior applications where the same camera type is used in several different areas and the scene illumination in each area is constant, specify the manually adjustable iris. This allows a manual iris adjustment appropriate for each particular area's illumination level at the time of installation. If the camera must operate in an area subject to a wide dynamic range of illumination levels (such as would be found outdoors), specify the automatically adjusted iris feature.

Cost Considerations

6-158. The cost of a CCTV system is usually quoted as cost-per-assessment zone. When estimating the total system cost, video-processor equipment costs and the video-transmission system's costs must be included. Other potentially significant costs are outdoor lighting system upgrades and the site preparation required to support the CCTV cameras. The CCTV systems are expensive compared to other electronic security subsystems and should be specified with discretion.

DESIGN GUIDELINES

6-159. The design and application of CCTV systems are quite complex and should be left to professionals who are abreast of the current state-of-the-art systems. Some of the general design guidelines include the following:

- **System familiarity.** Before designing an effective CCTV assessment system, the designer must be familiar with the ESS's sensor placement and the detection field's shape.
- **CCTV camera placement and lighting.** The placement of exterior cameras requires more attention than that of interior cameras because of weather and illumination extremes. The field-of-view alignment, illumination range, and balanced lighting are major design factors. Exterior CCTV design considerations include environmental housings, camera mounting heights, system types, and so forth. Indoor design considerations include the mounting location and tamper detection. The layout for indoor alarm-assessment cameras is subject to three constraints—
  - The camera's location should enclose the complete sensor detection field in the camera's field of view.
  - Lighting that is adequate to support alarm assessment will be provided.
  - Protection from tampering and inadvertent damage by collision during normal area operations will be provided.

Chapter 7

Access Control

Perimeter barriers, intrusion-detection devices, and protective lighting provide physical-security safeguards; however, they alone are not enough. An access-control system must be established and maintained to preclude unauthorized entry. Effective access-control procedures prevent the introduction of harmful devices, materiel, and components. They minimize the misappropriation, pilferage, or compromise of materiel or recorded information by controlling packages, materiel, and property movement. Access-control rosters, personal recognition, ID cards, badge-exchange procedures, and personnel escorts all contribute to an effective access-control system.

DESIGNATED RESTRICTED AREAS

7-1. The installation commander is responsible for designating and establishing restricted areas. A restricted area is any area that is subject to special restrictions or controls for security reasons. This does not include areas over which aircraft flight is restricted. Restricted areas may be established for the following:

- The enforcement of security measures and the exclusion of unauthorized personnel.
- Intensified controls in areas requiring special protection.
- The protection of classified information or critical equipment or materials.

DEGREE OF SECURITY

7-2. The degree of security and control required depends on the nature, sensitivity, or importance of the security interest. Restricted areas are classified as controlled, limited, or exclusion areas.

- A controlled area is that portion of a restricted area usually near or surrounding a limited or exclusion area. Entry to the controlled area is restricted to personnel with a need for access. Movement of authorized personnel within this area is not necessarily controlled since mere entry to the area does not provide access to the security interest. The controlled area is provided for administrative control, for safety, or as a buffer zone for in-depth security for the limited or exclusion area. The commander establishes the control of movement.
- A limited area is a restricted area within close proximity of a security interest. Uncontrolled movement may permit access to the item. Escorts and other internal restrictions may prevent access within limited areas.

- An exclusion area is a restricted area containing a security interest. Uncontrolled movement permits direct access to the item.

7-3. The security protection afforded by a restricted area pertains particularly to subversive-activity control; that is, protection against espionage, sabotage, or any such action adversely affecting national defense. Within this context, the designation "restricted area" is not applicable to an area solely for protection against common pilferage or misappropriation of property or material that is not classified or not essential to national defense. For example, an area devoted to the storage or use of classified documents, equipment, or materials should be designated as a restricted area to safeguard against espionage. An installation communications center should also be so designated to safeguard against sabotage. On the other hand, a cashier's cage or an ordinary mechanic's tool room should not be so designated, although the commander may impose controls to access. This may be a simple matter of posting an "off limits to unauthorized personnel" sign. The PM or the physical-security manager acts as an advisor to the commander. In his recommendations, he must consider evaluating the purpose of designating a restricted area and coordinating with the intelligence officer and the staff judge advocate (SJA).

7-4. A restricted area must be designated in writing by the commander and must be posted with warning signs according to AR 190-13. In areas where English is one of two or more languages commonly spoken, warning signs will be posted in English and in the local language (see Figure 7-1).

7-5. An installation may have varying degrees of security. It may be designated in its entirety as a restricted area, with no further restrictions; or it may be subdivided into controlled, limited, or exclusion areas with restrictions of movement and specific clear zones. Figure 7-2 depicts a simplified restricted area and the degrees of security.

CONSIDERATIONS

7-6. There are other important considerations concerning restricted areas and their lines of division. These considerations include the following:

- A survey and analysis of the installation, its missions, and its security interests. This can determine immediate and anticipated needs that require protection. Anticipated needs are determined from plans for the future.
- The size and nature of the security interest being protected. Safes may provide adequate protection for classified documents and small items; however, large items may have to be placed within guarded enclosures.
- Some security interests are more sensitive to compromise than others. Brief observation or a simple act by an untrained person may constitute a compromise in some cases. In others, detailed study and planned action by an expert may be required.
- All security interests should be evaluated according to their importance. This may be indicated by a security classification such as confidential, secret, or top secret.

RESTRICTED AREA

THIS (INSTALLATION OR ACTIVITY) HAS BEEN DECLARED A RESTRICTED AREA BY AUTHORITY OF (TITLE, COMMANDING GENERAL OR COMMANDING OFFICER) IN ACCORDANCE WITH THE PROVISIONS OF THE DIRECTIVE ISSUED BY THE SECRETARY OF DEFENSE ON 20 AUGUST 1954, PURSUANT TO THE PROVISIONS OF SECTION 21, INTERNAL SECURITY ACT OF 1950. UNATHORIZED ENTRY IS PROHIBITED.

ALL PERSONS AND VEHICLES ENTERING HEREIN ARE LIABLE TO SEARCH. PHOTOGRAPHING OR MAKING NOTES, DRAWINGS, MAPS, OR GRAPHIC REPRESENTATIVES OF THIS AREA OR ITS ACTIVITIES ARE PROHIBITED UNLESS SPECIFICALLY AUTHORIZED BY THE COMMANDER. ANY SUCH MATERIAL FOUND IN THE POSSESSION OF UNAUTHORIZED PERSONS WILL BE CONFISCATED.

Figure 7-1. Sample Restricted-Area Warning

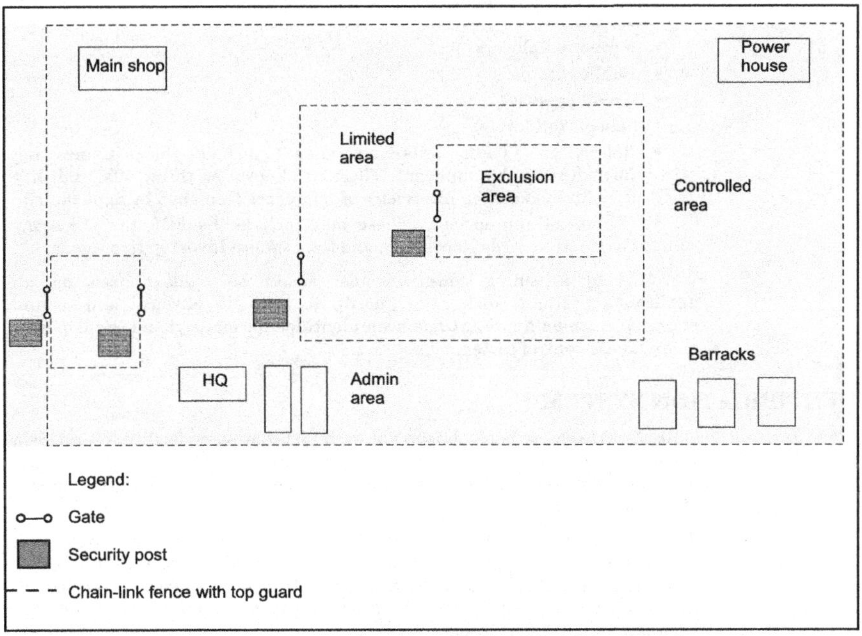

Figure 7-2. Schematic Diagram of a Simplified Restricted Area and the Degrees of Security

- Parking areas for privately owned vehicles (POVs) are established outside of restricted areas. Vehicle entrances must be kept at a minimum for safe and efficient control.
- Physical protective measures (such as fences, gates, and window bars) must be installed.

EMPLOYEE SCREENING

7-7. Screening job applicants to eliminate potential acts of espionage and sabotage and other security risks is important in peacetime and is critical during a national emergency. Personnel screenings must be incorporated into standard personnel policies.

7-8. An applicant should be required to complete a personnel security questionnaire, which is then screened for completeness and used to eliminate undesirable applicants. A careful investigation should be conducted to ensure that the applicant's character, associations, and suitability for employment are satisfactory. The following sources may be helpful in securing employment investigative data:

- State and local police (including national and local police in overseas areas).
- Former employers.
- Public records.
- Credit agencies.
- Schools (all levels).
- References. (These references should include those names not furnished by the applicant. These are known as throw offs, and they are obtained during interviews of references furnished by applicants.)
- Others as appropriate. (These may include the FBI, the US Army Criminal Records Repository, and the Defense Investigative Agency).

7-9. Medical screening considerations should be made (based on an applicant's position [such as a guard]) to evaluate physical and mental stamina. Once an applicant has been identified for employment, he is placed on an access-control roster.

IDENTIFICATION SYSTEM

7-10. An ID system is established at each installation or facility to provide a method of identifying personnel. The system provides for personal recognition and the use of security ID cards or badges to aid in the control and movement of personnel activities.

7-11. Standard ID cards are generally acceptable for access into areas that are unrestricted and have no security interest. Personnel requiring access to restricted areas should be issued a security ID card or badge as prescribed in AR 600-8-14. The card's/badge's design must be simple and provide for adequate control of personnel.

7-12. A security ID card/badge system must be established for restricted areas with 30 or more employees per shift. Commanders may (at their discretion) authorize a card/badge system in restricted areas for less than 30 people.

ID METHODS

7-13. Four of the most commonly used access-control ID methods are the personal-recognition system, the single-card or -badge system, the card- or badge-exchange system, and the multiple-card or -badge system.

Personal-Recognition System

7-14. The personal-recognition system is the simplest of all systems. A member of the security force providing access control visually checks the person requesting entry. Entry is granted based on—

- The individual being recognized.
- The need to enter has been established.
- The person is on an access-control roster.

Single-Card or -Badge System

7-15. This system reflects permission to enter specific areas by the badge depicting specific letters, numbers, or particular colors. This system lends to comparatively loose control and is not recommended for high-security areas. Permission to enter specific areas does not always go with the need to know. Because the ID cards/badges frequently remain in the bearer's possession while off duty, it affords the opportunity for alteration or duplication.

Card- or Badge-Exchange System

7-16. In this system, two cards/badges contain identical photographs. Each card/badge has a different background color, or one card/badge has an overprint. One card/badge is presented at the entrance to a specific area and exchanged for the second card/badge, which is worn or carried while in that area. Individual possession of the second card/badge occurs only while the bearer is in the area for which it was issued. When leaving the area, the second card/badge is returned and maintained in the security area. This method provides a greater degree of security and decreases the possibility of forgery, alteration, or duplication of the card/badge. The levels of protection described in TM 5-853-1 require multiple access-control elements as the levels of protection increase. In the case of the badge exchange, this system counts as two access-control elements.

Multiple-Card or -Badge System

7-17. This system provides the greatest degree of security. Instead of having specific markings on the cards/badges denoting permission to enter various restricted areas, the multiple card/badge system makes an exchange at the entrance to each security area. The card/badge information is identical and allows for comparisons. Exchange cards/badges are maintained at each area only for individuals who have access to the specific area.

MECHANIZED/AUTOMATED SYSTEMS

7-18. An alternative to using guards or military police (MP) to visually check cards/badges and access rosters is to use building card-access systems or biometric-access readers. These systems can control the flow of personnel entering and exiting a complex. Included in these systems are—

- Coded devices such as mechanical or electronic keypads or combination locks.
- Credential devices such as magnetic-strip or proximity card readers.
- Biometric devices such as fingerprint readers or retina scanners.

7-19. Access-control and ID systems base their judgment factor on a remote capability through a routine discriminating device for positive ID. These systems do not require guards at entry points; they identify an individual in the following manner:

- The system receives physical ID data from an individual.
- The data is encoded and compared to stored information.
- The system determines whether access is authorized.
- The information is translated into readable results.

7-20. Specialized mechanical systems are ideal for highly sensitive situations because they use a controlled process in a controlled environment to establish the required database and accuracy. One innovative technique applied to ID and admittance procedures involves dimension comparisons. The dimension of a person's full hand is compared to previously stored data to determine entry authorization. Other specialized machine readers can scan a single fingerprint or an eye retina and provide positive ID of anyone attempting entry.

7-21. An all-inclusive automated ID and access-control system reinforces the security in-depth ring through its easy and rapid change capability. The computer is able to do this through its memory. Changes can be made quickly by the system's administrator.

7-22. The commercial security market has a wide range of mechanized and automated hardware and software systems. Automated equipment is chosen only after considering the security needs and the environment in which it operates. These considerations include whether the equipment is outdoors or indoors, the temperature range, and weather conditions. Assessment of security needs and the use of planning, programming, and budgeting procedures greatly assist a security manager in improving the security posture.

CARD/BADGE SPECIFICATIONS

7-23. Security cards/badges should be designed and constructed to meet the requirements of AR 600-8-14. Upon issuing a card/badge, security personnel must explain to the bearer the wear required and the authorizations allowed with the card/badge. This includes—

- Designation of the areas where an ID card/badge is required.

- A description of the type of card/badge in use and the authorizations and limitations placed on the bearer.
- The required presentation of the card/badge when entering or leaving each area during all hours of the day.
- Details of when, where, and how the card/badge should be worn, displayed, or carried.
- Procedures to follow in case of loss or damage of the card.
- The disposition of the card/badge upon termination of employment, investigations, or personnel actions.
- Prerequisites for reissuing the card/badge.

VISITOR IDENTIFICATION AND CONTROL

7-24. Procedures must be implemented to properly identify and control personnel. This includes visitors presenting their cards/badges to guards at entrances of restricted areas. Visitors are required to stay with their assigned escort. Guards must ensure that visitors stay in areas relating to their visit; an uncontrolled visitor, although conspicuously identified, could acquire information for which he is not authorized. Foreign-national visitors should be escorted at all times.

7-25. Approval for visitors should be obtained at least 24 hours in advance (if possible). Where appropriate, the installation should prepare an agenda for the visitor and designate an escort officer. Measures must be in place to recover visitor cards/badges on the visit's expiration or when they are no longer required.

7-26. Physical-security precautions against pilferage, espionage, and sabotage require the screening, ID, and control of visitors. Further information about visiting requirements and procedures are found in ARs 12-15 and 381-20. Visitors are generally classed in the following categories:

- Persons with whom every installation or facility has business (such as suppliers, customers, insurance inspectors, and government inspectors).
- Individuals or groups who desire to visit an installation or facility for personal or educational reasons. Such visits may be desired by educational, technical, or scientific organizations.
- Individuals or groups specifically sponsored by the government (such as foreign nationals visiting under technical cooperation programs and similar visits by US nationals). Requests for visits by foreign nationals must be processed according to AR 380-10.
- Guided tours to selected portions of the installation in the interest of public relations.

7-27. The ID and control mechanisms for visitors must be in place. They may include the following:

- Methods of establishing the authority for admitting visitors and any limitations relative to access.

- Positive ID of visitors by personal recognition, visitor permit, or other identifying credentials. Contact the employer, supervisor, or officer in charge to validate the visit.
- The use of visitor registration forms. These forms provide a record of the visitor and the time, location, and duration of his visit.
- The use of visitor ID cards/badges. The cards/badges bear serial numbers, the area or areas to which access is authorized, the bearer's name, and escort requirements.

7-28. Individual groups entering a restricted area must meet specific prerequisites before being granted access. The following guidance is for group access into a restricted area:

Visitors

7-29. Before allowing visitors into a restricted area, contact the person or activity being visited. After verifying the visitor's identity, issue a badge, complete the registration forms, and assign an escort (if required). Visitors may include public-utility and commercial-service representatives.

Very Important Persons

7-30. The procedures for admitting very important persons (VIPs) and foreign nationals into restricted areas are contained in AR 12-15. Special considerations and coordination with the protocol office are necessary. A 24-hour advance notice is desirable for these requests, along with an agenda for the visit and the designation of an escort, if appropriate.

Civilians Working on Jobs Under Government Contract

7-31. To allow these personnel to conduct business in restricted areas, the security manager must coordinate with the procurement office. The security manager must also identify movement-control procedures for these employees.

Cleaning Teams

7-32. Supervisors using cleaning teams must seek technical advice from the physical-security office on internal controls for each specific building. This may include providing escorts.

DOD Employees in Work Areas After Normal Operating Hours

7-33. Supervisors establish internal controls based on coordination with the security manager. They also notify security personnel of the workers' presence, type, and duration of work.

ENFORCEMENT MEASURES

7-34. The most vulnerable link in any ID system is its enforcement. Security forces must be proactive in performing their duties. A routine performance of duty will adversely effect even the most elaborate system. Positive enforcement measures must be prescribed to enhance security. Some of these measures may include—

- Designating alert and tactful security personnel at entry control points.
- Ensuring that personnel possess quick perception and good judgment.
- Requiring entry-control personnel to conduct frequent irregular checks of their assigned areas.
- Formalizing standard procedures for conducting guard mounts and posting and relieving security personnel. These measures will prevent posting of unqualified personnel and a routine performance of duty.
- Prescribing a uniform method of handling or wearing security ID cards/badges. If carried on the person, the card must be removed from the wallet (or other holder) and handed to security personnel. When worn, the badge will be worn in a conspicuous position to expedite inspection and recognition from a distance.
- Designing entry and exit control points of restricted areas to force personnel to pass in a single file in front of security personnel. In some instances, the use of turnstiles may be advisable to assist in maintaining positive control.
- Providing lighting at control points. The lighting must illuminate the area to enable security personnel to compare the bearer with the ID card/badge.
- Enforcing access-control measures by educating security forces and employees. Enforcement of access-control systems rests primarily with the security forces; however, it is essential that they have the full cooperation of the employees. Employees must be instructed to consider each unidentified or improperly identified individual as a trespasser. In restricted areas where access is limited to a particular zone, employees must report unauthorized individuals to the security force.
- Positioning ID card/badge racks or containers at entry control points so that they are accessible only to guard-force personnel.
- Appointing a responsible custodian to accomplish control procedures of cards/badges according to AR 600-8-14. The custodian is responsible for the issue, turn in, recovery, and renewal of security ID cards/badges.

7-35. The degree of compromise tolerable in the ID system is in direct proportion to the degree of security required. The following control procedures are recommended for preserving the integrity of a card/badge system:

- Maintenance of an accurate written record or log listing (by serial number) all cards and badges and showing those on hand, to whom they are issued, and their disposition (lost, mutilated, or destroyed).
- Authentication of records and logs by the custodian.
- A periodic inventory of records by a commissioned officer.
- The prompt invalidation of lost cards/badges.
- The conspicuous posting at security control points of current lists of lost or invalidated cards/badges.
- The establishment of controls within restricted areas to enable security personnel to determine the number of persons within the area.
- The establishment of the two-person rule (when required).

- The establishment of procedures to control the movement of visitors. A visitor-control record will be maintained and located at entry control points.

SIGN/COUNTERSIGN AND CODE WORD

7-36. This method of verifying identity is primarily used in a tactical environment. According to the local SOP, the sign/countersign or code-word procedures must be changed immediately if compromised.

DURESS CODE

7-37. The duress code is a simple word or phrase used during normal conversation to alert other security personnel that an authorized person is under duress. A duress code requires planning and rehearsal to ensure an appropriate response. This code is changed frequently to minimize compromise.

ACCESS-CONTROL ROSTERS

7-38. Admission of personnel to a restricted area is granted to those identified and listed on an access-control roster. Pen-and-ink changes may be made to the roster. Changes are published in the same manner as the original roster.

7-39. Rosters are maintained at access control points. They are kept current, verified, and accounted for by an individual designated by the commander. Commanders or their designated representatives authenticate the rosters. Admission of persons other than those on the rosters is subject to specific approval by the security manager. These personnel may require an escort according to the local SOP.

METHODS OF CONTROL

7-40. There are a number of methods available to assist in the movement and control of personnel in limited, controlled, and restricted areas. The following paragraphs discuss the use of escorts and the two-person rule:

ESCORTS

7-41. Escorts are chosen because of their ability to accomplish tasks effectively and properly. They possess knowledge of the area being visited. Escorts may be guard-force personnel, but they are normally personnel from the area being visited. Local regulations and SOPs determine if a visitor requires an escort while in the restricted area. Personnel on the access list may be admitted to restricted areas without an escort.

TWO-PERSON RULE

7-42. The two-person rule is designed to prohibit access to sensitive areas or equipment by a lone individual. Two authorized persons are considered present when they are in a physical position from which they can positively detect incorrect or unauthorized procedures with respect to the task or operation being performed. The team is familiar with applicable safety and security requirements, and they are present during any operation that affords access to sensitive areas or equipment that requires the two-person rule. When application of the two-person rule is required, it is enforced constantly by the personnel who constitute the team.

7-43. The two-person rule is applied in many other aspects of physical-security operations, such as the following:

- When uncontrolled access to vital machinery, equipment, or materiel might provide opportunity for intentional or unintentional damage that could affect the installation's mission or operation.
- When uncontrolled access to funds could provide opportunity for diversion by falsification of accounts.
- When uncontrolled delivery or receipt for materials could provide opportunity for pilferage through "short" deliveries and false receipts.
- When access to an arms or ammunition storage room could provide an opportunity for theft. Keys should be issued so that at least two people must be present to unlock the locks required under the provisions of AR 190-11.

7-44. The two-person rule is limited to the creativity of the PM and the physical-security manager. They should explore every aspect of physical-security operations in which the two-person rule would provide additional security and assurance and include all appropriate recommendations and provisions of the physical-security plan. An electronic-entry control system may be used to enforce the two-person rule. The system can be programmed to deny access until two authorized people have successfully entered codes or swiped cards.

SECURITY CONTROLS OF PACKAGES, PERSONAL PROPERTY, AND VEHICLES

7-45. A good package-control system helps prevent or minimize pilferage, sabotage, and espionage. The local SOP may allow the entry of packages with proper authorization into restricted areas without inspection. A package-checking system is used at the entrance gate. When practical, inspect all outgoing packages except those properly authorized for removal. When a 100 percent inspection is impractical, conduct frequent unannounced spot checks. A good package-control system assists in the movement of authorized packages, material, and property.

7-46. Property controls are not limited to packages carried openly, but they include the control of anything that could be used to conceal property or material. Personnel should not be routinely searched except in unusual situations. Searches must be performed according to the local SOP.

7-47. All POVs on the installation should be registered with the PM or the installation's physical-security office. Security personnel should assign a temporary decal or other temporary ID tag to visitors' vehicles to permit ready recognition. The decal or the tag should be distinctly different from that of permanent-party personnel.

7-48. When authorized vehicles enter or exit a restricted area, they undergo a systematic search, including (but not limited to) the—

- Vehicle's interior.
- Engine compartment.
- External air breathers.
- Top of the vehicle.
- Battery compartment.
- Cargo compartment.
- Undercarriage.

7-49. The movement of trucks and railroad cars into and out of restricted areas should be supervised and inspected. Truck and railroad entrances are controlled by locked gates when not in use and are manned by security personnel when unlocked. The ID cards/badges are issued to operators to ensure proper ID and registration for access to specific loading and unloading areas.

7-50. All conveyances entering or leaving a protected area are required to pass through a service gate manned by security forces. Drivers, helpers, passengers, and vehicle contents must be carefully examined. The examination may include—

- Appropriate entries in the security log (including the date, operator's name, load description, and time entered and departed).
- A check of the operator's license.
- Verification of the seal number with the shipping document and examination of the seal for tampering.

7-51. Incoming trucks and railroad cars must be assigned escorts before they are permitted to enter designated limited or exclusion areas. Commanders should establish published procedures to control the movement of trucks and railroad cars that enter designated restricted areas to discharge or pick up cargo (escorts will be provided when necessary).

7-52. The best control is provided when all of these elements are incorporated into access-control procedures. Simple, understandable, and workable access-control procedures are used to achieve security objectives without impeding operations. When properly organized and administered, access-control procedures provide a method of positively identifying personnel who have the need to enter or leave an area.

TACTICAL-ENVIRONMENT CONSIDERATIONS

7-53. Access-control procedures during tactical operations may establish additional challenges for the commander. In some instances, the commander cannot provide a perimeter barrier (such as a fence) based on METT-TC. Commanders are still required to provide security measures for restricted areas, although they may not always have the necessary assets. Early-warning systems and the use of guards become crucial. A restricted area may become a requirement without prior notice during an operation. Figure 7-3 and Figure 7-4, page 7-14, are examples of temporary tactical restricted and exclusion areas.

7-54. Commanders must plan for these considerations when developing their budget. Funding must be requested and set aside to support physical-security requirements during tactical operations. Resources will not always be available; therefore, commanders must implement procedures that support access-control measures. Improvising will become common practice to overcome shortfalls concerning access-control equipment in the field.

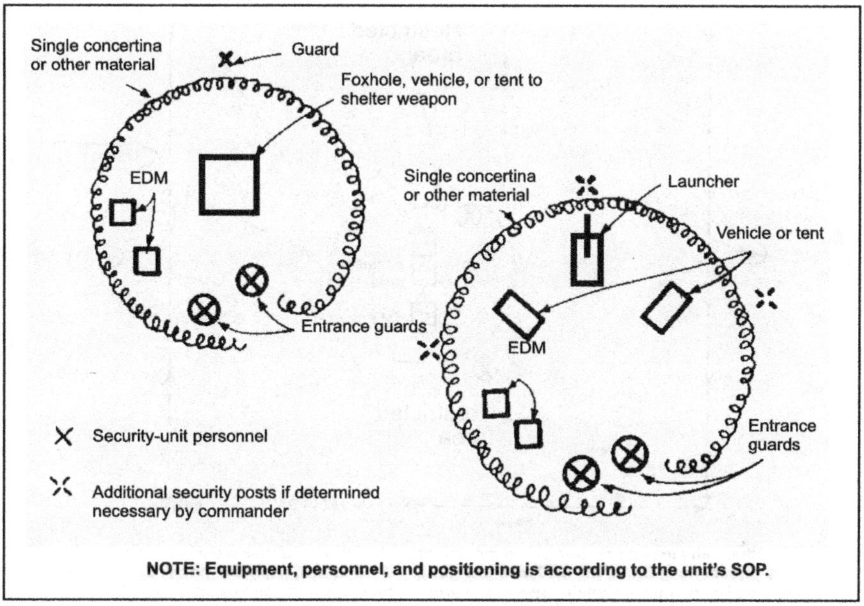

Figure 7-3. Sample Layout of Temporary Tactical Restricted Areas

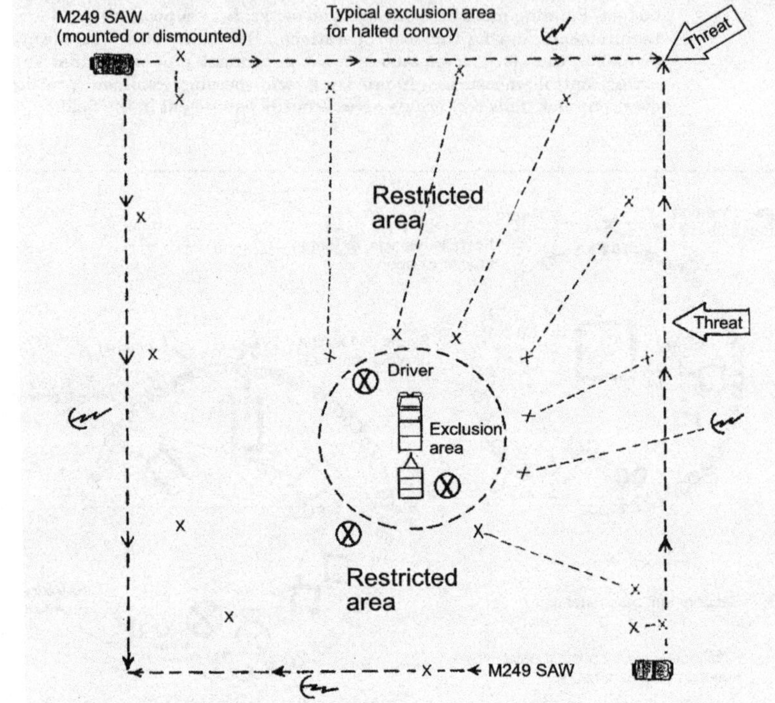

NOTE: Other vehicles and personnel, to include the LPs, form a defensive perimeter inside the primary directions of fire of the M60 machine guns. During attack, alternate machine-gun crews should be placed in the perimeter away from primary machine guns and be spaced so that they can be used as replacements and so that large gaps do not appear in the perimeter.

Legend:

X MP rifleman

Principal direction of fire

Temporary exclusion area (size depends on terrain and threat force)

Listening post

Figure 7-4. Sample Layout for a Temporary Tactical Exclusion Area

Chapter 8

Lock and Key Systems

Locks are the most acceptable and widely used security devices for protecting facilities, classified materials, and property. All containers, rooms, and facilities must be locked when not in actual use. Regardless of their quality or cost, locks are considered delay devices only. Some locks require considerable time and expert manipulation to open, but all locks can be defeated by force and with the proper tools. Locks must never be considered as a stand-alone method of security.

INSTALLATION AND MAINTENANCE

8-1. The USACE is responsible for installing locking devices in newly constructed facilities. Installation-level engineers are responsible for maintaining the locking devices. Physical-security personnel must work closely with engineer personnel to ensure that locks meet the standards and are installed according to applicable regulations. One source of assistance and information is the DOD Lock Program Technical Support Hotline at the Naval Facilities Engineering Services Center, Port Hueneme, California.

TYPES OF LOCKING DEVICES

8-2. The degree of protection afforded by a vault, a safe, or a filing cabinet may be measured in terms of the lock's resistance. Locking devices are listed in TM 5-805-8. Types of locking devices include key and combination locks.

8-3. ARs 190-11, 190-51, 50-5, and 50-6 prescribe specific types of locks for specific types of facilities. AR 380-5 prescribes standard facilities for storing classified material and contains guidance for different storage requirements.

KEY LOCKS

8-4. Key locks consist of, but are not limited to, the following:

- Cylindrical locksets are often called key-in-knob or key-in-lever locks. They are normally used to secure offices and storerooms. The locking cylinder located in the center of the doorknob distinguishes these locks. Some cylindrical locksets have keyways in each of the opposing knobs that require a key on either side to lock and unlock them. Others unlock with a key, but may be locked by pushing or rotating a button on the inside knob. These locks are suitable only for very low-security applications. Using these locks may require compensatory measures in the form of additional locks on containers within the room.
- Dead-bolt locks are sometimes called tubular dead bolts. They are mounted on the door in a manner similar to cylindrical locksets. The primary difference is in the bolt. When the bolt is extended (locked),

the dead bolt projects into the doorframe at least one inch, and it cannot be forced back (unlocked) by applying pressure to the end of the bolt. The dead-bolt lock has the potential for providing acceptable levels of protection for storerooms and other areas where more security is desired. It is recommended for use in military housing as an effective security measure in the installation's crime-prevention program. In situations where there is a window in or adjacent to the door, a double-cylinder dead-bolt lock (one that requires a key to open from either side) should be used.

- Mortise locks are so named because the lock case is mortised or recessed into the edge of the door. The most common variety of mortise locks has a doorknob on each side of the door. Entrance doors often have an exterior thumb latch rather than a doorknob. The mortise lock can be locked from inside by means of a thumb turn or by a button. Mortise locks are considered low-security devices since they weaken the door in the mortised area.
- Drop-bolt locks (often referred to as jimmy-proof locks) are normally used as auxiliary locks similar to dead bolts. Both the drop-bolt lock body and the strike have interlocking leaves similar to a door hinge. When closed, locking pins in the lock body drop down into the holes provided in the strike and secure the locking system. Since the lock body and the strike are interconnected with locking pins when closed, the lock essentially becomes a single unit and is extremely difficult to separate.
- Rim-cylinder locks are mounted to the door's inside surface and are secured by screws in the door face. These locks are generally used with drop-bolt and other surface-mounted locks and latches. They consist of an outer barrel, a cylinder and ring, a tailpiece, a back mounting plate, and two mounting screws. The tailpiece screws are usually scored so that the lock can be tailored to fit varying door thicknesses.
- Unit locks are ideal in heavily traveled facilities (such as hospitals or institutional buildings). These locks are a complete, one-piece unit that slides into a notch cut into the door's latch edge. The one-size cutout of the door edge simplifies the door preparation for the lock.
- Mechanical, push-button combination locks are digital (push buttons numbered 1 through 9) combination door-locking devices used to deny area access to any individual not authorized or cleared for a specific area. These locks are normally used for access control and should be backed up by door locking devices when the facility is unoccupied.
- Padlocks are detachable locks that are typically used with a hasp. Low-security padlocks, sometimes called secondary padlocks, are used to deter unauthorized access, and they provide only minimal resistance to force. Low-security locks are made with hardened steel shackles. Precautions must be taken to avoid confusing these locks with similar brass or bronze locks. The brass or bronze locks are commonly used but do not meet the security requirements of the hardened shackled locks. High-security padlocks may be used to secure AA&E. They provide the maximum resistance to unauthorized entry when used with a high-security hasp.

8-5. Some locks have interchangeable cores, which allow the same key system to include a variety of locks. Padlocks, door locks, cabinet locks, and electrical key switches can all be operated by the same key system. Because these cores are removable by a special key, this system allows for rapid rekeying of locks in the event that the key is compromised.

8-6. Locks are keyed in several different ways. When several locks are keyed differently, each is operated by its own key. When they are keyed alike, one key will open them all. Locks that are master-keyed are keyed differently, yet have one key that will open them all. Master-keying is done for convenience and represents the controlled loss of security. Master-keying is not used unless permitted by regulation.

COMBINATION LOCKS

8-7. Combination locks are available as padlocks or as mounted locks. They are low-security padlocks with combinations that are either fixed or changeable. Combination locks may be either mechanical or electronic. They are operated by entering a particular sequence of numbers. When the correct combination is entered, the lock's bolt is retracted. Combination locks used for securing classified material must meet Federal Specification FF-L-2740.

8-8. Although the lock is the most accepted and widely used security device, it is only a delay device and should never be considered as a positive bar to entry. A lock can (and will) be defeated. The best defense for locking devices is a good key-control program. Refer to AR 190-51, Appendix D, for standard key and lock procedures. Additional key and lock procedures for AA&E can be found in AR 190-11, Chapter 3.

Chapter 9

Security Forces

The security force for an installation or a facility provides the enforcement element in the physical-security program. This force consists of personnel specifically organized, trained, and equipped to protect the command's physical-security interests. It is a commander's most effective tool in a comprehensive, integrated, physical-security program. Vulnerability tests are periodically conducted to determine and ensure the state of readiness of security forces (see Appendix K).

TYPES OF SECURITY FORCES

9-1. On installations, security forces may be MP forces, security police, DOD civil-service security guards, or contract guards. Interior guard duties are performed by installation unit troops on a roster basis. MP forces normally perform security duties that require higher degrees of training and experience. These include—

- Security of restricted areas.
- Security of specific sensitive gates.
- Supervisory or coordinated roles with other military or DOD civil-service security guards.
- Responsibility for monitoring and responding to intrusion alarms.

9-2. An MP unit may perform the entire physical-security function alone based on METT-TC, the area, and the facilities. When an MP unit cannot assume responsibilities for all of the physical-security requirements in the command, other forces may be required. Additional forces may consist of the following:

- Personnel furnished by units of the installation's command on a daily or weekly basis. While this method has the single advantage of providing additional manpower, it has the disadvantages of rapid turnover and the lack of training. If this manpower is used, personnel should be assigned the least sensitive posts or patrols. For extended augmentation, units may be attached to MP units. The MP unit may also be augmented by reserve units or units in rotation.
- The combat-arms branches (especially the infantry) may attach their forces to MP units and may be designated as security guards assisting in the required operations.
- Military or paramilitary units of the host country may also be attached to or operate in coordination with MP forces. They may also be supplemented with national police of their own country.

- The installation's band may be a source of military force during wartime. (The band is assigned enemy-prisoner-of-war [EPW] duty as a wartime duty.) The band is doctrinally capable of providing security at the division tactical operations center (DTOC) and ASPs, assisting in the perimeter defense of the command post (CP), and operating the dismount point for the CP. It is capable of providing access control at the DTOC and the ASPs and augmenting or relieving security personnel on the defensive perimeter.

9-3. Civil-service security guards are uniformed civilian employees from a government agency. They are customarily trained and organized along military lines. The organization may be completely civil service or may be composed of civil-service personnel under military supervision. In either case, they are under the operational control of the PM or the security officer.

9-4. Labor-service personnel (local civilian personnel) have been organized and used successfully in theaters of operation. These types of units were organized after World War II and since that time have established enviable records in the physical-security field. They are distinctively uniformed, organized, and equipped. They have set and maintained the highest security standards, resulting in a minimal loss of property. While not military organizations, these units have successfully developed a high sense of duty and esprit de corps that has been reflected in their outstanding contributions to the physical security of installations in overseas commands.

AUTHORITY AND JURISDICTION

9-5. It is most important that the PM or the security officer determine (and instruct his security force in) the extent and limitations of the commander's jurisdiction in the field of law enforcement and investigations. Those jurisdictions include—

- Jurisdiction of place.
  - Military installations and facilities. Whether state or federal law or both are applicable on a military installation or facility depends largely on the nature of jurisdiction over the land involved. The amount of federal jurisdiction may vary between different areas of the installation or facility. The legal formalities of acquiring jurisdiction over land under the control of the Secretary of the Army are accomplished at DA level and according to the provisions of AR 405-20. Information and advice relating to jurisdictional questions should be referred to the local SJA.
  - Areas outside of military installations. Areas outside of military installations are generally subject to state and local laws; however, there are exceptions. Information and advice in this regard should be obtained through the local SJA.
  - Overseas areas. In overseas areas, jurisdiction varies according to the military situation and existing international treaties, contracts, and agreements. Guidance should be obtained in each instance from the commander and the SJA and set forth in appropriate command directives.

- Jurisdiction of personnel.
  - Jurisdiction of personnel generally follows the limitations of jurisdiction of the installation.
  - MP forces have jurisdiction and authority over personnel as described in AR 190-14 and related publications.
  - Authority for federal civilian employees assigned to security, police, and guard duties is derived from the installation's commanding officer. These personnel can have no more authority than he possesses and are subject to any limitations imposed thereon.
  - Security-force personnel may enforce all offenses under the Uniform Code of Military Justice (UCMJ), military regulations, federal laws and regulations, and state laws where applicable.
  - Security-force personnel may be given the same authority as MP forces over all personnel subject to military jurisdiction, including apprehension, detention, and search.
  - Civilian security-force personnel have no specific grant of authority over civilians other than the right of citizen's arrest.
  - The commander is the source of jurisdiction and authority for all other personnel assigned to security-force duties.

PERSONNEL SELECTION

9-6. Regardless of the use of structural, mechanical, or electronic equipment, the human element in security operations makes the difference between success or failure. Commanders and supervisors have a responsibility to ensure that security personnel who control access to restricted areas and classified activities are qualified based on criteria in AR 380-67. Personnel who perform physical-security duties must be disciplined and alert, have sound judgment, be confident and physically fit, and possess good interpersonal communication skills.

SECURITY CLEARANCE

9-7. Security-clearance criteria for security positions must be based on the security classifications of the information to which access will be granted. Security positions are normally designated as sensitive and require a secret security clearance. ARs 381-20 and 380-67 describe criteria and procedures governing security clearances. Appropriate civilian-personnel regulations should also be consulted when civilians are involved.

9-8. Positive evaluation of the reliability of all personnel must be made before they are entrusted with classified or sensitive information. (The Individual Reliability Program is prescribed in AR 190-56.) Follow-up action must be made on all personnel who are granted a security clearance to ensure that they continue to meet the criteria for their clearance. Personnel not meeting or adhering to the prescribed standards must have their security clearances revoked and thereby lose their access to areas containing classified information or material (see AR 380-67).

ORGANIZATION AND EMPLOYMENT OF FORCES

9-9. The organization of a security force will vary, depending on circumstances and the forces available. Forces consist of—

- Mobile patrols. A mobile detachment of ground, sea, or air forces dispatched to gather information or carry out a security mission.
- The response force. A mobile force with appropriate fire support (usually designated by the area commander) to deal with Level II threats in the rear area (Army). This is normally an MP function.
- Reserves. That portion of a force withheld from action or uncommitted to a specific course of action so as to be available for commitment at the decisive moment. Its primary purpose is to retain flexibility throughout an offensive action.
- Any combination of these three.

9-10. Instructions to the security force should be issued in writing. These instructions are normally in the form of general, special, or temporary orders. They should be carefully and clearly worded and include all phases of each assignment. They should be reviewed at least monthly to ensure that they are current. Categories of instructions of each are as follows:

- General orders are those orders that concern the security force as a whole and are applicable at all posts and patrols.
- Special orders pertain to a permanent post or patrol. Each permanent post or patrol should have special orders issued concerning the location, duties, hours manned, arms, ammunition, and other equipment required and the instructions for using force in enforcement and apprehension activities.
- Temporary orders are issued for a short period and cover a special or temporary situation. If it can be predetermined, such orders should indicate the period of time for which they are valid.

9-11. A security-force SOP that outlines policies, organization, authority, functions, and other required information should be prepared for required reading. Each security-force member should be held responsible for full knowledge and understanding of the contents of the SOP. Each installation PM, physical-security officer, or chief of a guard force should conduct periodic inspections and tests to determine each individual's degree of understanding of these instructions. Instructions should be provided in writing regarding the safeguarding and control of the SOP. Its contents may not be classified; however, the information could assist an intruder in breaching security.

HEADQUARTERS AND SHELTERS

9-12. The location of the security force's headquarters will depend on the size and layout of the installation or facility. The objectives are the efficient control of the security force and the adequate security of vital activities. On a small installation, there is frequently only one full-time entrance that may be supplemented by several part-time entrances. At these installations, the logical location of the headquarters would be at or near the main entrance. On larger installations, it might be better to locate the headquarters near the center of the cantonment area.

9-13. The security force's headquarters should be the control point for all physical-security matters for the installation and the monitoring point for protective alarm and communication systems. This office should have a reliable and independent means to contact nearby civil authorities. A list of key telephone numbers should be available for use in emergency operations.

9-14. Personnel shelters should be available to protect the guards from the elements. The design can be temporary or hardened and include adequate space for guard-force personnel only. The facility should have heat, ventilation, storage space for essential accessories, lighting that will not expose the occupant, and good visibility in all directions.

EXECUTION OF SECURITY ACTIVITIES

9-15. Security personnel must exercise good interpersonal communication skills when carrying out their duties with other employees. Bad employee relations can result if security personnel become impertinent and assume powers not rightfully theirs. Security personnel must understand the methods and techniques that will detect security hazards and assist in identifying violators and intruders.

9-16. Written reports or journals are recommended for security activities. These should be prepared by either the security force's supervisor or the personnel at the security post. These reports should record all activities, actions, and visits at the security post.

9-17. It must be strongly emphasized that security personnel will be used for security duties only and should not be given other routine functions except as directed by the commander or his representative. Security personnel should have no fire-fighting or similar duties regularly assigned. Such emergencies offer an excellent diversion to cover an intruder's entrance. Consequently, during such times, security personnel must be exceptionally alert when performing their duties. However, the security force may be cross-trained in other areas (such as fire fighting) so that they may be used when required and when circumstances permit (such as when they are off duty).

9-18. Personnel who are assigned to fixed posts should have a designated method of relief. The security force's shift supervisor should establish a relief schedule (about every two hours) according to local policies and the SOP. A simple but effective plan of operation should be worked out for the security force to meet every foreseeable emergency. Practice alarms should be conducted frequently to test the plan's effectiveness. Such plans should be designed to prevent a diversion at one point on the installation, drawing off the guards or distracting their attention from another section of the installation where unauthorized entry may be made. Routes and times for security patrols should also be varied at frequent intervals to preclude establishing a routine that may be observed by potential intruders.

TRAINING REQUIREMENTS

9-19. The extent and type of training required for security forces will vary according to the importance, vulnerability, size, and other factors affecting a particular installation or facility. The training program's objective is to ensure that all personnel are able to perform routine and emergency duties competently and efficiently.

BENEFITS OF PROPER TRAINING

9-20. Efficient and continuing training is the most effective means of obtaining and maintaining maximum proficiency of security-force personnel. Regardless of the selection process, new personnel seldom have all of the qualifications and experience necessary to do the job. In addition, new or revised job requirements frequently mean that personnel must be retrained. Training can bridge the void between ability and job requirement.

9-21. Supervisors need to remember that all personnel do not have the same training needs. It is a waste of valuable time to train an individual in a subject that he has already mastered. Past experience, training, acquired skills, and duty assignments should be evaluated for each person as an aid in planning an effective training program.

9-22. A good training program benefits both the installation and the security force. The task of supervising the security force is made easier, there is much less wasted time, fewer mistakes are made, and there is less friction with other agencies. A good training program helps to instill confidence through developing increased skill proficiency. The training program provides for more flexibility and better physical protection, fewer required personnel, and less time to learn duties. Training establishes systematic and uniform work habits.

BASIC TRAINING

9-23. As a minimum, personnel (including civil-service security personnel) who have not had security training should receive training in their security duties. This training includes—

- The care and use of weapons, if required. No person should be placed on security duty unless weapons training has occurred within the past 12 months. Weapons training must be according to AR 190-14.
- Areas of responsibility and authority of security personnel, particularly on apprehension, search and seizure, and the use of force.
- The location and use of first aid and fire-control equipment and electrical switches.
- Duties in case of emergencies such as alerts, fires, explosions, and civil disturbances.
- Common forms of sabotage and espionage activity.
- The location of hazardous and vulnerable equipment and material.

IN-SERVICE TRAINING

9-24. All newly assigned individuals are given special instructions for each post. When possible, their first assignment should be with an experienced person. Additional in-service training and periodic retraining to review basic material and procedures are continuous requirements.

9-25. Scheduling in-service training and classes to enable all of the security force or a complete shift to participate is often difficult. Therefore, the supervisor must exercise good judgment when scheduling training to ensure that each person has the opportunity to receive the training.

EVALUATION OF TRAINING

9-26. Testing designed to evaluate performance is a necessary step in the training program. These tests may be oral or written or may be a type of performance test. They should be administered annually to ensure that the entire force maintains high standards of proficiency. A testing program also helps to improve training by—

- Discovering gaps in learning.
- Emphasizing main points.
- Evaluating instructional methods.

9-27. Security training received by personnel at their units must be entered in unit training charts or records. The record serves to—

- Indicate individual degrees of skill.
- Establish priorities of instruction.
- Present a consolidated picture of the security force's training status.
- Help certify guard personnel.

SUPERVISION

9-28. A security supervisor is tasked with overseeing and directing the work and behavior of other members of the security force. To obtain maximum performance from each member of his force, the supervisor must have a complete understanding of leadership principles and be capable of applying them.

9-29. The supervisor is responsible for understanding the operations of all posts. Additionally, he is often responsible for selecting, inducting, training, and ensuring the productivity, safety, morale, and advancement of guard-force members.

9-30. To ensure an alert, presentable, and efficient security force, the leadership must provide consistent and intelligent supervision. To earn the respect and cooperation of the guard force, supervisors must be professional in their conduct. The security force's morale and efficiency is a direct reflection of the quality of its supervision.

9-31. The ratio of supervisory personnel to security personnel should be determined by the individual characteristics of each installation. At small installations, the ratio may be higher than at large installations.

9-32. There must be sufficient supervision to enable the inspection of each post and patrol. It is also essential that supervisors be in contact with security headquarters to control emergencies that may arise. Specific duties of a supervisor include the inspection and briefing of the relief shift and the inspection of posts, vehicles, and equipment during visits to posts and patrols.

SUPPLEMENTS TO SUPERVISION

9-33. Various means and devices may be used as supplements to personnel supervision. These include the following:

- **Recorded tour systems.** Personnel record their presence at strategic points throughout an installation by using portable watch clocks or similar devices. These are effective means of ensuring that such points are regularly covered. This system provides an after-the-fact type of supervision.
- **Supervisory tour systems.** A signal is transmitted to a manned central headquarters at the time the post is visited. These systems provide instantaneous supervision and a means of detecting interference with normal security activities and initiating an investigation or other appropriate action.

9-34. All personnel on security duty should be required to report periodically to headquarters by the usual means of communication. The frequency of such reports will vary, depending on a number of factors. Regularity should be avoided to preclude setting a pattern by which an intruder can gauge an appropriate time for entrance.

MANAGEMENT

9-35. The physical-security supervisor is responsible for managing and developing the security organization. A physical-security program is greatly enhanced by a well-developed educational program.

9-36. The physical-security supervisor acts as an advisor and assists in formulating policies for the installation's physical-security measures. The goal should be the best security within the restrictions of the commander's budget guidance. Physical-security planners must remember that anyone can provide adequate security with unlimited funds; however, this is not a realistic approach. There must be a constant endeavor to effect justifiable economy where possible without jeopardizing the physical-security program.

UNIFORMS

9-37. All security-force personnel are required to wear the complete prescribed uniform as outlined in their special orders. Deviations from the prescribed uniform should not be made except for items to protect the guard force's health, comfort, and safety. The duty uniform will be worn during all tours of duty and may be worn during off-duty hours only between the place of residence and the place of duty. Each member of the security force is required to maintain high standards of appearance.

VEHICLES

9-38. The security force should be furnished with sufficient and reliable vehicles to maintain patrol standards established by the installation commander. Vehicles assigned to the force should be equipped with two-way radios to obtain the greatest possible use of all personnel and vehicles.

FIREARMS

9-39. Before issuing weapons, the security force will be briefed on the use of force. Security-force personnel will be issued weapons as prescribed by AR 190-11 and the unit's SOP. The commander may prescribe other weapons for the security force based on needs and requirements. Weapons normally are loaded with live ammunition, except where prohibited for safety reasons. The use of privately owned weapons while on duty is not authorized. Weapons and ammunition issued to security-force personnel will not be removed from the installation except in the course of official duty. When not in use, weapons are secured in arm racks in storage rooms as prescribed by AR 190-11.

9-40. Weapons are inspected as necessary to ensure proper maintenance. A written report is prepared and filed on the discharge of any weapon except for authorized and supervised training. The patrol supervisor or an MP investigator prepares the report (DA Form 3975).

9-41. Ammunition supplies for the security force's use must be maintained in secured storage containers according to AR 190-11. Ammunition must be issued only under proper supervision for authorized purposes. Ammunition issued to members of the security force must be accounted for by individual members immediately upon completion of duty. Any ammunition unaccounted for will be the subject of a report of its disposition by the individual.

COMMUNICATIONS

9-42. The security force should be equipped with two-way radios. These may be vehicle-mounted and portable, or they may be telephones. A secure-voice capability should be used where possible. This equipment is considered essential for the efficient operation of the security force and the accomplishment of its assigned mission. Proper use and care by security personnel will enhance the equipment's usefulness and capability.

MISCELLANEOUS EQUIPMENT

9-43. Security managers or supervisors should obtain other equipment necessary to accomplish their security mission. Items in this category may include (but are not limited to) warning lights; sirens; spotlights; portable lights; flashlights; first aid kits; traffic-control devices; and items of wear for the health, comfort, and safety of security personnel. Some of this equipment may require local purchase.

MILITARY WORKING DOGS

9-44. The requirements for physical protection of installations or facilities within the US and overseas theaters of operation continue to increase. Manpower available for this purpose has always been (and probably will continue to be) limited. The MWD, properly trained and properly used, can enhance a physical-security program. See AR 190-12 and DA Pam 190-12 for information regarding the use of MWDs.

SUMMARY

9-45. A security force is the critical element of a successful physical-security program. It is as strong as its weakest member. A comprehensive training program is essential to a knowledgeable, disciplined, and alert security force. A well-trained security force will be prepared to respond to a security breach.

Chapter 10
In-Transit Security

In-transit security subjects the movement of cargo to different, and frequently, more demanding aspects of physical security. Cargoes may be moved via port, rail, pipeline, or convoy. Regardless of the mode of movement, commanders must aggressively apply the principles of physical security to their protection. Security forces must be provided at the most vulnerable areas of each cargo movement.

IN-PORT CARGO

10-1. Ports and harbors are prime targets for enemy and criminal activities. Perimeter areas of these facilities are more vulnerable because of the extensive distance and exposed beach or pier areas. Terminal areas may include fully developed piers and warehouses or may be an unimproved beach where logistics-over-the-shore (LOTS) or roll-on/roll-off (RORO) operations are conducted.

10-2. If a Theater Army Area Command (TAACOM) MP unit must provide security for cargo in a port, the main effort is to provide security from the perimeter of the port outward. Security measures focus on aggressive patrolling to detect, report and, if need be, combat enemy threats. Measures may include—

- Conducting route and area reconnaissance patrols.
- Developing police intelligence in the area of operations (AO).
- Controlling traffic in the area surrounding the port.
- Conducting mounted or dismounted patrols (with MWDs, if available) around the port's perimeter.
- Establishing an access-control/ID section.
- Watching for diversions of supplies out of the port.
- Providing a response force to react to incidents inside the port's perimeter.
- Providing observation and early warning of threat ground and air attacks.

10-3. When providing security for cargo, the focus is on providing a security overwatch for the cargo as it moves from the port to the combat area. Inside a port's perimeter, access to cargo is limited by—

- Operating random mounted or dismounted patrols (with MWDs, if available).
- Using combined patrols as a response force for incidents inside the perimeter.

- Controlling access to the most restricted areas.

10-4. On occasion, the MP may have to safeguard highly critical cargo inside a port's perimeter. The type and degree of security provided is based on logistical security information. Some examples are the—

- Types and values of the cargo stored.
- Vulnerability of the cargo to a land threat.
- Likelihood of theft, diversion, pilferage, or sabotage by military personnel, local workers, black marketers, or enemy agents.
- Location and nature of the port facilities.
- HN agreements.
- Degree of entrance and exit controls.

10-5. Safeguarding the most critical cargo waiting to be transferred to land transport is the priority. The following measures help to safeguard stored cargo:

- Establishing access-control procedures.
- Searching bundles and packages being taken from the area.
- Examining trip tickets and documentation of cargo vehicles.

10-6. If the restricted area is a pier or other maritime environment, access from the water must be controlled as well as from the land. Entry on the landward side of a pier can be limited with fencing, pass control, and aggressive patrolling; but the part of the pier that protrudes over the water is accessible from the sides and from below. Methods for securing the pier along its water boundaries include—

- Patrols (both walking on the pier and in small boats).
- Protective lighting.
- Log booms.
- Nets.
- Buoys or floats.
- Anchored or pile-mounted navigational aids and signaling devices.
- Barges.

10-7. While most of the barriers described above will stop or impede access to facilities from boats or swimmers, nets are among the most effective. Well-marked, partially submerged objects are also effective; however, there may be legal prohibitions against placing barriers that may constitute a hazard to navigation. These barriers should be placed only after coordination with and approval by the appropriate legal and HN authorities. Sometimes it is best to close off the waterside of a pier. A floating boom will keep small boats out. Suspending a cable or a chain-link net from the bottom of the boom will deny access underwater.

10-8. At least two security zones must be established on a facility's waterside—the reaction zone and the keep-out (exclusion) zone. Security forces in these zones notify vessels, craft, and swimmers that they are entering restricted waters and should alter their course. Security forces may stop and search intruders if necessary. Security zones should extend at least 1,000 meters from the nearest protected asset; however, in some port areas this large security zone is not possible. In such cases, other measures (such as boat patrols) must be increased to mitigate the possibility of attack.

10-9. A reaction zone extends from the high-water mark to a distance beyond the maximum range of anticipated waterborne threats. Security forces will stop and challenge intruders inside the reaction zone.

10-10. The keep-out zone is the zone closest to the protected assets. It extends from the asset to the maximum range of anticipated threat weapons. Security forces should prevent the entry of all unauthorized craft or vessels into this zone. The tactical response force (in this case, a boat) may be used. In addition to organic security, forces may be provided by HN or contracted personnel.

10-11. To keep the cargo secured while transferring from one transport method to another, the traffic moving in and out of cargo-handling areas must be controlled. MP forces can—

- Set up a single access-control point.
- Erect field-expedient barriers. Truck trailers or other large vehicles can be used to constrict the traffic flow if permanent barriers are not in place.
- Limit entry to mission-essential personnel, vehicles, and equipment (as designated by the port authority).

10-12. A holding area should be provided if gates are used by vehicles other than cargo vehicles. Cargo vehicles can pull into the holding area while they are being checked. The holding area should be large enough to handle the volume and size of traffic. A wooden deck or platform at, or slightly higher than, the level of the truck bed can be used to facilitate checking. The platform must be at least as long as the vehicle (such as an empty flatbed trailer). Such a platform makes it quicker and easier to observe and check cargo.

10-13. Cargo is less likely to be diverted if a close watch is kept on cargo documentation and container safety. Containerized cargo is less likely to be stolen or sabotaged. However, containers must be watched closely as they are filled and sealed. Cargo can be pilfered before the seal is applied. An unsealed container can be moved to a stacking area; or someone may apply a false seal, break the seal later, remove the cargo, and then apply a legitimate seal.

10-14. At access-control points—

- Inbound and outbound containers should be inspected. Signs of damage or unserviceability should be observed.
- Containers must be inspected for the presence of seals or locks and hinges. Their serviceability should also be checked.
- The document's transport number, container number, and seal number should be checked to ensure that they match those numbers on the transportation control-and-movement document. (Check the seals by handling them, not simply by a visual check.)
- Containers with valid documents only should be allowed to pass inbound or outbound through the control point.

RAIL CARGO

10-15. Because a train's movement is determined directly by the condition of the tracks, cargo moving by rail is particularly vulnerable to attack. The destruction of switches, signals, or the track may be a delaying harassment; or it could trigger a major catastrophe. Since railroads can be such high-value targets, the commander may task MP or other US forces to provide on-board security for critical cargo.

10-16. Most train crews consist of four or five people who control the train—the engineer, a conductor, a fireman, a senior brakeman, and a brakeman or a flagman. The conductor is the train commander unless a transportation railway service officer is assigned to the train. The train commander is responsible for the train's operation and security. He makes all decisions affecting the train. The security force's commander is responsible for the cargo's security. The train crew and the security force watch for and report any discrepancies or interruptions to normal procedures at any time during the movement. Information about the movement is usually sent along the movement route by the chief dispatcher through a telephone circuit.

10-17. A four- to six-person security force is usually enough to secure railway shipments of sensitive freight, but additional security forces may be needed for moving critical cargo. In addition to a military security force, the shipper or loading agency may send specially trained personnel with highly sensitive cargo. The number of MP in a train security force depends on the—

- Sensitivity of the freight.
- Priority of need for the freight.
- Terrain over which the train will pass.
- Length of the train.
- Duration of the trip.
- Degree of threat.

10-18. Security forces prepare and maintain a record (by car number) of guarded cars in the train. Security forces can ride in—

- A specific car that requires protection.
- The caboose.
- A security-force car. (If only one security car is used, it should be near the center of the train; if more than one is used, cars should be spaced to provide the best protection for the train.)

10-19. The security force on a train must keep a constant check on car doors, seals, wires, and locks to detect tampering. The following instances must be noted and reported immediately:

- Irregularities in procedures.
- The presence or actions of unauthorized persons.
- Deficiencies or incidents that occur.

10-20. When planning rail-cargo security, the time schedule for the rail movement must be obtained. A map reconnaissance of the route should be provided, detailing bridges and tunnels that are especially vulnerable.

10-21. Security-force actions should be planned at scheduled stops or relief points, and forces should be deployed according to these plans. Locations of MP units and other friendly forces should be plotted along the route, and their radio frequencies and call signs should be noted. An intelligence report covering the route should also be obtained. This report should indicate sites where sabotage may occur, attacks may be expected, or thefts and pilferage are likely.

10-22. The shipper is responsible for the security of all carload freight until it is turned over to the Transportation Railway Service and the loaded cars are coupled to a locomotive for movement. The shipper or field transportation officer should complete the freight waybill or the government bill of lading. This report shows the car number, a brief description of contents, the weight of the load, the consignor, the consignee, the origin, and the destination. In addition, it may show special instructions for the movement or security of the car and its contents. Careful documentation is essential for—

- Securing the shipment.
- Locating cars with critical cargo.
- Ensuring that priority movement is authorized.

10-23. Transportation officers are responsible for the completeness, correctness, and proper handling of waybills. Each car must have a waybill; this allows cars to be detached or left behind should they become defective en route. If this occurs, a team from the security force must remain with the cargo until they are relieved.

10-24. Railway cars are sealed after loading. A seal shows that a car has been inventoried and inspected. The standard method of sealing a railway boxcar door (in addition to padlocks or wires) is with a soft metal strap or a cable seal that contains a serial number. Maintaining rigid accountability of all seals is necessary to prevent the undetected replacement of an original seal with another. While sealing does not prevent pilferage, a broken seal is a good indicator that the car and its contents have been tampered with. Train security forces or operating crews can easily check the seals on cars when the train stops. Broken seals should be reported immediately to help pinpoint the time and place of a possible theft. When vehicles are shipped by railcar, sensitive and high-value items must not be secured in the vehicles. Container-express (CONEX) and military-van (MILVAN) containers are ideal for shipping these and other small items on flatcars since they greatly reduce the chance of pilferage. These containers must be locked and sealed and, if possible, placed door to door for additional security.

10-25. When operations permit, cars containing highly pilferable freight, high-priority cargo, or special shipments are grouped in the train to permit the most economical use of security forces. When flatcars or gondolas are used to transport sensitive or easily pilfered freight, security forces should be placed where they can continuously observe and protect these cars.

10-26. When the train is stopped, security forces should dismount and check both sides of the train, verifying that seals, locks, and wires are intact. They must report a broken seal immediately to help pinpoint the time and place of the theft.

10-27. If the security force is relieved by another security force while en route, a joint inspection of the cars is conducted. The relief force signs the record being kept on the guarded cars. Consignees assume responsibility for the security of loaded freight cars at the time they arrive at their destination. When the trip is complete, the receiver or his agent will inspect the cars. The security force obtains a receipt for the cars, which is then attached to the trip report. The trip report should include—

- Dates and times the trip started and ended.
- Any additional information required by the local SOP or command directive.
- Recommendations for correcting deficiencies or for improving future security on trains.

10-28. Because unloading points are highly vulnerable to pilferage and sabotage, cars should be unloaded as soon as possible to reduce the opportunity for loss. MP forces are normally not available for the security of freight in railway yards. For more information regarding rail cargo, see FM 55-20.

PIPELINE CARGO

10-29. Pipeline systems are widely used in a theater of operation to transport bulk petroleum products or other liquids. Such systems are open to a number of security threats from the point of entry to the point of final delivery. Pipeline systems are composed of storage and dispersing facilities, pump stations, and extended pipelines. They also include discharging facilities for tankers at ports or other water terminals.

10-30. The type and extent of risk to a pipeline varies with the level of conflict in the AO. In a communications zone, the chief hazard is likely to be pilferage. Pipelines can be tapped by loosening the flange bolts that join sections of pipe or by cutting holes in the hose line. The risk rises if gasoline is scarce and expensive on the civilian market. Sabotage is a security hazard during all levels of conflict. It is committed by any method such as simply opening pipe flanges, cutting a hose line, or setting fires and causing explosions to destroy portions of the line.

10-31. In areas of conflict, the likelihood of sabotage and interdiction increases. Pipeline systems are vulnerable to air attacks, especially aboveground sections of the pipeline, pump stations, and storage facilities.

10-32. Security forces should be deployed in the best manner to provide coverage to the most vulnerable portions of the pipeline that are at the greatest risk to enemy, terrorist, partisan, and ground attack. Patrols should be set up to screen isolated areas and remote pumping stations. Sensors should also be considered, along with aerial security. Security patrols will—

- Detect, report, and respond to attacks on or sabotages of the pipeline.
- Monitor critical parts of the pipeline on a routine but random basis.
- Monitor ground sensors and other intrusion-detection devices. These are often used at pump stations and elsewhere along the pipeline to detect and identify threats to the system.

- Check line-pressure devices in pipeline and pumping facilities. These devices monitor the flow and detect breaks in the line, which may indicate pilferage of gasoline (or other petroleum products).

10-33. Dedicated security forces are rarely sufficient in number for the surveillance of an entire pipeline system. All available supporting forces (in the course of their normal duties) should observe and report items of intelligence for further investigation. Examples of suspicious activities in the pipeline area might include the unusual presence of commercial tanker trucks, the appearance of gasoline drums or cans, or an increased use of motor vehicles in fuel-scarce areas. Other resources available to the commander for coordination and support include HN and MP elements responsible for the AO, as well as the security officer of the petroleum group or battalion.

CONVOY MOVEMENT

10-34. As convoy movements are tactical in nature and are discussed in detail in FM 19-4, they will be briefly discussed here. When moving by convoy, consideration should be made for the following:

- Congested traffic areas.
- Travel during night hours when traffic is reduced rather than travel during daylight hours when traffic congestion is heaviest.
- National holidays. Traffic may be three times heavier than on a normal day. Also, if you are moving a convoy overseas on a national holiday, the HN people may not be receptive to your action and the result may be unwanted reactions on their part.
- The use of a marked HN police vehicle in conjunction with the convoy. The HN people are more receptive to an activity when it is represented by one of their own. Additionally, the HN police may be able to diffuse a potential crisis.
- Security of the convoy. Security of the convoy is foremost important both during movement and stops. During extended or overnight stops, special consideration must be given to securing the loaded vehicles.

Chapter 11

Inspections and Surveys

Inspections and surveys are valuable tools to a commander's physical-security program. These tools collectively measure and identify the readiness of a commander's physical-security program. The survey provides the installation commander with an overall security posture of the installation.

INSPECTIONS

11-1. Physical-security inspections are conducted at DA installations, activities, and facilities by trained physical-security inspectors. Some facilities on an installation may be exempt from inspection due to their mission. These facilities are inspected under the guidance of regulations and directives unique to those activities. Inspection personnel will be trained and will conduct inspections according to AR 190-13.

11-2. A physical-security inspection is a recorded assessment of physical-security procedures and measures implemented by a unit or an activity to protect its assets. The inspection is recorded on DA Form 2806-1-R (see AR 190-13 for using the form).

COORDINATION

11-3. Liaison and coordination should be established with other agencies on the installation before an inspection. The director of facility engineers can provide information to benefit the overall security program. Other agencies, such as MI (threat analysis) and local law-enforcement agencies, may have input essential to the security program.

SECURITY LIBRARY

11-4. A security library is necessary to help personnel prepare for and conduct an inspection. This library may include—

- The mission and history of each activity to be inspected.
- Previous inspection reports.
- A copy of the most current risk analysis.
- The SOPs and ARs specific to physical security.

ENTRANCE INTERVIEWS

11-5. Entrance interviews are usually required before the actual inspection. During the interview, the inspector establishes a rapport with the unit representative. The inspector identifies the following during the interview:

- All members of the inspection team.

- An overview of the last inspection.
- Areas to be inspected and the order of inspection.
- A review of waivers, work orders, and exceptions.
- Changes to the unit's mission (if any).

CONDUCTING INSPECTIONS

11-6. The inspection should be conducted from the outside to the inside of the facility, activity, or area with regard to the following:

- Observation of the facility will be conducted during all hours of the day.
- Interviews of managerial and operational personnel will be performed.
- Security forces should be inspected so as not to disrupt the mission (if possible).
- An assessment should be made of security-force training, especially if security-force knowledge proves inadequate.
- Inspection of entry and movement control by the guard force should not hinder operations.
- All communications systems used by the guard force should be thoroughly inspected. The guard force should have two reliable and efficient means of communication, one of which is a radio.
- Inspections should be conducted according to regulations appropriate for the facility.

EXIT INTERVIEWS

11-7. Exit interviews should be conducted as soon as possible after the inspection. The commander should be informed of any deficiencies or compliments noted. A rating on the inspection's results will not be provided during the exit interview. The approving authority, not the inspector, will determine the inspection rating. The rating will be forwarded to the unit along with the final report.

11-8. Recommendations will be made according to regulations. Written reports should be forwarded through channels in a timely manner according to the PM's SOP. The commander's report of actions taken will be required and reviewed by the PM's staff.

SURVEYS

11-9. A physical-security survey differs from an inspection in that a survey covers a formal assessment of an installation's physical-security program. Each survey includes a complete reconnaissance, study, and analysis of installation property and operations. The survey provides the commander with an assessment of the installation's overall security posture. It consists of the threat and the mission, and it advises the commander on the installation's physical-security program's strengths and weaknesses.

PHYSICAL-SECURITY SURVEY

11-10. The physical-security survey is a formal recorded assessment of an installation's physical-security program. See AR 190-13 for further information on this type of survey.

SECURITY-ENGINEERING SURVEYS

11-11. While a security-engineering survey is largely an engineer function, it must be coordinated with physical-security personnel to be successful. A security-engineering survey is the process of identifying (by means of an on-site survey) engineering requirements associated with facility enhancements for physical security and antiterrorism, including an IDS installation. This type of survey should be conducted when planning new construction, renovations, or upgrades to existing facilities where there are likely to be physical-security requirements. A security-engineering survey may also be requested by the PM or an equivalent security officer to evaluate existing security. This survey—

- Identifies assets to be protected.
- Identifies threats to these assets and the level of protection required to protect them.
- Identifies the protective measures.
- Determines the cost of the protective measures.

Appendix A

Metric Conversion Chart

This appendix complies with current Army directives which state that the metric system will be incorporated into all new publications. Table A-1 is a conversion chart.

Table A-1. Metric Conversion Chart

Metric to English			English to Metric		
Multiply	**By**	**To Obtain**	**Multiply**	**By**	**To Obtain**
Length					
Centimeters	0.0394	Inches	Inches	2.54	Centimeters
Meters	3.28	Feet	Feet	0.0305	Meters
Meters	1.094	Yards	Yards	0.9144	Meters
Kilometers	0.621	Miles (stat)	Miles (stat)	1.5609	Kilometers
Kilometers	0.540	Miles (naut)	Miles (naut)	1.853	Kilometers
Millimeters	0.039	Inches	Inches	25.40	Millimeters
Area					
Square centimeters	0.1550	Square inches	Square inches	6.45	Square centimeters
Square meters	10.76	Square feet	Square feet	0.0929	Square meters
Square meters	1.196	Square yards	Square yards	0.836	Square meters
Volume					
Cubic centimeters	0.610	Cubic inches	Cubic inches	16.39	Cubic centimeters
Cubic meters	35.3	Cubic feet	Cubic feet	0.0283	Cubic meters
Cubic meters	1.308	Cubic yards	Cubic yards	0.765	Cubic meters
Milliliters	0.0338	US liq ounces	US liq ounces	29.6	Milliliters
Liters	1.057	US liq quarts	US liq quarts	0.946	Liters
Liters	0.264	US liq gallons	US liq gallons	3.79	Liters
Weight					
Grams	0.0353	Ounces	Ounces	28.4	Grams
Kilograms	2.20	Pounds	Pounds	0.454	Kilograms
Metric tons	1.102	Short tons	Short tons	0.907	Metric tons
Metric tons	0.984	Long tons	Long tons	1.016	Metric tons

Appendix B

Sample Installation Crime-Prevention Handbook

This appendix provides guidance on planning, organizing, directing, and controlling installation crime-prevention programs. It provides guidance on developing an installation program, criminal analyses to identify crimes, guidance on which crimes to address, command and individual countermeasures for particular crimes, and program-evaluation procedures.

SECTION I — INSTALLATION CRIME-PREVENTION PROGRAMS

B-1. In the past few years, the Army has shifted an increasingly larger percentage of its manpower from combat-service-support activities to combat organizations. This change means that fewer MP personnel are available to support a larger number of units. To meet this challenge, it is necessary to reevaluate the way we do business and to emphasize those programs or procedures that have the greatest impact on our installation crime rates. Crime prevention is one program that can have a major impact on installation crime rates at a relatively minor cost in both dollars and manpower. It takes less effort to discourage a criminal from perpetrating a crime or to teach a soldier to avoid becoming a victim than it does to investigate a crime, identify the offender, prosecute him, and punish him. In addition, a proactive approach to law enforcement can help maintain the high quality of service life that can improve the retention of first-term soldiers.

B-2. The Army is a large organization that performs a variety of activities in many different environments. Crimes that are major problems on one installation may be totally absent from others. For example, most military installations have a significant number of robberies while most depots have none. Because of this, any rigid, centrally controlled program—no matter how carefully thought out—is bound to be inappropriate in many locations. Therefore, DA has elected to provide only the most general guidance and to allow commanders to develop crime-prevention programs that address their local problems.

CRIME-PREVENTION WORKING GROUPS

B-3. The installation is the smallest practical level for implementing crime-prevention programs. If these programs are developed and implemented at a lower level, then crime is often not eliminated but is merely displaced from units with good programs to units with less effective programs. Also, crime does not affect personnel only when they are in their place of duty. In many cases, a company commander's troops are victimized in areas over which he has little control. Unit commanders are responsible for implementing many anticrime measures; however, the selection of overall program goals, the ID of appropriate countermeasures, and quality control should be done at the installation level.

B-4. Crime prevention must always be recognized as a commander's program rather than as an MP program. MP personnel have the expertise to analyze data, identify major problems, and develop lists of possible countermeasures. They should perform these functions in support of an installation crime-prevention council appointed by the installation commander and composed of representatives of all of the installation's major organizations and activities. The advantages of using this type of system are—

- It provides representatives of all major segments of the post population with a forum where they can identify criminal problems that are of the greatest concern to them.
- It allows the representatives of all major commands to review the available options to counter a crime and to select the level of resource commitment that is compatible with their missions and internal priorities.
- It helps ensure that the resources of the entire community, rather than only those of the MP force, are mobilized to attack the problem.
- It is easier to obtain the support of the whole population if its representatives are instrumental in the development of the program.

CRIME-PREVENTION OFFICERS

B-5. The installation's crime-prevention officer is normally a senior NCO or an officer who has a solid background as an MP investigator or a physical-security inspector (PSI). He supports the installation council by performing a crime-data analysis to identify problem areas, drafting programs for the council's consideration, inspecting the implementation of council-mandated measures, and coordinating the efforts of unit crime-prevention officers in the implementation of the crime-prevention program.

B-6. As a member of the PM's staff, the crime-prevention officer develops the law-enforcement section of the crime-prevention program, develops and maintains the written crime-prevention plan, and coordinates crime-prevention programs with civilian police agencies and community groups.

B-7. Crime-prevention officers are also appointed in each organization down to the company level. At this level, written crime-prevention plans are not required; however, SOPs are established. The crime-prevention officers serve as their organizations' focal points for coordinating installation crime-prevention plans; they supervise the implementation of the installation's program within their organizations.

CRIME-PREVENTION PROGRAM DEVELOPMENT

B-8. The starting point for developing a crime-prevention program must be a thorough analysis of criminal activity on the installation. This identifies significant criminal problems that are susceptible to crime-prevention efforts. Crimes that are most susceptible to crime-prevention measures are those for which a high probability of reoccurrence exists. Crimes such as murder normally are not repetitive and are poor candidates for inclusion in the crime-prevention program. Since it is seldom practical to attack all criminal problems simultaneously, they should be prioritized based on their impact on the command's ability to perform its mission and their impact on installation personnel. Next, the whole range of countermeasures that can be used to combat each problem must be identified (see Figure B-1). Table B-1, page B-4, identifies (by offense) programs that have been successful in countering specific criminal problems. Sections III and V of this appendix contain discussions of the strengths, weaknesses, and applicability of the countermeasures listed in Table B-1. Once developed and prioritized, the list of criminal problems and possible countermeasures must be presented to the installation crime-prevention council for action. The council should decide which crimes will be addressed and which countermeasures will be used for each crime. The council must then identify specific objectives for its anticrime campaigns.

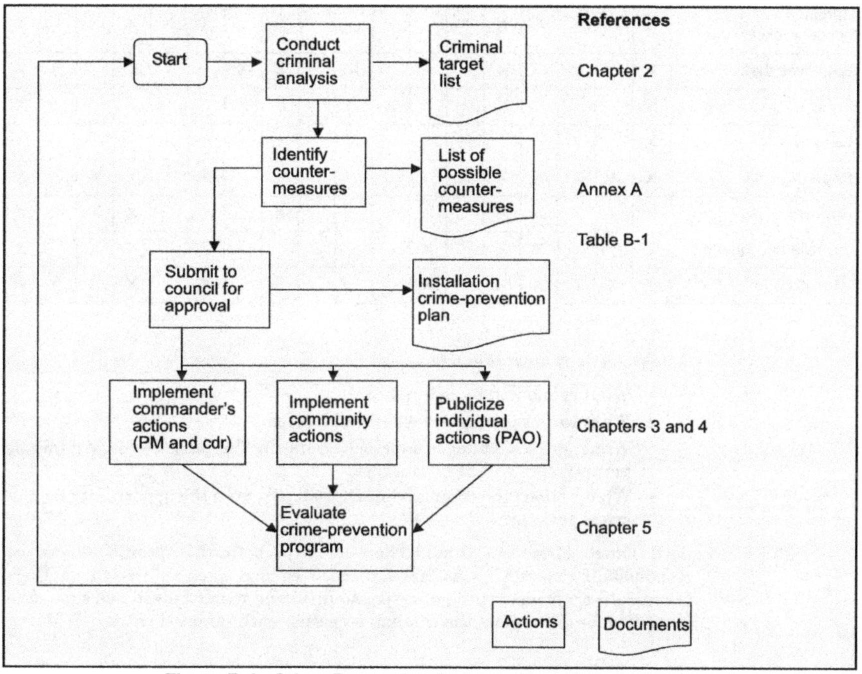

Figure B-1. Crime-Prevention Program-Development Cycle

Table B-1. Offenses Countermeasure Matrix

Crime	Command/Law-Enforcement Countermeasures												Community Programs					
	Crime Hot line	Lighting	Environmental Changes	Patrol/Surveillance	Publicity Campaigns	Residential Security Survey	Juvenile Programs	Fraud Programs	Employee Theft	Arson Programs	CCTV	Warning Signs	Neighborhood Watch	Operation ID	Neighborhood Walks/Escorts	Mobile Patrols	Project Lock	Education
Arson				X	X					X			X			X		X
Auto theft	X	X			X								X		X	X	X	X
Burglary/ housebreaking	X	X	X	X	X	X							X	X	X	X		X
Employee theft	X				X				X		X	X		X				X
Fraud	X				X			X				X						X
Larceny	X		X			X			X					X				X
Rape	X	X	X	X	X	X							X		X	X		X
Robbery	X	X	X	X	X						X		X		X	X		X
Juvenile delinquency	X			X	X		X						X		X	X		X
Vandalism	X	X		X	X		X						X		X	X		X

B-9. Objectives must identify—

- What crime will be reduced.
- What target population will be addressed.
- What specific changes and behaviors on the part of the victims or perpetrators will be encouraged.
- What actions the command must take to reduce the opportunity for the crime to occur.

B-10. Once objectives have been clearly defined, specific areas of responsibility should be assigned to each council member (based on their organization's primary area of responsibility) and major milestones should be identified for developing the campaign against each targeted crime.

TRAINING

B-11. The prerequisite skills for successful performance as an installation crime-prevention officer are best developed through on-the-job experience as the supervisor of MP investigations or physical-security inspections. More important than any technical skill is the cultivation of a frame of mind that instinctively examines each case to determine not only what occurred, but also how the crime could have been prevented. Technical skills (such as criminal-data analysis) that may not have been developed as an MP investigator or physical-security supervisor are presented in courses taught by several civilian agencies. These classes should be used to the fullest extent possible.

CIVILIAN CRIME-PREVENTION ORGANIZATIONS

B-12. There are many civilian crime-prevention organizations at the national, state, and local levels. Many of these organizations have produced crime-prevention material (including posters, radio spots, and leaflets). Material and programs sponsored by civilian agencies should be used to support Army crime-prevention efforts. However, when material from a source outside of DOD is used, a copyright release must be obtained. Normally, it is necessary to get a release for each separate item that is used. If there is any doubt as to the necessity of securing a copyright release, the crime-prevention officer should refer the matter to the local SJA.

SECTION II — CRIMINAL ANALYSIS

B-13. Criminal analysis is a system for identifying trends and patterns where they may exist. It is a routine, ongoing function for the PM and battalion- and brigade-level staffs. Criminal analysis is the foundation upon which the installation force-protection program is based. Moreover, criminal analysis is an integral component of the police intelligence-operations function and is applicable across the operational continuum. An effective criminal analysis establishes the following:

- Crimes having a significant impact on the installation.
- The segments of the population being victimized.
- The ID of criminals/perpetrators.
- The most common time of occurrence.
- The areas that experience the highest number of incidents.
- Offense information (such as types of weapons or victims' actions that contribute to the offense).
- Information critical to an installation's VA.
- Information essential in formulating a successful patrol-distribution plan.
- Police information and criminal intelligence fused with tactical intelligence.

B-14. With this type of information, specific countermeasures are developed to reduce the opportunity for a crime to occur or to remove the incentives for perpetrators. Without an effective criminal analysis, the overall security effort is unfocused. Moreover, the installation/base patrol-distribution plan may be skewed. When this occurs, broad countermeasures are implemented and meaningful results are lost.

B-15. A professional analysis of criminal data is essential for protecting soldiers, units, and installations. The Military Police Management Information System (MPMIS) provided the initial software capable of assisting in the development of criminal-analysis information. As this system is modernized and replaced by the Military Police Automated Control System (MPACS), the MP Corps and the CID will have interface with the Army Battle Command System (ABCS). The analysis of police, criminal, and tactical information results in the development of police intelligence (PI). The PI, which is coordinated with the Assistant Chief of Staff, G2 (Intelligence) (G2) or the Intelligence Directorate (Joint Command) (J2), provides relevant information and intelligence (RII) that greatly contribute to the commander's critical information requirements (CCIR). The RII provided vertically and horizontally ensures that the common operational picture is properly shared.

SOURCES OF INFORMATION

B-16. The basic source for identifying crimes that warrant examination is the installation's Law Enforcement and Discipline Report (DA Form 2819). The information in this report can be used to plot the amount of and seasonal variations for each major type of crime. This will tell little about conditions that produce the crime, but it is useful in identifying crimes that can be eliminated from detailed analyses due to low frequency rates.

B-17. The Law Enforcement and Discipline Report identifies which major category of crime should be targeted, but it does not tell which type of crime is causing the most problems. For example, It may indicate that robbery is a problem, but it will not discriminate between robbery of commercial establishments (such as the post exchange [PX], the Class VI store, or banks) and muggings of individuals. Since countermeasures for these two types of robbery are different, it is necessary to collect additional information. The best sources for this information are the military-police report (MPR) and the CID report of investigation (ROI). If there are fewer than 200 cases in the past year for a particular type of crime, they should all be examined. With a larger annual caseload, a random sample that is large enough to give results of plus or minus 5 percent accuracy should be examined.

B-18. General factors must be identified for any type of crime. These are the types of victims, the perpetrators, geographical data, and chronological data.

VICTIMS

B-19. It is essential to determine (as precisely as possible) the segment of a post population that is being victimized. Junior soldiers live in different areas on post than officers and senior NCOs. They patronize different clubs and, for the most part, work in different areas. Information programs must use multiple sources to ensure that the same message reaches the different portions of the population. If the specific population segment that is being victimized is not identified, it is possible to spend a large amount of resources and have little impact on the targeted crime because the message is not getting to the people who need it.

PERPETRATORS

B-20. As with the victim profile, data on perpetrators is essential to ensure that countermeasures are targeted against the correct population. When on-post auto thefts are committed by civilians with legitimate reasons for entering the installation, it may not make sense to increase security (as a countermeasure) at post entrances. Countermeasures are developed based on specific profiles identified by case type and the information described above. Successful countermeasures depend on correct criminal analyses. An analysis and profile of perpetrators will help to focus MP/CID countermeasures.

GEOGRAPHICAL DATA

B-21. The MPR and the ROI should contain a street address or a description of the incident's location (such as "in a parking lot on the east side of building 1409"). This type of description fulfills requirements for identifying the location of the incident in court. It is further enhanced by identifying the normal duty hours and the type of activity that takes place in the location (for example, "in a parking lot on the east side of the NCO club (building 1401), normal operating hours 1800-0400" or "in a parking lot on the east side of building 1408, a troop billets for Company A, 15th Cav"). This type of information may help to develop a list of specific types of areas where a particular crime is occurring, making it easier for the PM to provide increased MP patrols in these areas. For example, if there have been several incidents in which vehicles have been broken into and stereos removed, it would help to know that 90 percent of them occurred in the parking lot by troop billets.

CHRONOLOGICAL DATA

B-22. As with geographical data, the more specific the time-of-occurrence pattern is for a crime, the easier it is to apply sufficient resources to affect the crime rate. For each crime having a significant impact, determine the following:

- Major seasonal variations.
- Monthly variations. Is there a concentration of crimes immediately before or after payday?
- Weekend occurrences. Is there a concentration of incidents on weekends? Each day represents about 14.25 percent of the week. Concentrations of crime higher than that for any particular day may be significant.
- Time of day. Is there a particular period that accounts for a disproportionate share of the incidents?

CRIME-SPECIFIC FACTORS

B-23. In addition to the factors that should be examined for all crimes, the following crime-specific factors are useful in analyzing specific offenses:

- Housebreaking and burglary.

  - The type of building that was attacked (family-housing unit, troop barracks, PX, and so forth).
  - Whether the facility was occupied or unoccupied.
  - The point of entry (door, window, and so forth).
  - The method of entry (unsecured door, forced door, forced window, and so forth).
  - The property that was stolen. (Was it marked?)
- Robbery.
  - The number of perpetrators.
  - The perpetrator's method of operation.
  - The type of weapon used.
  - The type of robbery (street mugging, residential robbery, or commercial robbery).
  - The victim's injuries.
  - Actions by the victim that contributed to his being targeted.
- Larceny.
  - The type of property taken.
  - Whether the property was secured or unsecured.
  - The perpetrator's method of operation.
- Auto theft.
  - The type of vehicle stolen (POV, motorcycle, and so forth).
  - Whether the vehicle was secured or unsecured.
  - Whether the vehicle was recovered and where it was recovered.
  - Whether the vehicle was stripped of parts.
  - The perpetrator's method of operation.
- Forgery.
  - The type of document that was forged.
  - How the document was obtained.
  - The type of ID used in passing the forged document.
  - The con games or techniques used.
- Rape and sex offenses.
  - The perpetrator's method of operation.
  - The relationship between the victim and the perpetrator (blood relatives, acquaintances, or strangers).
  - The degree of force used.
- Aggravated assault or murder.
  - The relationship between the victim and the perpetrator.
  - The motivation.
  - The weapon used.

B-24. Personnel who have tried to use MPRs and ROIs as a sole source of data know that in many cases the required information is not contained in sufficient detail to be useful. To correct this situation, investigators and inspectors must be trained to recognize conditions that contributed to the crime as well as the information needed to identify and prosecute the offender. As a minimum, the PM ensures that all law-enforcement personnel know general and specific crime factors that are required to analyze each type of crime. He must also provide feedback through the law-enforcement operations staff when incomplete reports are received. If this is done, high-quality data can be collected without generating additional reports to collect criminal information for analysis.

B-25. The MPRs and ROIs are not the only sources of information on which to base criminal analyses. Physical-security inspection results, summaries of common deficiencies noted by the inspector general, data from the SJA claims section, and information summaries from reports of survey on lost government material can all produce worthwhile information on conditions that lead to the commission of crimes. Additionally, a review of civilian criminal statistics and civilian police information may contribute to the pattern analysis. The reviewing authority must understand which information is relevant. The MP commanders ensure that the CCIR are published in operations orders (OPORDs), operation plans (OPLANs), and SOPs. This effort ensures that MP forces at every level recognize relevant information.

INDIVIDUAL CRIMINAL ANALYSIS

B-26. A criminal analysis is most effective when it is applied to the class of criminal offenses with a high probability of recurrence. Single-incident crimes do not lend themselves to analyses. Most crimes against persons do not usually benefit from analyses, with the notable exceptions of rape, robbery, and related combinations of offenses (such as kidnapping and rape; robbery and attempted murder; burglary and rape; and burglary, robbery, and kidnapping). Analyzing isolated criminal offenses has some value (such as gaining knowledge of where these offenses are most likely to reoccur). However, this knowledge is usually difficult to use effectively for preventive purposes.

B-27. There are universal factors available for analysis for most crimes. The availability of these factors varies greatly between criminal types and specific reported offenses.

B-28. In addition to the universal factors, there are an almost infinite number of factors that may be considered specific to a particular criminal class or type. These crime-specific factors are data elements that are usually recorded during the reporting of a particular type of offense and are used for analysis purposes.

B-29. Crime-specific factors provide information that can be used by the analyst to connect crimes with similar characteristics. Information regarding physical evidence may have considerable value in the analysis of several criminal types (such as burglary and auto theft). Therefore, the suitability of different crimes to analyses depends on the general factors, specific modus operandi (MO) factors, and physical evidence. With these considerations in mind, different criminal types and classifications are examined to determine their applicability.

RESIDENTIAL HOUSEBREAKING AND BURGLARY

B-30. Although housebreaking and burglary are two of the most difficult crimes to prevent, they are the most suitable for analyses. The typical housebreaker or burglar establishes an MO pattern based on successful past offenses. Usually, a burglar will continue to commit similar crimes until he is apprehended.

B-31. The available information for evaluation must consider the specificity, accuracy, and value of the information received. Informational factors are presented in the order of relative importance. It should be emphasized that the factors listed in Figure B-2 are for analysis purposes; the order for investigative solutions may vary.

COMMERCIAL HOUSEBREAKING

B-32. The analysis of commercial housebreaking is in many respects easier than that of residential housebreaking or burglaries. More pertinent information is available to the analyst for analyzing commercial housebreaking (which is more specialized and will exhibit more specific MO characteristics). The analyst normally has specific information regarding the point and method of entry, the victim/target description, and the property-loss description. The time factors for commercial housebreaking may be of less value in a commercial-housebreaking analysis than in a residential-burglary analysis since most commercial housebreakings occur at night or during weekends.

B-33. The commercial housebreaker is generally more mobile than the residential burglar. An analysis of commercial housebreaking is not as

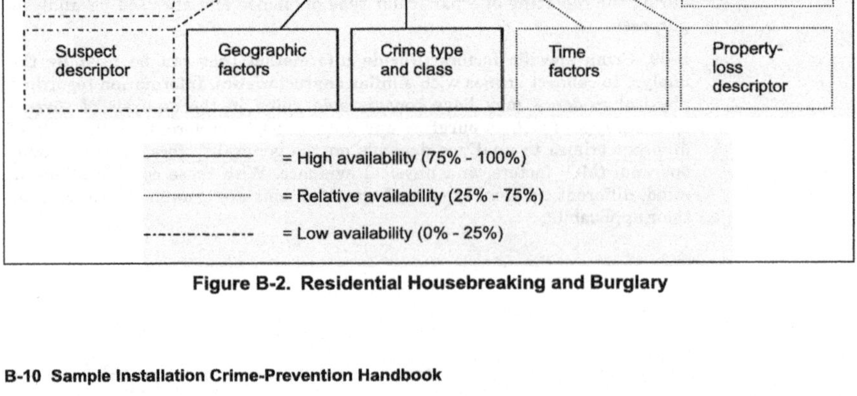

Figure B-2. Residential Housebreaking and Burglary

restricted in geographical area as an analysis of residential burglaries. The commercial burglar may travel a considerable distance to attack a particular type of business. The residential burglar is less discriminating because he has a greater number of potential targets.

B-34. The analysis of commercial housebreaking should be systematically directed to examining the information factors shown in Figure B-3. A commercial housebreaker may be a juvenile criminal, an addict, a professional thief, a soldier, or an enemy agent. Each of these types provides a different set of factors to analyze.

ROBBERY

B-35. Robberies are well suited for analysis. This class of criminal offender usually operates in a given geographical area. The commercial robber generally seeks a sizeable amount of cash; while targets of the street robber frequently include credit cards, checks, and other valuables in addition to cash.

B-36. The presence of physical evidence is more probable in street robberies than in commercial robberies because the criminal offender frequently discards evidence (such as purses and wallets) after completing the offense. The mugger frequently uses surprise as part of his MO in order to reduce his chance of apprehension by physical ID.

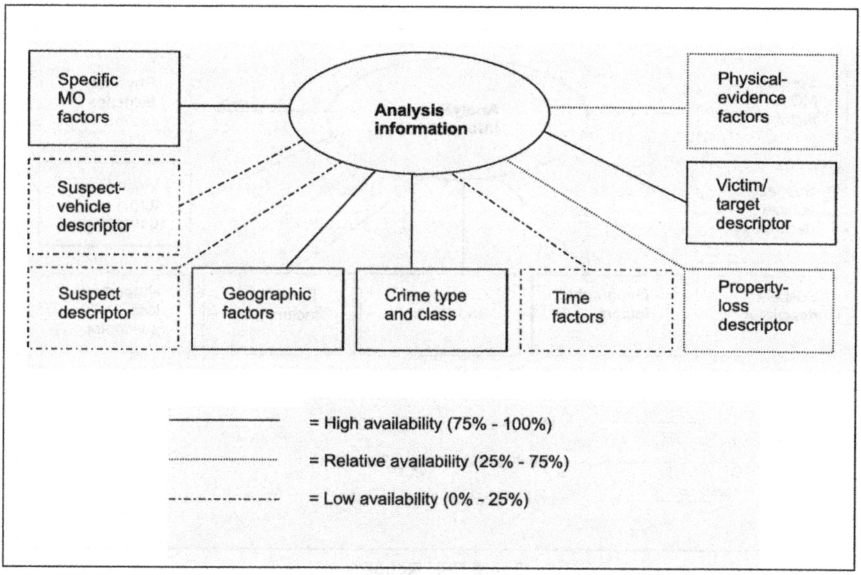

Figure B-3. Commercial Housebreaking

B-37. The probability of universal and crime-specific informational factors for a typical robbery is a systematic method for examining the case. See Figure B-4 for the analysis of a robbery case.

AUTO THEFT

B-38. Auto theft is probably the best-suited crime for analysis. The analyst frequently has more information available on any given auto theft than any other crime. Suspect descriptive information is usually lacking in auto larceny cases. However, the greater availability of crime-specific information makes auto theft extremely receptive to an analysis.

B-39. Auto-theft offenders include joyriders, professional car dismantlers, and wholesalers of stolen vehicles. In many cases, these offenders will establish and maintain a particular MO until apprehended.

B-40. Helpful elements in analyzing auto-theft cases are the availability of information, the presence of physical evidence, and the ability of the stolen property to be traced. Of special importance are the geographic factors, suspect-vehicle descriptors, and property-loss descriptors. In analyzing an auto theft, the analyst usually has information concerning two geographic locations for analysis—where the vehicle was stolen and where it was recovered. The suspect-vehicle descriptors, property-loss descriptors, and victim/target descriptors are the same (see Figure B-5). This greatly augments the analysis of auto thefts. In an examination of specific MO factors in an auto

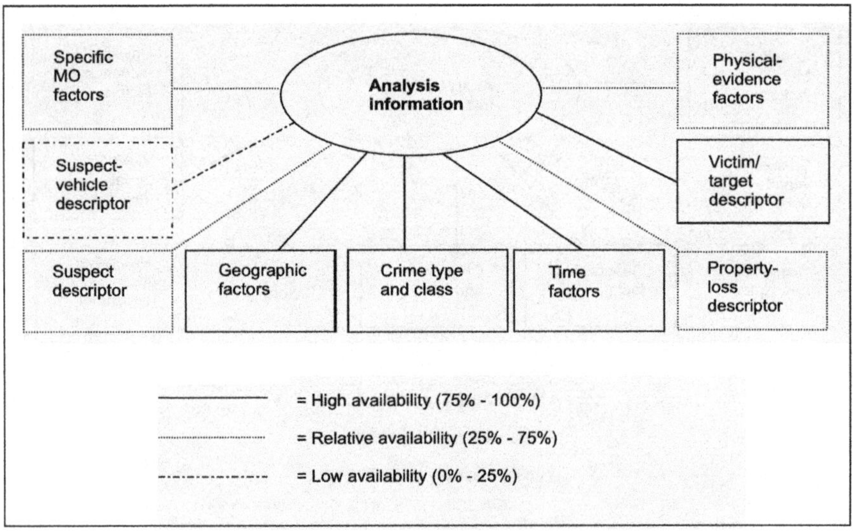

Figure B-4. Robbery

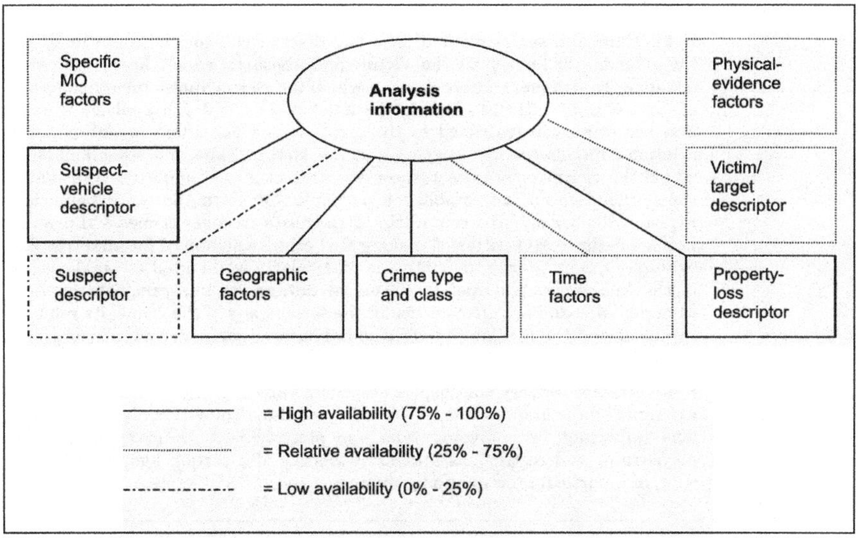

Figure B-5. Auto Theft

theft, the analyst will place emphasis on the vehicle's condition when recovered. Factors of primary importance in analyzing auto-theft cases are the vehicle's recovery location, the vehicle's make and model, and the degree of stripping. Coordination with local civilian police or HN police may reveal potential buyers of stolen automobiles.

B-41. Larceny and theft cases are not usually well suited for analysis. However, benefits are derived from analyzing larceny offenses when specific factors are selected, such as auto accession thefts.

B-42. The problem with trying to analyze general theft cases results from the large volume of reported offenses, the large number of possible crime types, and the many possible MO patterns. To analyze theft cases, the analysis operation must first restrict the number of cases and crime classification to a workable level. This may be done by classifying general theft cases into special categories and analyzing them by considering only special MO factors. Logical classifications include thefts of autos, auto accessories, bicycles, items on shipping docks, and tools and equipment.

B-43. Many of the general theft cases are helpful in analyzing other types of crimes. For example, an increasing trend in stolen automobile parts may have common perpetrators involved in auto theft and stripping. When particular patterns appear, a specific in-depth analysis can be conducted. It is essential that information developed during the analysis process be shared with the G2/J2, the SJA, the CID, and others (as authorized).

RAPE AND SEX CRIMES

B-44. Rape and sex crimes fall into two distinct categories—those in which the offender is known to the victim and those in which he is unknown (stranger to stranger). Those cases in which the victim knows the offender are of limited informational value to the analyst. The in-depth analysis of rape and sex crimes is restricted to those cases in which there is no apparent relationship between the suspect and the victim. Rape and sex offenses in which the victim knows the suspect are often crimes of opportunity. Usually, these offenders do not establish a particular MO. On the other hand, the rapist or the sex offender committing stranger-to-stranger crimes will usually provide definite MO patterns, making this crime well suited for analysis. For example, the burglary/rape motive is particularly well-suited for analysis, as is the kidnapper/child molester. The advantages to analyzing stranger-to-stranger rape and sex offenses lies in the seriousness of the crime, its relative rareness, and the availability of information for a particular offense or group of offenses.

B-45. As with robbery, the suspect descriptors are important to the analyst for examining rape and sex offenses (see Figure B-6). The victim descriptors are also important in analyzing rape and sex offenses. In many cases, the perpetrator will restrict his attacks to victims of a certain age, a particular race, or a particular occupation grouping.

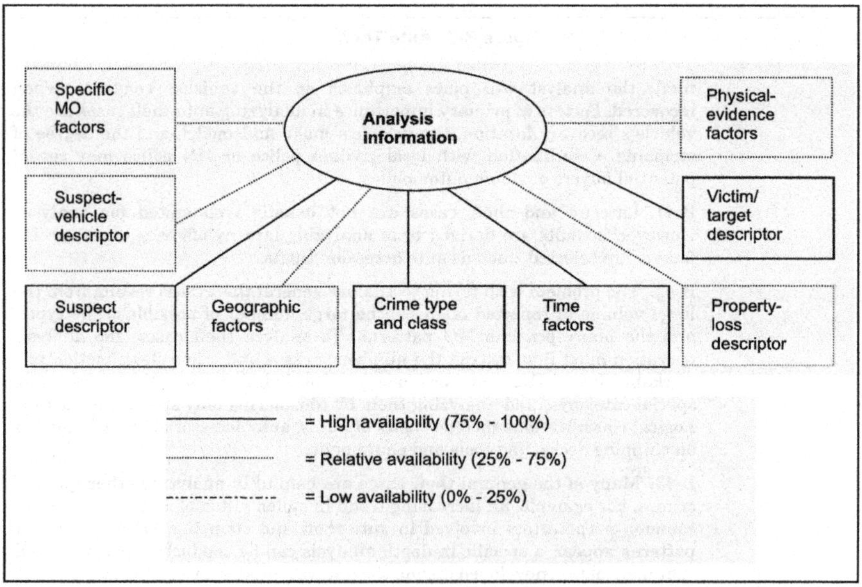

Figure B-6. Rape and Sex Crimes

FORGERY AND FRAUD

B-46. Forgery and fraud cases can provide correlative information for other reported criminal offenses such as burglary and strong-armed robbery. The analysis of forgeries is also aided by the fact that most forgers are repeaters who have established definite MO patterns.

B-47. An analysis of forgery and fraud cases can generate information that is disseminated to area merchants in the form of check-warning bulletins. The use of these bulletins can enhance community relations.

B-48. The primary problem in analyzing forgery and fraud cases is the immediate availability of information. In some cases, a delay of between three or four days to several months occurs between the time the offense was committed and the time it was reported.

ASSAULT AND MURDER

B-49. Most criminal offenses constituting assault and murder do not lend themselves to analysis. The comparative rarity of these crimes involving complete strangers makes it difficult to plot and predict assaults and murders. A tremendous amount of time and research is necessary to predict assaults and murders effectively. This data may identify specific areas with high rates of violent crime. This allows the commander to identify unsafe areas and warn soldiers to stay away from them.

CRIMINAL-ANALYSIS PROCEDURES

B-50. To organize the flow of information within a PMO and to facilitate analyses in support of crime-prevention efforts, the staff reviews a list of offenses. This list contains offenses that lend to analyses. The analyses provide the azimuth for crime suppression and countermeasures (for example, focused investigation/surveillance and patrol distribution). When final action on an MPR or an ROI covering one of these offenses is completed, it is forwarded to the staff for review.

B-51. The staff uses preprinted work sheets containing the factors of particular interest for an offense. They annotate and review the specific criminal factors. If some of the information is not available, the MP or CID investigator continues the investigation. The period covered by each work sheet will depend on the volume of cases at the particular installation. For a low-incidence-rate crime (such as robbery), one work sheet will suffice for the entire calendar year. For a more frequent crime (such as larceny), it may be necessary to use a different work sheet for each month or quarter. A summary sheet can be used to keep track of the data from all of the reporting periods in the year.

B-52. The data-collection sheets give specific information on individual types of crime. This information, together with general crime trends that are identified by comparing the data from DA Form 2819 for the current period to earlier periods and the geographic data from the crime-occurrence map, gives most of the information that is required to develop tightly targeted crime-prevention programs. While this system requires additional forms (the data-collection sheets and the crime-occurrence map) to be developed for use by the

MP Intelligence Officer (US Army) (S2), it does not increase the patrol's workload.

B-53. In addition to the information recorded in the forms, handwritten notes should be kept on items or trends that become apparent to the analyst during the review. For example, as a result of reading the cases on robbery to extract information for the collection form, the analyst may notice that a high proportion of the victims were attacked while crossing an unlighted athletic field near a troop billet area. This should be recorded so that it may be used in publicity campaigns and so that it may be passed along to patrols who routinely patrol the troop billet areas.

B-54. A geographic analysis is performed to determine information not available in the original data-element source (for example, a crime report). When a pin is used to indicate a crime location on a map, the relationship to other reported crimes of the same type becomes apparent to the analyst. Mapping is usually the source for a geographic analysis.

B-55. Mapping-analysis techniques involve the use of a map to depict the actual geographical relationships between particular criminal events according to prescribed data elements. A pin map displaying the actual locations of burglaries over a period of time is an example of such a technique.

B-56. In selecting a mapping technique, several things must be considered. These include the number of data elements that are to be recorded in the mapping, the retrievability of stored data, and the number of maps to be maintained.

B-57. The number of data elements to be recorded on a map can dictate the technique to be used. For example, if the case number is to be recorded in conjunction with the location of the offense, a dot (color-coded, self-adhesive paper disk) map or a flag-pin map can be used, whereas a typical color-coded pin map cannot. If particular MO factors are recorded, various color-coded and marked pins can be applied to identify specific MO patterns.

B-58. In addition, the length of time a particular map or set of maps is maintained should be considered. No set rules are established regarding a map's maintenance time; however, a number of agencies develop annual statistics based upon maps. Generally, maps are maintained for short-term (quarterly) and long-term (annually) periods for each type of crime. This decision should be based on the volume of criminal activity, physical-space limitations, the specific mapping technique adopted, and the number of maps to be maintained. As geospatial-terrain data becomes available, mapping techniques can be fully integrated into the MPACS.

B-59. Stored map data can be easily retrieved for comparing data. Pin maps can be photographically recorded and digitized for software applications. Computers and scanning equipment can be used to store the data for later retrieval. Color-coded paper dots placed on acetate overlays covering area maps facilitate the retrieval of recorded data. Digitized photographs of the maps are excellent sources of evidence for trials and court-martials. Moreover, the developed maps may contribute to the CCIR.

B-60. The number of maps to be maintained must be based on physical-space limitations, staffing, the types of crimes selected for analysis, and the maintenance period for each map. There are five types of crime suited to geographical analysis (burglary, robbery, housebreaking, auto theft, and sex crimes). Other crimes (such as theft of government property) may be depicted as a geographical analysis and contribute to RII. Housebreaking and burglaries can be placed on the same map by using different colored pins for each type of crime. The incorporation of numbered pins of different colors can be used to designate the type of premise attacked. Auto-theft and recovery maps may be kept on the same map with recorded information on thefts of vehicle parts.

B-61. The number of maps maintained should reflect the needs of the agency according to the volume of reported crimes and the different data elements or informational factors to be recorded. Therefore, the more detailed the analysis, the more maps are required. The actual number of maps maintained must be left to the discretion of the PM or the commander.

B-62. Periodically, a summary of the information that is collected, countermeasures, and operational/tactical procedures are briefed to the installation commander. Whether this is done monthly, quarterly, or semiannually will depend on the volume of reported crime and the commander's preference.

CRIMINAL-ANALYSIS SUMMARY

B-63. Criminal analysis is an important element to the MP function of PI. The techniques mentioned in this appendix are not all-inclusive. The PM/CID commander establishes formal procedures for integrating data and developing PI. It may require the formation of an ad hoc team, or it may be accomplished with computers and the MPACS. Criminal analysis requires close coordination with the G2/J2, the public affairs office (PAO), civilian police, and others based on METT-TC.

SECTION III — COMMAND AND LAW-ENFORCEMENT COUNTERMEASURES

B-64. This section discusses actions that commanders or PMs can take to reduce crime on Army installations. In many cases, the actions described have applications other than strictly crime prevention. However, only the crime-prevention aspects are discussed.

B-65. Most of the data available on the effectiveness of law-enforcement measures in reducing crime comes from studies conducted by civilians. While most of the findings of these studies are applicable to crime-prevention programs on Army installations, differences in population, the degree of control that can be exercised by authorities, and other environmental factors may dictate that the civilian recommendation be modified before implementation on military installations.

CRIME HOT LINES

B-66. A crime hot line is a dedicated crime-reporting telephone number located at the MP or security desk. This hot line allows anyone in the community to make an immediate report of an observed crime or suspicious activity. An effective crime-reporting program is a deterrent to crime and enhances law-enforcement responses to such incidents. Considerations in implementing a reporting program include the following:

- It should be publicized that personnel reporting incidents are allowed to remain anonymous if they desire. Some individuals will report their observations only if they know they can remain anonymous. If deemed feasible, neighborhood-watch blocks could be provided designated block numbers that could be used when reporting crimes or other suspicious activities. This system would allow neighborhoods to receive feedback on the disposition of the reported incident.
- The hot line's phone number should be easy to remember. This could include a number where extension digits are all the same, are in ascending or descending order, or spell out a word (such as 4357 [HELP]). Sticker labels listing the number could be placed on the phone with other emergency numbers for quick reference.
- The program should be well-publicized.
- Members on the installation should be educated on the desired procedures for reporting incidents.
- Each call should be documented. Records should be maintained concerning the results of these calls to evaluate the program's effectiveness.

CRIME PREVENTION THROUGH ENVIRONMENTAL DESIGN

B-67. Crime-prevention practitioners are recognizing the importance of considering design and physical planning in crime reduction. The crime-prevention officer has an opportunity to influence the design of facilities through the installation planning board. However, to be effective, he must understand a number of concepts about the relationship between the physical design of buildings and crime occurrences. These include the concepts of territoriality, natural surveillance, and defensible space.

TERRITORIALITY

B-68. Historically, a single-family home on its own piece of land and somewhat isolated from its neighbors (but often by as little as a few feet) has been considered to be the family's territory. The single-family home sits on a piece of land buffered from neighbors and the public street by intervening grounds. At times, symbolic shrubs or fences reinforce a boundary. The positioning of lights in windows that look out on the grounds also act to reinforce the claim.

B-69. Unfortunately, as the population has grown and the need for housing has increased, the trend toward developing single-family units has been paralleled, if not surpassed, by the development of row houses, apartment buildings, and various high-rise structures. Architects, planners, and designers involved in developing structures have not paid a great deal of attention to crime control or the need for an individual or a family group to identify with its home in a manner that might affect crime. Therefore, most families living in apartment buildings consider the space outside their apartment door to be distinctly public. In effect, they relegate responsibility for all activity outside the immediate confines of their apartment to public authorities. A question is whether environmental design can be used to extend the boundaries of these private realms, subdividing public space outside quarters so that more of the common space comes under the resident's influence and responsibility.

B-70. Through extensive research of efficiently functioning housing developments, a number of mechanisms have been identified that may be used in the design process (or may be added after construction). These mechanisms encourage the residents of multifamily dwellings to identify more with the ground or area around their immediate home site and to assume responsibility for its protection. Presented below is a brief discussion of a number of the mechanisms.

- **Site design.** If the grounds around a set of quarters can be directly identified with a particular building and the residents of that building take a personal interest in the use or upkeep of that area, they will play a role in protecting it. Through proper site design, a recreational area adjoining a building may be used as a buffer zone by providing play equipment for young children and seating areas for adults. The fact that children play and adults sit in these areas serves to increase the residents' concerns with the activities taking place there. Strangers are usually recognized and their activities come under observation and immediate questioning.
- **Street design.** Research has shown that by the placement, enclosure, or rerouting of streets and traffic the nature of a particular area can be changed and the crime rate reduced. For example, a particular portion of a street might be closed to vehicular traffic, and play equipment and seats may be added. In a number of areas where this technique has been used, it has been found that most residents know or at least recognize people up and down the block and strangers on the street are identified. Similar approaches that involve rerouting traffic, using one-way streets, or blocking off streets has lowered the crime rate in some areas.
- **Symbolic barriers.** The types of barriers that planners may use in laying out an area include open gateways, light standards, low walls, and plantings. Both physical and symbolic barriers serve the same purpose—to inform an individual that he is passing from a public to a private space. Symbolic barriers identified by residents as boundary lines serve as defining areas of comparative safety. Many places warrant the use of symbolic barriers, including transition points between a public street and the semipublic grounds of a building; an area between a building's lobby and its corridors; or hallways on particular floors of a building.
- **Internal design.** Although economics may sometimes enter the picture, a building's interior may be designed for specific groupings of apartment units and shared entrances. These factors may cause the residents of these apartments to develop a concern for the space immediately adjacent to their dwelling. For example, on each floor of an apartment building, two to four families might be required to share

a common corridor area. The apartment doors would be grouped around that common corridor, and access to elevators or stairs might be screened by a glazed partition. The net effect would be that the floor's residents would adopt the corridor as a collective extension of their dwelling unit and would take an increased interest in its maintenance and use.

- **Facilities and amenities.** The location of particular facilities (such as play and sitting areas and laundry facilities) will tend to give an area a high intensity of use and support the idea of territoriality. The presence of residents involved in various activities (children at play and people chatting or engaged in other types of activities) allows for casual surveillance by concerned members of the family and screens out possible intruders.

B-71. Reducing the number of apartment units grouped together to share a collectively defined area and limiting the number of buildings that comprise a housing project are important factors for creating an environment that residents will help to protect. Research has documented the fact that housing projects comprised of fewer high-rise buildings (two to four) have lower crime rates than projects containing a larger number of buildings. Based on this finding, it is argued that there appears to be much less freedom of movement in the public spaces of the smaller high-rise projects. Unlike buildings and large developments, every building of a small grouping usually has an entrance directly off a public street. These dwellings more closely resemble middle-income, high-rise developments and look more private.

B-72. As a crime-prevention officer, you may not be in a position to directly use these techniques. However, your familiarity with these approaches and the value of their use in the crime-prevention process are important elements in your arsenal of tools to create public involvement in reducing crime. In particular, the purpose of outlining these tools is not to equip you to be a designer, but rather to equip you to communicate with those who are involved in that profession. The discussion that follows will further enhance your ability to converse with designers.

NATURAL SURVEILLANCE

B-73. Experience has shown that the ability to observe criminal activity may not be adequate to stimulate an observer to respond with assistance to the person or property being victimized. The decision to act depends on the presence of motivational conditions, including—

- The degree to which the observer has developed a sense of personal and property rights that are being violated by the criminal act.
- The degree to which the observer feels that the event is within his area of influence.
- The observer's ability to clearly identify whether the act is unusual for the particular area.
- The observer's identification with either the victim or the property being vandalized.
- The degree to which the observer believes he can effectively alter the course of events he is observing.

B-74. Based on these conditions, a number of mechanisms have been identified that can be used to design the grounds and internal areas of apartment units, housing developments, and other residential areas to facilitate natural monitoring of activities taking place. By providing opportunities for surveillance through the positioning of windows in relation to stairs, corridors, or outside areas, continual natural observation will be maintained and crime will be deterred. If such steps are taken, the security of observed areas will be understood by the potential criminal, making him think twice before committing a crime.

B-75. The first of these natural surveillance mechanisms involves the positioning of service areas and access paths leading to apartment buildings to facilitate surveillance by residents and authorities. For example, buildings might be designed so that their entries face and are within 50 feet of a street, so that well-lit paths lead to the front door or the lobby, and so that the lobby is arranged to afford good visibility from the street. Other related steps focus on the strategic placement of windows, fire stairwells, lobby lights, and mailboxes so that they can be easily viewed from the street. Elevator waiting areas on each floor can also be designed so that they can be seen from the street level. Research has proven that if steps such as these are taken, residents will be more likely to become involved with protecting the facility, MP patrols will be in a better position to observe what is going on, and criminals will be discouraged from vandalizing the site.

B-76. A second technique that might be used to increase surveillance is to design facilities so that people within them will naturally view commonly used paths, entries, and play and seating areas during their normal household activities. This concept also focuses on the strategic placement of windows, lighting, and open areas so that natural surveillance by residents is improved.

B-77. Another mechanism involves the subdivision of housing areas into small, recognizable, and identifiable groupings that improve visual surveillance possibilities. Research has shown that in housing developments where the surveillance of a neighbor's outside activities was possible, residents were found to be very familiar with everyone's comings and goings. The overall effect was to cement collective identity and responsibility through social pressure.

DEFENSIBLE SPACE

B-78. Defensible space is a term for a range of combined security measures that bring an environment more under the control of its residents. A defensible space is a residential environment that can be used by inhabitants for the enhancement of their lives while providing security for their families, neighbors, and friends. The physical mechanisms suggested to create safety and improve upkeep (as part of the defensible-space concept) are self-help tools wherein design catalyzes the natural impulses of residents rather than forcing them to surrender their shared social responsibilities to any formal authority.

B-79. Research has revealed investigative techniques that might be used to modify existing housing areas to make them more secure. The following methods may require alteration or adaptation to the particular situation on your installation:

- Widening major pathways and using colored decorative paving.
- Differentiating small private areas (front lawns) outside each dwelling unit from the public path with low, symbolic walls.
- Adding public-seating areas in the center of public paths far enough from private-dwelling units to eliminate conflicts of use but close enough to be under constant surveillance by residents.
- Designing play areas as an integral part of open space.
- Adding new and decorative lighting to highlight various paths and recreation areas at night and extending the residents' surveillance potential and feeling of security.
- Adding seats and path networks to recreational facilities where large, central court areas exist. This increases the interest and usability of the areas.
- Redesigning parking and play areas around buildings to create the illusion that the buildings are grouped where natural opportunities exist.
- Modernizing building entrances to create breezeways into building courts and to accommodate a telephone intercom for opening entry doors to the lobby.
- Providing video surveillance of public grounds and central paths by security of public monitors.
- Installing audio surveillance capabilities in elevators and at the doors of residences.

CRIME-PREVENTION MODEL

B-80. The model for crime prevention through environmental design is based on the theory that action must be taken to counter crime before it occurs. The critical element in this model is the environmental-engineering component. It provides both direct and indirect controls against criminal activity by reducing the opportunity for crime through science and technology and the use of various urban planning and design techniques. The model explains what environmental engineering is and how it supports crime prevention. With this information, you may be in a better position to understand and respond to questions and discussions on how urban design and planning can have an impact on the installation's criminal element.

THE ENVIRONMENTAL INFLUENCE ON CRIMINAL BEHAVIOR

B-81. The basic theory that supports crime prevention through environmental design is that urban environments can influence criminal behavior in two ways. First, the physical surroundings in which people live have an effect on each individual. These physical characteristics include noise, pollution, overcrowding, and the existence and unmonitored spreading of refuse and other unsightly waste. The second element that must be dealt with in the environmental-engineering formula concerns the social characteristics of the community that provide individuals with social relationships to which they must respond. Characteristics such as alienation, loneliness, anxiety, and dehumanization are seen as keys to criminal behavior.

B-82. In terms of these environmental characteristics, buildings are all too often constructed to be dangerous, with corridors and passageways hidden from public view. Elevators, basements, and storage and washroom areas are also laden with danger due to their design. Various large-scale housing developments are not secure in that they are often isolated from the main flow of traffic (both human and automobile) and are closed to public use and public view.

B-83. With regard to altering the social characteristics of the community and their relationship to criminal behavior, it should be recognized that behavior is future-oriented, not past-oriented. A man steals so that he can have a car or money in the future, not because in the past he experienced psychic trauma, a broken home, poverty, or delinquent associates. Criminal behavior can be explained directly in terms of the consequences of behavior and in terms of noncriminal variables such as poverty, race, or social class. Criminal behavior is viewed as a problem to be dealt with and not symptomatic of other problems (such as poverty, mental conflict, class conflict, unemployment, or undereducation). To change criminal behavior, it must be dealt with directly by removing the environmental reinforcement that maintains the behavior. The approach advocated is to change the environment to which the individual responds.

ACTION APPROACHES TO CRIME PREVENTION THROUGH PHYSICAL PLANNING

B-84. The primary focus of crime prevention has been on what architects, planners, and other nonpolice professionals can do in terms of various physical-planning strategies to reduce criminal opportunity. Experienced MP personnel have long recognized that certain physical conditions can contribute to the rate and nature of crime. They have also developed a capability to identify high crime-risk locations by noting such factors as poor lighting and weak points of entry as potential targets. The critical job is to identify specific areas concerning physical planning and design that can be responded to and actions can be taken against on the installation.

B-85. Attempting to reduce crime or the fear of crime by regulating physical environments is easier said than done. In fact, although crime prevention can be built into almost every aspect of community planning, it is often ignored for a number of reasons. For example, fragmentation of responsible agencies is a key problem. In addition, crime has historically been looked upon as the exclusive responsibility of MP forces; not of those in charge of education, housing, or health and welfare. Yet, with an understanding of crime prevention combined with the knowledge that design techniques can change the opportunity for criminal behavior, PSIs will be able to talk the language of the planner and the designer and to be able to advise them from a police perspective.

B-86. It is notable that a number of civilian police agencies have become involved in the physical-planning process and have achieved notable results from their work. For example, the Fremont, California, Police Department has been involved in a planning process and maintains that law enforcement should become an integral part of the master- or comprehensive-plan review to screen all redevelopment plans for safety and crime hazards. Working with other units of municipal government as well as architects and designers, the department drew up a set of model guidelines for the evaluation of projects. The model included evaluation criteria dealing with such subjects as the accessibility of buildings to patrol units; traffic flow and off-street parking provisions; and the location and regulation of cul-de-sacs, playgrounds, common greens, fences, and security entrances. In addition, working with agencies such as the American Institute of Architects, the National Public Works Association, the Association of Public Utilities, and others, the department identified a number of subjects that are of specific concern to police officers and that should be considered in the design and planning stage. As a result of these efforts, the following list of design concerns was developed by the department:

- Building setbacks (front, side, and rear).
- Wall construction, interior and exterior (industrial, commercial, and residential).
- Door construction, setbacks and security (industrial, commercial, and residential) (including carports, garages, and sliding-glass doors).
- Windows and skylights, setbacks, heights (from ground), show-window displays, and the type of frame or pane.
- Stairs (stairwells and staircases).
- Balconies.
- Utility boxes.
- Fences, walls, hedges, screens, setbacks, heights, and louvers.
- Parking (public and private).
- Lighting (industrial, commercial, and residential).
- Streets, sidewalks, and walkways (locations, slopes, curvature, grades, and the length of a block).
- Alleys (blind and through alleys).
- Visibility of valuables (people, safes, cash registers, and personal property).
- Signs (street signs and signals, traffic signs and signals, and advertising signs).
- Accessibility; approach, entrance, and exit (pedestrian, vehicular, services, residential, commercial, and industrial).
- Public utilities and easements (gas, water, telephone, and electrical).
- Public areas and facilities (public restrooms, parks, bus stops and shelters, playgrounds, recreation halls, and so forth).
- Street trees and shrubbery (types, heights, and locations).

B-87. With this information, you can improve the security aspects of the community's physical-planning process. There is a probability that the work and recommendations of the National Advisory Commission on Criminal Justice Goals and Standards may help you in your efforts to get involved in the design process. More specifically, the Commission noted that "every police agency should participate with local planning agencies and organizations, public and private, in community physical planning that affects the rate or nature of crime or the fear of crime."

B-88. The future role of MP forces in crime prevention through physical planning will depend on the PSI's initiative. The perspectives and knowledge of what is happening in this field, combined with a working knowledge of the language, should equip PSIs to sell this approach as part of the overall program. It is important to point out that others who represent professions other than crime prevention are aware of the relationships between urban planning and crime. Remember, they deal in concepts, approaches, and ideas that have not involved the realities that law-enforcement personnel face.

SPECIALIZED PATROL TACTICS AND SURVEILLANCE

B-89. Specialized patrol operations use a variety of tactics in attempting to control identified crime problems. The most common tactics include uniformed tactical patrols and suspect and area surveillances. The following paragraphs discuss these tactics in terms of their target crimes, operation requirements and characteristics, and established or perceived levels of effectiveness.

B-90. The appropriateness of a given tactic depends on the characteristics of a particular crime problem. The selection of specialized patrol tactics should be made on the basis of a careful and continuous analysis of crimes. Most crimes can be addressed by more than one tactic. Several tactics might be tried in an effort to find the best one, and it is quite possible that the most effective approach to a given crime problem will include the combination of several tactics.

UNIFORMED TACTICAL PATROLS

B-91. A uniformed tactical patrol is the most traditional and widely used form of specialized patrol. It is a simple, straightforward approach to specialized patrol that involves the same procedures and techniques used by MP officers on routine patrol. These include constant visible movement throughout an area to generate a sense of police presence, careful observation of street activity, vehicle and pedestrian stops, and citizen contacts. The difference between uniformed tactical patrols and routine patrols are that uniformed tactical patrols use these tactics in an intense, concentrated fashion. MP officers are relieved of the responsibility for responding to routine calls for services so that they can devote their full time and attention to patrol, thus intensifying its impact. In addition, uniformed tactical operations typically deploy a number of MP officers in target areas, thereby increasing the level of patrol in these areas.

B-92. Uniformed tactical patrols can be used to control virtually any type of suppressible crime (for example, crimes that can be viewed from locations where the police have a legitimate right to be and those that can be potentially affected by police operations). These suppressible crimes include street robberies, purse snatches, vehicle thefts, burglaries, and housebreakings. Uniformed tactical patrols can also have an impact on other types of crime as officers use observation, field interrogation, and citizen contacts to develop information on the locations, activities, vehicles, and associates of suspects.

B-93. The primary purpose of uniformed tactical patrol is deterrence. This tactic is based on the assumption that highly visible, active patrols will deter potential offenders. By increasing the perceived probability of apprehension, conspicuous patrol is thought to reduce the likelihood that crimes will occur. If the deterrence should fail, heightened patrol coverage is believed to increase the probability of the immediate apprehension of the suspects.

B-94. Uniformed tactical patrols are often used to saturate an area that is experiencing a particularly serious crime problem. Although it has been widely used for years, saturation patrol has never been clearly and adequately defined. Exactly what level and intensity of patrol constitutes saturation has never been determined, nor have the effects of different levels of patrol been clearly established. It is difficult to prescribe the level of uniformed tactical patrols that should be used to disrupt a crime pattern in a particular area. This should be determined through an analysis of the size and characteristics of the area of concentration of each potential target crime pattern coupled with an assessment of manpower availability.

B-95. Some patterns can be effectively handled by a very small increase in the patrol level. For example, in Portland, Oregon, Operation Crime Reduction Involving Many People (CRIMP) had a significant impact on street robberies and assaults in the city's skid-row area by initiating two-officer, uniformed foot beats in the area. This was sufficient to saturate the primary locations of the target crimes during the high-crime hours, and it led to a substantial reduction in these crimes with little apparent spillover into adjacent areas.

B-96. Other departments have used saturation patrols on a much larger scale. For example, in the mid-1950s, the New York City Police Department attempted to saturate an entire precinct by assigning over 200 additional officers to the precinct's patrol force. Foot beats covering extremely small areas were arranged in straight lines so that an officer could keep the entire street area of his beat in view at all times. The four-month experiment led to a dramatic reduction in crime in the precinct. Compared with the same period in the previous year, street muggings fell by 89.9 percent, burglaries where entry was made from the front of the building dropped by 78 percent, and so on. The only crime category that was not affected was the relatively private crime of murder. Since crime displacement was not examined in the experiment, its true impact remains unknown. The experiment strongly suggests that massive additions of police manpower can have a substantial effect on crime. The problem is that most departments do not have the ability to conduct even a limited version of this experiment.

B-97. The amount of resources required for saturation patrols can vary tremendously, and there is no definite way of determining how much additional patrol is needed. This can best be decided on a problem-by-problem basis, using experience and evaluations of past efforts as a guide. As a rule, patrol augmentation should be sufficient to affect rather quickly the perceptions of would-be offenders concerning the level of police activity in a particular area. It appears that for such a strategy to be effective, sufficient resources should be applied in a limited area to ensure a true saturation effect.

B-98. Uniformed tactical patrols can use various modes of transportation. Patrol cars are most often used; however, foot patrols can be effective in congested service districts and bikes have been used to good advantage in high-density residential areas. The mode of transportation should be selected based on visibility, mobility, and access to citizens.

B-99. Some specialized units deploy MP officers in unmarked police cars. This is done in an effort to effect a balance between overt and covert operations, hopefully gaining many of the advantages of both. Unmarked cars may also be readily available since in many departments investigators work primarily during the day, which leaves their vehicles free for specialized patrol in the evening and early morning watches. This approach has serious drawbacks. First, unmarked MP cars are somewhat less visible than marked cars, yet they are still easily recognizable as police vehicles to large segments of the public (especially when the officers in them are in uniform). Second, the use of these cars in uniformed tactical patrols could lead to the sacrifice of some of the deterrent effects of high visibility without realizing the apprehension benefits of truly covert patrols.

B-100. Several patrol techniques have been tried to increase the visibility of uniformed patrol and to enhance the sense of police presence. A tandem patrol uses two marked patrol cars that follow each other at intervals of one-half to several blocks. Two units can also patrol on parallel streets one block over or in an alley. Another approach combines foot and vehicle patrols in an effort to increase visibility. Officers park their marked cars in conspicuous locations in high-crime areas and then are transported to other high-crime areas where they patrol on foot. The frequent repetition of this procedure is seen as a way of multiplying patrol visibility.

B-101. Unless a target area is more or less completely saturated (as in the New York experiment), MP patrols should move in a random, unpredictable pattern. This will make it difficult for potential offenders to plan their crimes on the basis of observation of patrol activities. One department found that its uniformed-tactical-patrol operation was actually helping burglars to commit break-ins. Interviews with confessed burglars indicated that they were aware of a visible patrol passing through the neighborhoods at regular intervals and they planned their crimes accordingly.

B-102. In addition to increasing the level and visibility of patrol in high-crime areas, uniformed tactical patrols often use aggressive patrol tactics involving frequent vehicle and pedestrian stops. MP patrols stop, question, and perhaps search citizens when there are reasonable grounds for suspecting that the citizens may have committed, may be committing, or may be about to commit a crime. Since the concept of reasonable suspicion is vague, MP officers have a considerable amount of discretion in conducting field interviews. Field interviews that do not result in either immediate arrest of the citizen or in alleviation of the officer's suspicions are usually documented in field-interview cards or spot reports. Field interviews serve to generate information about the activities of probable suspects; more importantly, they make the suspects aware that the police know of their presence in a given area, regard them as suspicious, and are watching them closely. This is expected to reduce the likelihood that they will commit crimes, at least in the area in which the tactical force is working.

B-103. The extensive use of field interviews is often highly controversial. Tactical units that emphasize field interviews have been accused of being storm troopers who constantly harass citizens and abuse their rights. There is sometimes a certain amount of truth to the allegations of harassment, especially when the activities of a particular type or group of suspects are being closely monitored. Not surprisingly, individuals who are stopped and questioned frequently are likely to complain, particularly if they have reason to be concerned about close police scrutiny. A recent study of field interviews in San Diego found that when conducted at moderate levels, field-interview activities do not have a major effect on police and community relations. The majority of citizens in all three study areas accept field interviews as legitimate and properly conducted police activity. The majority of all citizens who were the subjects of contacts felt that the contact was justified and properly conducted. The study found that the suspension of field interviews in a particular area was associated with a significant increase in the level of suppressible crime and when reinstated in the area, these crimes declined. The results of the San Diego study provide confirmation for the widely held belief that field interviews can have an important deterring effect on suppressible crime without doing irreparable damage to police and community relations.

B-104. The potentially negative impact of field interviews on certain segments of the community can be held to a minimum if the interviews are conducted in a professional manner. Citizens should be informed of the reasons why they are being stopped. They should be detained for as little time as possible and should not be subjected to verbal or physical abuse. There is also no need to stop everything that moves.

B-105. While experience and a limited amount of research indicate that uniformed tactical patrols can have a positive impact on the level of suppressible crime in areas, the overall effectiveness of this tactic is a controversial and much-debated issue. The principal concern is that rather than reducing crime, uniformed tactical patrols may simply lead to its displacement from one area to another or from one period to another.

B-106. When conducting uniformed tactical patrols, a PM activity should carefully monitor crime trends for indications of possible displacement effects. This is an important aspect of evaluating the impact of tactical operations, and it will provide information to guide future deployments and tactical decisions.

AREA SURVEILLANCE

B-107. Covert patrol and surveillance of high-crime areas can be used to make apprehensions for crime problems. These problems include those for which there are no suspects who warrant personal surveillance, the suspects are too numerous to permit personal surveillance, and there are too many potential targets to conduct either physical or electronic stakeouts. Examples of these types of problems would be a rash of residential burglaries or auto thefts in a particular area.

B-108. This tactic simply involves the covert patrol of a particular area and the observation of suspicious or unusual activities and occurrences that might indicate the likelihood of a crime. Suspicious individuals are not stopped but are watched until they either commit an offense or the officers' suspicions are removed.

B-109. The list of various techniques that can be used in an area surveillance is virtually endless. The following are some techniques that have been effectively used by specialized patrol units:

- Mingling with citizens at the crime scene to pick up information on possible suspects.
- Maintaining rooftop surveillance of a shopping center's parking lot to locate larcenies from vehicles.
- Surveilling housing areas by posing as maintenance workers.
- Following likely crime victims such as elderly citizens leaving a bank.
- Surveilling rooftops for unusual activity from aircraft or higher buildings. Binoculars are used to facilitate surveillance, and rooftops are marked so that street units can be dispatched to check out suspicious circumstances.

B-110. As in all types of plain or uniformed patrol, care should be taken to ensure that area surveillance is truly covert. Rental vehicles that can be changed frequently provide an excellent, though expensive, means of covert transportation.

B-111. Police on covert patrol should be dressed to blend in with the environment in which they are working, and they should have apparently legitimate, nonpolice-related reasons for being where they are. Several specialized units have found that surveillance teams composed of one male and one female officer can work in many situations without arousing suspicion. An apparently married or romantically involved couple lingering in a park, meandering slowly down the street, or sitting together in a parked car would generally appear less suspicious than two male MP officers doing the same things. Finally, it should be noted that in some small neighborhoods where residents know each other well, covert surveillance may be difficult, if not impossible, since the presence of any stranger arouses immediate curiosity and suspicion.

SUMMARY

B-112. These tactics represent the basic approaches that specialized patrol operations normally take in trying to control suppressible crime. Some of the tactics (such as uniformed tactical patrols) are directed primarily at crime deterrence, while others (such as suspect surveillance) are used to achieve apprehensions for target crimes. The tactics are most commonly used independently of one another. However, there are some indications that the combined use of several tactics in an integrated operation might be an effective way of coping with particular types of crime. Especially promising is the coordinated use of highly visible and covert patrols. A visible patrol force could be deployed to a particular area to deter crime there and direct it toward other areas in which MP forces using covert tactics are working. To date, efforts to direct criminal activity to areas or targets where MP forces are set up to make apprehensions have only been tried on a sporadic basis. However, this appears to be a promising approach to crime control and warrants greater attention in the future. It can be viewed as the creative use of crime displacement.

B-113. Many of the tactics mentioned previously could not be effectively used by MP patrols that are responsible for handling routine calls for service. It is obvious that the effectiveness of decoy operations, physical stakeouts, and suspect surveillance would be destroyed if the MP officers had to interrupt these activities to handle calls for service. It is also generally unwise to have MP officers in civilian clothes respond to calls for service.

B-114. The importance of crime analyses in identifying crime problems and suggesting appropriate tactics deserves repetition. Specialized patrol units should be deployed to address clearly identified crime patterns, and they should choose their tactics on the basis of an analysis of these patterns and a comprehensive knowledge of the area of occurrence. The nature and characteristics of a particular crime pattern should be the driving force behind the selection of tactics. None of the tactics of specialized patrol can be effectively used unless crime patterns have been identified and analyzed.

PUBLICITY CAMPAIGNS

B-115. Public-information campaigns are an essential part of every crime-prevention effort. The installation PAO can provide assistance in—

- Informing the public of the magnitude of the local crime problem.
- Disseminating information on crime circumstances.
- Generating the interest and enthusiasm necessary to initiate and sustain community crime-prevention programs.

B-116. The PAO should be a member of the installation's crime-prevention council and must be involved from the start in planning crime-prevention efforts. There are many information vehicles at the installation level available to carry the message to the public, including—

- Digital marquees.
- Radio.
- CCTV.
- Installation and unit newspapers.
- Posters and leaflets.
- Commanders calls and similar meetings.
- Town-hall meetings.

B-117. Not all media are equally effective in reaching a particular segment of the post population. To ensure that the media campaign is effective, identify the segment of the population requiring the information and the exact message to be communicated. The more specific the target audience, the more effective the media campaign will be.

B-118. Whatever media is used, it is important to provide coordinated input to the PAO well in advance of the desired publicity campaign date to allow for writing, rewriting, and publishing. For material to be published in a magazine, a minimum lead time of six months is normally required. For installation newspapers, a lead time of a week may be sufficient. Check with the PAO to determine the correct lead times for local publications.

B-119. When addressing civic groups or school audiences, films are often more effective than straight lectures because of the animation and other special techniques used to illustrate critical points. Several Army films are available as are a large number of commercial films.

RESIDENTIAL-SECURITY SURVEYS

B-120. A security survey is an in-depth, on-site examination of a physical facility and its surrounding property. The survey is conducted to determine a residence's security status, to identify deficiencies or security risks, to define the protection needed, and to make recommendations to minimize criminal opportunity. Because of several common characteristics, one expert has likened the security survey to the traditional criminal investigation. This comparison hinges primarily on the facts that both techniques are systematic in nature; are aimed at identifying the method of a criminal act; and are, in effect, more an art than a science. It should be recognized, however, that the survey has two other advantages. First, it can be conducted before the commission of a crime; second, it can offer protection against, rather than just remedial action after, criminal victimization.

THE RESIDENTIAL-SECURITY SURVEY AS A CRIME-PREVENTION TOOL

B-121. A residential survey is not a substitute for an aggressive neighborhood watch. It supplements these efforts and should be established after Neighborhood Watch and Operation ID programs are established.

B-122. The security survey is the primary tool used in crime prevention to recognize, appraise, and anticipate potential loss in residential areas. It is often defined as the backbone of a local crime-prevention program. In practice, it combines the security experience, training, and education of the local crime-prevention officer and focuses on a single element—the analysis of a residential facility.

B-123. The inherent value of surveys has been proven by nearly 300 civilian police departments across the country that have established crime-prevention bureaus or units. An even broader endorsement of the survey technique, however, was provided by the National Advisory Commission on Criminal Justice Standards and Goals, which stated that, "Every police agency should conduct, upon request, security inspections of businesses and residences and recommend measures to avoid being victimized by crime."

B-124. In short, the security survey is a tool that informs a homeowner of the particular areas in which his home is susceptible to criminal victimization along with steps that can be taken to reduce and minimize that potential. Further, the survey is a tangible document that reflects the efforts of the MP force not only to be responsive to community needs, but to get the community more directly involved in the criminal-justice process.

THE CRIME-PREVENTION SURVEY—A PROCESS OF INVESTIGATION

B-125. An important factor in understanding the residential-security survey is that it must be considered as an ongoing process. While a particular survey will result in specific recommendations, each survey will provide a foundation for future action. In combination, these surveys will provide a database that can be used in the analysis of the community's crime problems and guide action toward the resolution or reduction of the problem on a community-wide basis.

B-126. As a starting point, five steps must be used in carrying out the actual survey, while four additional steps must be remembered afterward. These steps are—

- Analyzing the overall environment (neighborhood, block, and so forth).
- Assessing the general vulnerability of the premises.
- Defining the specific points of vulnerability.
- Recommending specific security procedures.
- Including specific remedial hardware recommendations.
- Urging the implementation of the recommendations.
- Conducting a follow-up to ensure that recommendations have been implemented.
- Keeping crime statistics to evaluate the survey's effect and the implementation of recommendations.
- Conducting a second survey of the premise's statistical analysis to determine the alteration in criminal activity in the areas surveyed.

B-127. This process is the continued involvement and participation of the police with the community. Once the survey is completed, the job is not finished. In fact, if this posture is assumed, it may later be learned that the recommendations were not implemented and that the work was done in vain. This can easily lead to a loss of community-wide public confidence in the programs. Similarly, even in those cases when recommendations are implemented, additional crime might be experienced. Attention to this fact, coupled with immediate follow-up, will be essential to avoid losing the confidence gained from the original survey. Through prompt actions, the proposal of additional tactics may become the final step needed to substantially reduce criminal opportunity. In short, the security-survey process is not a one-shot operation and it is not to be looked upon as a simple, straightforward task. It is a continuing, difficult, rigorous, yet effective approach to reducing crime that has not been systematically applied by police agencies in this country.

THE ROLE OF POLICE IN IMPLEMENTING CRIME-REDUCTION PROGRAMS THROUGH SECURITY SURVEYS

B-128. At a minimum, there are two ways to encourage people to improve their personal security. First, you can organize, conduct, and participate in broad-based public-information and -education programs that make use of such media as radio, television, and the press. Second, you can organize, undertake, and follow up on a series of person-to-person security surveys. Clearly, both of these techniques have their advantages. The security survey has a unique quality that does not exist in the public-education program. It provides an added incentive on the part of citizens to implement recommendations because of the personal relationships and respect established during the actual survey by a crime-prevention officer.

B-129. For example, consider a homeowner listening to someone on television recommending improvements of security by installing better locks or alarm systems. You might ask yourself these questions: How secure am I? Are the locks that I have on my doors adequate? If they are inadequate, what kind of locks should I install? Do I really need an alarm system? What can the MP officers do for me in terms of making recommendations about security? In short, you would be aware of the possible need for improved security, but you would not know enough of the specifics to warrant real action. Because home security is a complicated matter requiring careful analysis and the installation of carefully tailored systems, implementation must be approached in a personal way. Certainly, public-education programs can assist in some aspects of crime prevention (such as prevention against schemes, auto theft, and personal protection), but they are not as effective in causing improved security as is personal contact.

B-130. Moreover, for the first time, police are placed in a position where they can actually provide a different kind of advice or service that can be offered in an environment where a crisis has not yet occurred. This is a truly unique opportunity for crime-prevention officers.

CONDUCTING THE SURVEY

B-131. To develop a proper perspective of the types of crimes that a PSI will most frequently be trying to reduce, a review of the cases is necessary to get a broader feel for the actual conditions in your area. During this review process, pay specific attention to photographs in the files. Study crime scenes in an effort to identify the type of security device that was defeated. In particular, if a door was used as a point of entry, note whether it was of adequate construction; if the door frame was broken or separated; if hardware, such as strike plates and door trim was inadequate; and so forth. In addition, review photos to determine if lock cylinders were pulled or if door chains fastened to trim moldings were simply pulled away to permit easy entrance. If photographs of the crime scenes are not available, visit as many crime scenes as possible. While doing so, photograph security risks you can study later and use as examples in future presentations to community groups.

B-132. By becoming familiar with the MO of persons committing such crimes as burglary and larceny, you will be better-equipped to understand potential risk situations and to point them out to potential victims. Quite surprisingly, many of the cases investigated were invited by some obvious crime-risk hazard that was overlooked by a resident. Also, many additional crime risks existed at a particular crime scene that a burglar could have exploited. Such vulnerability might be an indication of other crime targets within the community to which you should pay particular attention in your survey work.

B-133. To be an effective security surveyor, it is necessary to develop an intimate knowledge of the crime factors in your community. You do not have to become a statistician; however, the more you know and understand about crime problems, the methods used in your community, and security devices or systems that were defeated, the better you will be equipped to analyze the potential crime when surveying a home.

B-134. In addition to the general types of crimes that occur in your community, it is necessary to develop an understanding of the details of particular types of offenses. For example, with regard to residential housebreaking, you should be familiar with the types of burglaries and approaches used in particular sections of the community.

B-135. On the surface, it might appear as if this would be a monumental task. However, in terms of housebreaking, you might pull the files for the last quarter and carry out the following steps:

- Tally the number of times entry was made through the front door, rear door, or through a window.
- Identify the approach used for entry (kicking the lock, throwing a shoulder through the door, jimmying the lock, and so forth).
- Determine how often a window was broken or removed or a mechanism was used to force the latch (when entry was made through a window).
- Attempt to determine whether security devices were used in residences (such as alarms, special lighting, or other systems).
- Identify the burglar's general escape route (down a back alley, through a schoolyard, and so forth).

B-136. In developing an understanding of the MO of crimes, statistics that illustrate exactly what is happening in your community will be valuable tools. Not only will you be able to use this information in explaining crime risks while you are surveying the site, but it will be invaluable in making public presentations.

B-137. Only when you have developed the ability to visualize the potential for criminal activity will you have become an effective crime-scene surveyor. It is important that when you arrive on a survey site, you are prepared to give a property owner sound advice on the type of security precautions he should consider.

B-138. In summary, to be a good crime-prevention surveyor, you have to be a good investigator. You must understand the criminal's method of operation and the limitations of standard security devices. In addition, you must be knowledgeable about the type of security hardware necessary to provide various degrees of protection. (Specific information on conducting surveys is contained in Chapter 11).

JUVENILE CRIME PREVENTION

B-139. Law enforcement has long recognized the importance of establishing personal contact with youth. This need was first met with the "Officer Friendly" concept. The goal of this concept was to enhance the image of police among younger children and attempt to negate the unfavorable image other segments of society provided children. This concept has evolved into today's Drug Abuse Resistance and Education (DARE) program which has spread to most law-enforcement agencies.

B-140. Crime-prevention officers can have an impact on juvenile crime by establishing a positive interaction with juveniles. Many areas suffer a high crime rate of which over 50 percent is attributable to juveniles. Nationally, it is known that youth 17 years of age and under (who make up just 16 percent of the population) commit 42 percent of the crimes that cause injury or loss of property. As the punitive aspects of the juvenile justice system today are of questionable value, it is logical that crime-prevention officers should develop programs aimed at juvenile crime.

B-141. The basic problem in establishing any juvenile crime-prevention program is the destruction of preconceived ideas by both youths and law-enforcement personnel. This obstacle can only be overcome by the building of an honest rapport with individual youths. Just as a crime-prevention officer must sell crime-prevention procedures to the public, he must also sell himself and his program to the youth.

B-142. One method of establishing the needed rapport with youth reverts to a similar system used in the Officer Friendly concept. Classroom presentations in high schools and junior high schools can be used to destroy the preconceived ideas held by youth regarding police and law enforcement in general. Once an MP officer enters the classroom, he discovers that the majority of students obtain their ideas of police one of two ways—

- From television shows that portray every officer as a slow-witted moron (such as the old "Car 54" show).
- From their peers at school who relate every negative aspect of law enforcement.

B-143. However the students gain their information, it seldom resembles fact and must be changed to effectively establish a juvenile crime-prevention program.

B-144. In making classroom presentations for crime-prevention purposes, the crime-prevention officer must display honesty at all times. When confronted by uncomfortable questions, the MP officer should render the law-enforcement's point of view and make a truthful explanation. Obviously, some answers will not always be well received, but they will establish the credibility of his presentation. Further, when making these presentations, many areas dealing with specific areas of crime prevention will arise (such as rape prevention, property engravement, residential and apartment security, vehicle theft, and burglary prevention).

B-145. In dealing with schools, the crime-prevention officer can establish many positive contacts with individuals inside the school system that will prove to be of great value. Individuals such as administrators, counselors, and faculty members offer the crime-prevention officer assistance in many different ways. Perhaps the most valuable assistance is in the area of establishing a juvenile crime-prevention program that is based on peer influence.

B-146. A crime-prevention program aimed at peer influence is essential to a crime-prevention unit. Peer influence is directly related to the Determination Theory of Delinquency Causation. Simply stated, the theory holds that delinquency occurs as a result of external influences on youth. While attempting to control all external influences on youths is an impossible task, the control of peer influence is possible and perhaps the most important external influence to be controlled.

PEER-INFLUENCE CRIME-PREVENTION PROGRAM

B-147. A peer-influence crime-prevention program can be established after the crime-prevention officer establishes himself in the junior/senior high schools as described earlier. The basic concept of the program is the negation of undesirable peer influence on predelinquent youths and the establishment of a positive peer influence.

B-148. The peer-influence program functions within the school district. A problem youth is detected at the junior high school level by his teacher, counselor, or the crime-prevention officer. When it is believed that the peer-influence program could serve a youth, his counselor arranges a conference with him. During the conference, the counselor attempts to determine the basic problem experienced by the youth and has him complete an application for admittance into the program.

B-149. When the application is complete, the counselor contacts the youth's parent or guardian and explains the program. If the parent agrees to allow participation, the application is sent to the parent for his signature.

B-150. When the completed application is returned to the counselor, it is forwarded to a senior high school counselor for review. If necessary, the senior high school counselor may contact the junior high school counselor personally to discuss the youth in more detail. The high school counselor then selects a suitable senior high school student who has been accepted for the peer-influence program and matches the student with the junior high school youth.

B-151. Once the match has been made, the senior high school student is given a briefing on the junior high school student by the crime-prevention officer. The senior youth is instructed to contact the junior youth (preferably at the junior youth's home) and introduce himself to the youth and to the youth's parent or guardian. After the initial contact has been made, the senior youth is instructed to plan activities with the junior youth and spend as much time as possible with him.

B-152. Each senior youth that volunteers for the program is required to complete an application. This application gives personal background information and references that are thoroughly checked by the crime-prevention officer. The youth's academic record is examined, but prime consideration is given to citizenship factors rather than grades. The crime-prevention officer personally interviews each applicant.

B-153. When a senior youth is selected for the program, he is given a basic outline to follow. He is warned not to preach to the junior youth but to lead him by allowing him to observe the senior youth's actions at different activities. The senior youth must devote the necessary amount of time to limit the junior youth's opportunity to continue the negative associations he had in the past.

B-154. While the peer-influence program is simple in nature, the results can be remarkable. During a three-year period in Irvin, Texas, approximately 600 students participated in the program. From this number, seven arrests were made from the group with one juvenile accounting for four of the seven arrests. As about 30 percent of the juveniles participating were legally judged delinquent, the relapse rate was officially established as 0.1 percent.

B-155. While juveniles judged as delinquent can participate in the program, the crime-prevention officer should concentrate his efforts on those juveniles identified as predelinquent by counselors or faculty members. Again, the personal contact of the crime-prevention officer and the school faculty is essential to the program's success. A personal rapport should be cultivated and maintained.

B-156. Operationally, the peer-influence program can be administered by the collective efforts of school personnel and MP officers who are committed to the program. Once the program is established, it will probably not require the services of an officer on a full-time basis (depending on the number of students involved). However, constant contact is necessary; the more time that can be devoted, the more success will be apt to follow.

CRIME-PREVENTION PRESENTATIONS FOR ELEMENTARY AND JUNIOR HIGH SCHOOLS

B-157. As the traditional American family structure has experienced a significant decline over recent years, more and more emphasis is being placed on schools to provide guidance for children. As a crime-prevention officer who is frequently invited to speak to students, you can play an important role in assisting the school to provide this guidance. To stress the importance of the above statement, many experts in child behavior believe that the school is second only to the family in molding a child's socialization and behavior. Many people feel that the school is increasingly replacing much of the family socialization process. For this reason, schools are the most logical environment to begin a juvenile crime-prevention program.

B-158. Crime-prevention officers can have an impact on the most common types of juvenile crime by preparing specific lectures for particular grade levels in the elementary and junior high schools. Classes discussing shoplifting, vandalism, and bicycle theft are well received by students and have a wide-reaching effect on the community as a whole due to the children relaying the information to their parents.

B-159. A number of films are currently on the market that assist in making crime-prevention presentations in the classroom. In addition to specifically addressing juvenile crimes, the films directly relate to the peer-influence causation of crime.

OVERVIEW OF SELECTED JUVENILE CRIME-PREVENTION PROGRAMS

B-160. In addition to the programs described above, there are a variety of other juvenile programs. You will notice that some of the programs are designed to offer activities for the juvenile while at the same time attempting to provide one or more of the following:

- Positive peer influence.
- Education.

- Police/juvenile cooperation.
- An understanding of law-enforcement and MP duties.

B-161. One program deals with the enforcement aspect, with the goal being truancy reduction. Also mentioned are programs designed for child safety and victimization reduction.

DARE Program

B-162. DARE is a drug-abuse educational program that gives children and parents the skills to recognize and resist the subtle and overt pressures that cause them to experiment with drugs, alcohol, and violence. This unique program uses uniformed law-enforcement officers to teach a formal curriculum to students ranging from kindergarten to twelfth grade. DARE gives special attention to fifth through ninth graders to prepare them for entry into middle/junior high and high school where they are most likely to encounter pressures to use drugs or alcohol.

Police Explorer Programs

B-163. Explorer programs are run in conjunction with the Boy Scouts of America and a sponsoring police agency. The programs offer a wide range of flexibility in that they may be structured along the lines of a police cadet program with all activities centered around law-enforcement training. This type of explorer program usually seeks juveniles who have an active interest in pursuing a law-enforcement career. Being an explorer in this type of program will offer the student the opportunity to view the duties of a police officer and to take part in minor police activities (such as traffic direction and observation riding in patrol cars). The selection and training of the students for this type of program strictly depend on the sponsoring agency.

B-164. Other police-sponsored explorer programs may be designed to provide the members with other nonpolice activities (such as hiking, camping, and canoeing). At the same time, this type of program will offer the student the chance to meet with police officers and learn more about police functions, though this is a secondary purpose. Selection for this program would not be as stringent as the above-mentioned program. A local council office of the Boy Scouts can provide additional information on this program.

Alateen

B-165. The Alateen program is run in cooperation with the Alcoholics Anonymous organization. This program's intended purpose is to provide assistance for a child's emotional needs if he or she is faced with family problems caused by alcoholism of one or both parents. Your local chapter of Alcoholics Anonymous can provide more information on this program.

Truancy-Enforcement Program

B-166. Various cities have successfully adopted truancy-enforcement programs that focus attention on identifying and apprehending youths that are on unexcused absence from school. The primary goal behind this program is to prevent or reduce the incidence of daytime residential burglary. The San Angelo, Texas, Truancy Enforcement Program selected a target area in an effort to see how effective the program would be in reducing daytime burglaries. The initial evaluation of the program showed positive results. The San Angelo program uses uniformed crime-prevention officers in marked police cars patrolling around the school and known hangouts. All street personnel in the patrol division are also encouraged to check school-age individuals. Those students under 17 years of age who are suspected of truancy are taken to the school office while those over 17 years of age are detained only long enough to obtain their name and other identifying information, which is then turned into the school office. The assistant principal in charge of attendance completes the investigation and takes the appropriate action. The action varies (depending on the circumstances) but can include counseling with the student, the teacher, and the parents or, in extreme cases, suspension for a short period.

B-167. The emphasis behind these programs is to inform youths of the facts of the law. It was shown that many of the problems with the youthful offenders would be eliminated if they were presented the facts of the law to which they are subject. The courses can be adapted to fit the needs and desires of the individual school system, and the extent of the areas covered will depend on the commitment and availability of the local MP officers. Various civilian school districts have incorporated these law-enforcement courses as part of their regular curriculum.

Criminal Victimization Reduction

B-168. Programs designed toward this goal normally are structured to reduce the possibility of a child becoming a victim of a child molester. Most commonly, officers give a verbal presentation to groups of children using visual aids that help to act as a positive reinforcement to learning. Some effective methods have been the use of "talking traffic lights," "talking bicycles," and "talking motorcycles." These items are usually equipped with a tape-recorded message that can be controlled by the officer conducting the program. The talking traffic light uses a full-size traffic light that flashes red or green according to the type of person that is being described.

VANDALISM

B-169. Today's vandals often attack their own territory. School vandalism—the illegal and deliberate destruction of school property—is committed by students themselves. They break so many windows that in large districts the funds spent annually on replacing broken windows could pay for a new school. Vandals destroy about $3 million worth of school-bus seats annually, and they commit enough arson to account for 40 percent of all vandalism costs.

B-170. School vandalism outranks all other assaults on private and public property. At the end of the 1973 school year, the average cost of damages from vandalism was estimated at $63,031 per school district. That figure could have paid the salaries of eight reading specialists or could have financed a school breakfast program for 133 children for one year. A typical school's chance of being vandalized in a month are greater than one in four, and the average cost of each act of vandalism is $81. Yet, these figures do not include the hidden costs of school vandalism—increased expenses for fencing, intrusion and fire detectors, special lighting, emergency communications equipment, and vandal-resistant windows. Every dollar spent in replacing a window or installing an alarm cannot be spent on education.

B-171. School vandalism can also have enormous social cost. The impact of an 89-cent can of spray paint used to cover a wall with racial epithets far exceeds the monetary cost of removing the paint. An abusive word scrawled across a hallway wall can destroy student morale, disrupt intergroup relations, undermine the authority of an administration, or even close the school. Incidents with high social costs damage the educational process as much as those with high monetary costs. Today's vandal is not a hardened, war-scarred veteran. Instead of grizzled whiskers, he sports peach fuzz. He is almost literally the boy next door. In fact, the typical vandal differs quite dramatically from the typical juvenile delinquent.

B-172. It is significant that vandals fall into a well-defined and relatively narrow age group. They are usually early adolescent males who are highly subject to group pressures and transitory impulses. It is not at all unusual for adolescents to act out whatever is controlling them at the moment—rage, boredom, pent-up energy, or the sheer joy of "wreckreation." While there are conditions that may predispose or provoke a youth toward vandalism, the problem seems to be almost human nature. Few among us have never written on a sidewalk or scrawled initials on a school desk. Vandalism cuts across all strata of society, all geographic regions, and all racial lines.

B-173. The causes of vandalism remain obscure. Though research addressing the "why" of vandalism is growing, it has yet to yield clear-cut answers. Among the motivating factors often cited are anger, frustration, hostility, bitterness, alienation, futility, inequality, restricted opportunity, emotional pain, failure, prejudice, revenge, and the need for attention. Although much of the research is convincing, the fact remains that many vandals do not appear to be among the most angry, frustrated, hostile, alienated, or needy youth. Only a small fraction of the youngsters who fall into that category actually commit acts of vandalism. So, while most experts agree that vandalism is not totally senseless, they do not claim to fully understand its causes. In fact, vandalism is often not understood by vandals themselves. Many vandals report that they do not know why they did it. Many others, according to case reports, offer the unsolicited observation that destruction is fun. Still others express satisfaction and exhilaration. Few consider themselves criminals. For the time being, we can conclude only that motives for vandalism are diverse. But the whys notwithstanding, the vandal profile suggests that our task is, in large part, to anticipate and redirect the impulses of young teenagers.

B-174. Schools are by no means the helpless victims of early adolescence. Many school factors, most of which are amenable to change, influence the amount of vandalism that schools experience. The following characteristics are typical of schools that suffer high property damage or loss:

- Vandalism is higher when there is poor communication between the faculty and the administration (such as when the principal fails to define policy or makes policy decisions unilaterally).
- Hostility and authoritarian attitudes on the part of teachers toward students often result in students "taking it out" on the school.

- Limited contact between teachers and students reduces student involvement with the school and increases the likelihood of vandalism.
- Schools characterized by intense competition for leadership positions suffer greater property damage and loss.
- The chances for vandalism increase when the students do not value their teachers' opinions of them.
- Schools at which students strive to get good grades experience more vandalism.
- Parents of students in high-damage schools express less favorable attitudes toward their schools than do other parents.
- The school is a convenient target for vandalism when it is close to students' homes.
- The chances for vandalism are increased when grades are used as a disciplinary tool.
- Damage is greater in larger schools where there is more property to destroy. This correlation between school size and vandalism prevails regardless of whether the school is located in an urban, suburban, or rural setting.
- Fewer offenses occur when rules are well understood by students and are consistently and firmly enforced by teachers and administrators.

Vandalism Prevention

B-175. If the special problems of early adolescence, often intensified by social or personal pressures, interact with school conditions to produce vandalism, then preventive measures must address the nature of both the child and the school. Furthermore, prevention must include both physical and human responses. At present, most vandalism-prevention or -reduction programs rely on physical security—bigger and better electronic alarm systems, patrol guards, dogs, tamperproof locks, and window grilles. These techniques help, but they address only 20 percent of the problem—those incidents involving breakage. These incidents usually occur when school is not in session and in the absence of witnesses. The techniques have little effect on the day-to-day trashing of the school or on the disruptive acts aimed at the school's routine (bomb threats, the setting of fires, and false fire alarms) that are committed during school hours. The most sophisticated physical and electronic barriers are not sufficient to keep vandals from what they consider an attractive target. In fact, it has been argued that alarms and armed guards, besides lowering student and staff morale, often themselves become a challenge, inviting rather than deterring vandals. Vandalism prevention requires not a narrow or piecemeal approach, but a varied and comprehensive effort that includes both physical and human components geared to the school's specific problems. Furthermore, an effective long-term program must involve the community, parents, neighbors, police, and civic groups as well as students, teachers, and school administrators.

B-176. Schools are an easy target for vandals. Most are public, secular, and often unoccupied. Most will remain public and secular; but they need not remain unoccupied, unprotected, or unobserved. The following are techniques that have made some schools less vulnerable to vandals. These are especially effective against problems occurring during nonschool hours.

B-177. **Occupy the School.** Employ a custodial force around the clock. In most schools, the entire custodial force works at one time, leaving the school at night. As an alternative, custodians can be assigned staggered shifts so that the school is always occupied. Twenty-four-hour custodians are particularly appropriate in schools suffering sporadic property damage that demand more than a roving patrol but less than permanent security guards.

B-178. Invite police to use the school buildings at night. Police can be issued keys to the schools in their patrol areas so that they can use school offices to write their reports. This places a police officer in the school when it might otherwise be unoccupied, and it places a police car in front of the school.

B-179. Bring the community into the school. The school is an excellent place for recreational programs; health clinics; adult-education classes; counseling centers; community gatherings; and Boy Scout, Girl Scout, and Parent-Teacher Organization (PTO) meetings. The presence of people in the school building not only reduces the opportunity for vandalism, but also stimulates community and student interest in the school.

B-180. **Watch the School.** Use school neighbors as eyes and ears. Ask nearby homeowners to watch the school and report suspicious activities. Emphasize careful observation and rapid reporting, but discourage direct involvement in any situation observed. Such programs work best if they are organized but based on informal involvement, if they are accompanied by overall involvement of parents and community with the school, and if they offer some sort of prestige to participants.

B-181. **Employ Roving Patrols.** A uniformed patrol used instead of or in conjunction with an alarm system can deter vandalism. The individuals who patrol should establish rapport with neighborhood youths and open communication with community leaders. They should also vary their patrol patterns.

B-182. Hire student patrols during the summer and on weekends. The school district or community can provide its youth with part-time or summer employment and, at the same time, curb vandalism by paying students to patrol the school grounds during weekends, holidays, and summer vacations. These students should be paid an adequate wage and considered an integral part of the school's security force.

B-183. **Control Access to the School by Using an Alarm System.** Alarms are the most expensive vandalism control measure a school can use. While they can detect vandals, they cannot apprehend them; they can merely signal the alarm-system monitor, which may be miles away. They can, however, be an efficient deterrent and should be considered as part of any comprehensive plan to control vandalism. If alarm systems are linked with a surveillance camera, the chances of apprehending intruders are greatly increased.

B-184. **Design the School With Vandalism Prevention in Mind.** The following designs for preventing vandalism can be implemented when building a school:

- Limit ground-to-roof access.
- Eliminate low, overhanging roofs.
- Avoid unnecessary exterior fixtures.
- Plant trees that cannot be climbed near buildings.
- Consider raising as much of the school building as possible from ground level.
- Build the school at some distance from residential areas. While it should be located near the homes of most of those it serves, it will suffer less property damage if there is a buffer zone between it and surrounding residential areas.
- Design the school with plenty of defensible space so that the normal flow of school traffic allows continuing, casual surveillance of the premises.
- Use vandal-resistant surfaces. Use harder surfaces in damage-prone areas. For walls, use epoxy paint or glazed tiles that are easily and inexpensively replaced or repaired; use small wall panels and keep replacement panels in stock; and place permanent signs, building names, and decorative hardware at a level that cannot be reached from the ground. Replacing damaged areas immediately shows a sense of pride in the appearance and helps to eliminate copycat acts of vandalism.
- Plan windows carefully. Avoid windows that are vulnerably placed. Use small panes of glass to simplify replacement; use thick, tempered glass, thick acrylic, or Plexiglas® for windows in heavily traveled or hangout areas. Avoid useless windows in student stores, administrative offices, and industrial-arts storage areas.
- Plan entries with multiple uses in mind. Install flexible internal gates to block off specific areas or corridors when necessary. Provide separate exterior entries for community and student use. Inside the building, create areas for informal gatherings near entrances and exits by installing soft-drink machines and benches.
- Locate or relocate the playground's access roads to provide better surveillance by roving patrols.
- Consider outdoor lighting. Opinions on this issue are divided. Many schools report a decline in vandalism after installing hardened exterior night lighting. Others report that elimination of all night lighting reduces vandalism, presumably because young adolescents are afraid of the dark. If lighting is used, it should be directed away from windows to keep vandals from seeing the process of destruction or its outcome.
- Channel graffiti. Graffiti artists will usually select light, smooth surfaces rather than dark, rough surfaces. Therefore, school officials can channel graffiti onto one or two walls designed to withstand such treatment. Students or maintenance staff can paint most walls at regular but not too frequent intervals. One wall can be officially designated a "legitimate" graffiti wall, though this approach removes some of the challenge inherent in informal graffiti.

B-185. **Make the Target Less Attractive.** The school is not only an easy target but also an attractive one. It represents enforced learning, discipline, and mandatory attendance to students who are, simply by virtue of their age, rejecting adult standards and striving to achieve independence from adult control. Students are additionally provoked if the school functions in an impersonal, undemocratic, repressive, or manipulative manner, further increasing the likelihood of vandalism.

B-186. **Revise the Curriculum.** The list of characteristics associated with vandals and vandalized schools indicates that property damage and loss are higher when competition for rewards is intense, the availability of rewards is limited, or the distribution of rewards is unfair. All of this suggests that school policy and atmosphere have a direct bearing on school vandalism. Below are changes in school governance that can help remove the features that make a school an attractive target for vandals. These procedures, especially in combination, have proven effective against all forms of vandalism, including those that most commonly occur while school is in session.

- **Alternative schools.** Though originally designed to perform educational functions, alternative schools have proven effective in reducing school violence and vandalism. They provide an option to students who are not benefiting from the regular academic classroom. These schools operate within or alongside the traditional school. They are characterized by low student-to-teacher ratios, intense individual counseling, and extensive use of the community as a learning resource. They offer an alternative to suspension, personalize the learning environment, and provide successful experiences to students who have performed poorly in the past.
- **Law-related education and police-school liaison programs.** In many communities, the police department has assigned school resource or liaison officers to local junior and senior high schools. These officers may on occasion assume policing duties, but their primary function is to counsel students and teach law-related courses. These activities acquaint students with the positive role that law plays in our society and personalize the relationship between the "cop on the beat" and "the kid on the corner."
- **Action learning.** This term refers to apprenticeship programs as well as training in practical aspects of adult life. The former allows students academic credit for community work (such as tutoring and assisting physicians, lawyers, or other professionals). The latter provides instruction in skills such as checkbook balancing, comparative shopping, and applying for a job. Both address the boredom and frustration that are linked to truancy, violence, and vandalism.
- **Reward distribution.** The school's reward structure is related to school crime. Although the school may offer attractive incentives, most students receive very little in the way of rewards. Many of those who lose out still care about the competition. They become frustrated, and they vent their anger on the apparent source of their problems, the school. It should be possible to spread the rewards around without compromising performance standards, perhaps by recognizing improvement as well as achievement or by acknowledging forms of achievement other than scholastic, athletic, and social.

B-187. **Change Administrative Policy and Organizational Structure.** According to the National Institute of Education's Safe-School Study, the principal's leadership is a critical factor in reducing or preventing school crime. Strong principals who are visible, available, and responsive to students and staff appear to be most successful in eliminating violence and vandalism. Those who are less successful are often described as unavailable and ineffective.

B-188. The Safe-School Study also found that the exercise of discipline through clear enunciation and even-handed enforcement of rules is associated with a low incidence of school crime. To increase the likelihood that students will find school a fulfilling experience, many districts are establishing minischools—schools small enough to allow the individual student to feel significant. Similarly, several large schools are currently functioning on a house basis—the school is divided programmatically into several smaller units with which students can more easily identify.

B-189. **Involve the Students.** Establishing a vandalism fund uses peer pressure to the school's advantage. The community or the school district puts aside a certain amount of money and announces that the funds will be used to cover the costs of vandalism. Any money left over reverts to the students to be used as they choose. This plan works because it educates students about the costs of vandalism, allows them to see the positive results of curbing property damage and, most important, gives them full responsibility for the problem.

B-190. Several school districts have established voluntary student security advisory councils that conduct workshops and small group discussions focusing on vandalism and violence prevention. These committees increase awareness of the school's problems, generate recommendations for action, and give students opportunities to participate in school decision making.

B-191. Older students can be helpful in influencing younger students. In several communities, junior and senior high school students visit fourth- and sixth-grade classrooms where they show a film about vandalism and then lead a discussion on the causes and consequences of vandalism.

B-192. School beautification projects involve students in the care of the school building and grounds in an attempt to increase their pride in the responsibility for the school. The more effective projects are selected by the students themselves and continue throughout the year. The projects focus on marginal students rather than school leaders.

B-193. **Involve the Parents.** An open-door policy allows parents to go directly to their child's classroom whenever they wish without securing a special visitor's permit from the office. This simple policy offers parents concrete evidence that they are indeed welcome at school.

B-194. In some schools, parents serve on antitruancy committees along with teachers and students. They visit youngsters and their families in an effort to resolve truancy problems. In other schools, parents serve as hall monitors, supervise extracurricular activities, and help in classrooms. Parents are also the school's best source of guest speakers and contacts for work-study or apprenticeship programs. Some school districts have initiated faculty men's clubs to acquaint fathers with the male teaching staff and to encourage them to assume more responsibility for their children's progress in school.

B-195. **Involve the Community.** In some communities, students and law-enforcement, school-district, and city personnel sponsor daylong forums on vandalism. The forums introduce citizens to the causes and costs of vandalism and give them an opportunity to voice their concerns and initiate preventive programs.

B-196. Police departments can initiate public-relations programs within schools and youth-service agencies. In addition, they can enlist the help of youth in preventing vandalism through police-sponsored groups, such as the Police-Youth Service Corps, which recruits adolescents from high-crime areas to work as public-safety aides. Similarly, law students from neighboring universities can be brought into the schools at minimal or no cost to discuss the legal implications of vandalism.

Picking up the Pieces

B-197. The preventive measures listed on the preceding pages can function as part of a long-range, proactive response plan. However, they do not address the question of immediate response. What should the school do, right away, about 20 broken windows, a cherry bomb in the toilet, or recurring racial graffiti on the wall? The answer is to repair the damage immediately. Quick repair keeps perpetrators from admiring their handiwork, lessens the epidemic effect of vandalism, and minimizes any social impact the act may have.

B-198. Initiate formal record-keeping procedures, and ensure that they are followed. Schools faced with serious problems should begin recording all acts of vandalism. They should also consult law-enforcement personnel about when police should and should not be called. When a school begins to have problems, it should work with the juvenile justice system so that the two institutions can coordinate their efforts with regard to school-age offenders.

B-199. Careful record keeping allows a school to plot the incidence of vandalism to find out precisely where and when each type of offense is occurring. For example, using incident analyses, the National Institute for Education's Safe-School Study found that—

- Fire and bomb threats most often occur on Tuesdays.
- School-property offenses tend to occur with greater frequency toward the end of each semester, especially in November and December.
- Break-ins and school-property offenses occur most often on weekends and Mondays.

B-200. This type of information is invaluable in planning a vandalism reduction and prevention plan.

Restitution

B-201. Restitution is a set of legal and administrative procedures through which the school receives payment from vandals for damages they cause. While it seems reasonable to require payment for damages, restitution does not appear to be worth the effort. In the first place, most vandals are not caught. In the Los Angeles School District (which has an aggressive restitution program), only 30 percent of the offenders are ever identified.

From this 30 percent, most restitution is paid before matters get to court. Going through lengthy legal processes to obtain the rest is simply not cost-effective. However, a parent faced with the possibility of a court case may make a greater effort to keep his or her child out of trouble.

CONCLUSION

B-202. The goal of crime prevention is to reduce crime through public awareness and education. The skills that crime-prevention officers acquire can eventually benefit all segments of our communities. It is logical then for crime-prevention officers to attack one of the major crime problems in our society—juvenile delinquency. Efforts of crime-prevention units will certainly be lacking if the juvenile-crime problem is not given a high priority and if juvenile-crime programs are not established.

FRAUD

B-203. Two types of fraud can be affected by crime-prevention efforts—fraud against the government and fraud against individuals in the Army. Since the countermeasures for each of these categories of fraud are different, they will be discussed separately.

FRAUD AGAINST THE GOVERNMENT

B-204. Fraud of this type is a loss to the Army due to manipulation of systems from within the government with criminal intent. Typical examples of fraud are—

- The diversion or theft of government property by falsifying documents such as purchase orders, shipping documents, and so forth.
- False claims for temporary-duty (TDY) pay, for reimbursement for losses due to the movement of household goods, or for reimbursement for material reported as stolen.
- Overcharges or underproduction on contracts with the Army.
- Bribery.
- Kickbacks to secure purchase orders.
- Use of one's official position for personal gain.

B-205. This list is not all-inclusive. There are many more ways to defraud the government. Some of the factors that make it easy to perpetrate a fraud are—

- The lack of independent verification of records, transactions, and reports.
- The lack of adequate supervision.
- Unrealistic budgeting and acquisition requirements.
- Failure to correct deficiencies identified by existing systems.
- Concentration, at the operational level, of responsibility and authority for an entire process in one individual.

B-206. There are built-in measures to discourage fraud for most Army material-control systems. These programs are routinely examined by the inspector general (IG) along with other command inspectors. A crime-prevention officer's main function is to encourage the reporting of fraud by ensuring that the community understands what type of activities should be reported and by ensuring that the numbers are posted in work areas where personnel are in a position to detect fraud.

FRAUD AGAINST INDIVIDUALS

B-207. The second type of fraud is con games or consumer fraud. This crime is perpetrated by professional con men, unscrupulous businesses, and amateurs who see a chance to make a fast buck. Most of the time, the operator offers a "something for nothing" deal. By the time the victim realizes that he has been duped, the con man is long gone. Frequently, the victim is so embarrassed that he has been duped that he does not report the crime. The only way that a crime-prevention officer can help prevent this crime is to publicize common con games and illegal business practices and to encourage victims to report them. When a case of fraud occurs, the specific con technique that was used should be advertised widely to prevent others from being fooled by the same ploy.

Repair Fraud

B-208. Repair frauds are simple to execute but difficult to detect. Some unscrupulous repairmen do not fix the problem, but they charge you anyway. Some use inferior parts; others charge you for work that you did not expect or need. Some also do "insurance" work; they will repair one thing but ensure that something else will soon go wrong. The following are methods that you can use to protect yourself from repair fraud:

- Shop around. Ask friends, neighbors, or coworkers for references. When you find repairmen you trust, stick with them.
- Do not try to diagnose the problem yourself unless you are an expert. The mechanic may take your advice, even if it is wrong. If you know exactly what the problem is, do not tell the mechanics; wait and see if their recommendations agree with your diagnosis. That way you will know whether needless repairs are suggested.
- Try to get several detailed written estimates before any work is done. Compare job descriptions and materials to be used. Be sure to ask if there is a charge for an estimate.
- Ask for the old parts to ensure that replacements were really installed.
- Ensure that you get a guarantee on any work done.
- Make sure that the work was done before you pay. Take your car for a test drive. Plug in the refrigerator. Test the television.
- Ask your local consumer affairs office about the laws in your state and what specific protection it gives regarding professional services to be licensed or certified.

Home-Improvement Fraud

B-209. Home repairs and improvements can be costly. Be cautious if somebody offers to do an expensive job for an unusually low price, if a firm offers to make a free inspection, or if the workers "just happened to be in the neighborhood." These are the favorite tricks of dishonest home-repair firms.

Some firms offer a price you just cannot resist. Once you sign the contract, you learn why—they never deliver the service. Others send door-to-door inspectors to do a free roof, termite, or furnace inspection. These free inspections will turn up plenty of expensive repairs. Some fly-by-night companies will offer to do the work on the spot. When they leave, you may be left with a large bill and a faulty repair job. To avoid home-improvement fraud, try the following:

- Get several estimates for every repair job. Compare prices and terms. Check to see if there is a charge for estimates.
- Ask your friends for recommendations, or ask the firm for references and check them.
- Check the ID of all inspectors.
- Call the local consumer affairs office or the Better Business Bureau to check the company's reputation before you authorize any work.
- Be suspicious of high-pressure sales tactics.
- Pay by check, never with cash. Arrange to make payments in installments—one-third at the beginning of the job, one-third when the work is nearly completed, and one-third after the job is done.

Land Fraud

B-210. Real estate can be a great investment. The enterprising real estate salesperson knows how anxious you are to find just the right property, especially for an investment or a retirement home—a nice warm climate, not too crowded, or a new development. Some dishonest agents will promise you anything—a swimming pool, country club, or private lake—to get your name on the contract. Even if the sales agent promises you luxury, they may not guarantee the basics such as water, energy sources, and sewage disposal.

B-211. Most land developers offering 50 or more lots (of less than 5 acres each) for sale or lease through the mail or by interstate commerce are required by law to file a Statement of Record with the Housing and Urban Development Administration (HUD). This document tells you almost everything you need to know about your future home—legal title, facilities available in the area (such as schools and transportation), availability of utilities and water, plans for sewage disposal, and local regulations and development plans. All of this information must be given to you in a property report prepared by the developer. Always ask to see this report before you sign anything.

B-212. If the developer does not give you a copy of the property report for the lot you are considering, you can obtain it from HUD for a fee. Send requests to: Department of Housing and Urban Development, Office of Interstate Land Sales Registration, 451 Seventh Street South West, Washington, DC 20410.

Door-to-Door Sales

B-213. Beware of the following door-to-door sales tactics:

- "Nothing like it in the stores!" This is a true statement. The vacuum cleaners in the stores are probably of better quality and come with a better warranty.

- "Won't find this price anywhere." This is also a true statement. The prices in the stores are probably lower.
- "Easy credit!" This is another true statement. They do not care what your credit rating looks like. Once you sign for the purchase, paying for it is your problem. Be wary of low monthly payments. Find out the total amount you will pay over the life of the loan, then subtract the actual cost of the item itself. The difference is what you will pay in interest. Your bank, your credit union, or a local legal-aid society can tell you if the interest rate is fair.

B-214. Be cautious of these words, and be firm if the salesperson pressures you to buy. If you do get trapped, you are protected by a Federal Trade Commission regulation. When you make a purchase in your home totaling $25 or more, the salesperson must give you a written contract and two Notice of Cancellation forms. You have three days to change your mind and use one of these forms to cancel your contract.

Charity Fraud

B-215. Charity fraud does a lot of harm. The swindler takes advantage of a person's good will and takes his cash—money that was meant for people in need. You can ensure that any money you give gets into the right hands. Remember these pointers when somebody asks you for a donation:

- Ask for ID, either the organization's or the solicitor's. Find out what the charity's purpose is and how the funds are used. Ask if contributions are tax deductible. If you are not satisfied with the answers, do not give money.
- Give to charities that you know. Check out those you have never heard of or others whose names are similar to a well-known charity.
- Do not fall for high-pressure tactics. If solicitors will not take no for an answer, tell them no anyway, but do not give them your money.
- Be suspicious of charities that only accept cash. Always send a check made out to the organization, not to an individual.
- If a solicitor reaches you by telephone, offer to mail your donation. Shady solicitors usually want to collect quickly.

Self-Improvement Fraud

B-216. Con artists know that everyone wants to look better, feel better, and be a better person. Selling worthless plans and cures is one of the easiest ways for them to make a "quick buck." The following ads can look tempting:

- "Miracle reducing plan."
- "Look like a model in only five days."
- "Learn to speak Spanish while you sleep."
- "You can have a new, dynamic personality."

B-217. What can you do? Be careful! Read the small print. Know what the product contains. You should check with your doctor before you embark on any diet or exercise program.

Medical and Health Fraud

B-218. Most of us do not know much about medicine; that is why we go to doctors. It is also why we fall for miracle cures and other phony health products and services. Patent medicines, health spas, and mail-in lab tests should be warning signs for the potential consumer. A laboratory in Texas advertised nationally that it had perfected a fail-safe urine test for cancer. More than 15,000 tests were made at $10 each before authorities stopped this fraudulent company.

Unsolicited Merchandise

B-219. Cagey con artists will send you a gift in the mail—a tie, a good luck charm, or a key chain. What do you do with it if you did not order it? If you are the kind of person they are looking for, you will feel guilty and pay for it, but you are not obligated to. If you—

- Have not opened the package, mark it "return to sender." The post office will send it back at no charge to you.
- Open the package and do not like what you find, throw it away.
- Open the package and like what you find, keep it, free of charge. This is a rare instance where the rule of "finders, keepers" applies unconditionally.

B-220. Whatever you do, do not pay for it. Look at your gift as an honest-to-goodness way of getting something for nothing. Do not get conned if the giver follows up with a phone call or a visit—by law the gift is yours to keep.

Mail Fraud

B-221. The following are examples of mail fraud:

- **The contest winner.** "You've won! This beautiful brand-name sewing machine is yours for a song. To claim your prize, come to our store and select one of these attractive cabinets for your new machine. Bring this letter with you and go home with a new sewing machine for next to nothing." Treat an offer like this carefully. Shop around before you claim your prize. Chances are, the cost of the cabinet will be more than the machine and cabinet are worth.
- **The missing heir.** You have just received a very official looking document. The sender is looking for the rightful heirs to the estate of someone with your last name. It could be you. To find out, just send $10 for more information. There may be thousands of people with your last name, and letters like these are often mailed nationwide. Even if there really was claimed estate, it is highly unlikely that you would be an heir. Save your money; why help a swindler get rich?

B-222. These are just two examples of mail fraud. Many of the other frauds described in this section can be handled through the mail. When they are, the US Postal Service can launch a full-scale investigation. If you think you have been cheated in a mail-fraud scheme—

- Save all letters, including envelopes.

- See if your neighbors or business associates received the same material.
- Contact your local postmaster who can direct you to your regional Postal Inspector's Office.

SUMMARY

B-223. The following are recommendations to avoid being duped:

- Do not believe something-for-nothing offers. You get what you pay for.
- Be suspicious of high-pressure sales efforts.
- Take your time. Think about the deal before you part with your money.
- Get all agreements in writing. Insist that agreements are made in "plain English," not "legalese."
- Read all contracts and agreements before signing. Have a lawyer examine all major contracts.
- Compare services, prices, and credit offers before agreeing to a deal. Ask friends what their experiences have been with the firm or service in question.
- Check the firm's reputation with your local consumer affairs office or the Better Business Bureau.

INTERNAL THEFT

B-224. Employee theft is a major problem in most large organizations. It is estimated that one-third of all business bankruptcies are a result of theft. Of course, in the Army there is no danger of the organization going out of business; however, waste and internal theft can divert significant amounts of critical resources from mission-essential activities. Unlike many businesses, the Army has recognized that internal losses can be a problem, and most areas have adequate controls mandated by regulation. As crime-prevention officers, we should monitor reports of survey, physical-security inspection reports, and results of crime-prevention inspections to ensure that control measures are being followed. When it is apparent that the correct measures are not being followed, the problem must be identified for the senior commander's action.

B-225. Research has shown that an organization's atmosphere is just as important as management controls and physical security in preventing employee theft. To control this problem, the following items are essential:

- Employees, both civilians and soldiers, must believe they are part of a professional organization that expects superior performance from all of its members. When second-rate work is accepted, the lax attitude carries over into property-control procedures and losses increase. Standards must be high, but fair.
- Leadership must set the example. If leaders, supervisors, and managers take advantage of their positions to use government material or services for their own benefit (even in minor ways), their employees will also feel justified in diverting Army resources to their own use.

- The organization must show a genuine concern for the problems of its personnel. If an employee feels he has been treated unfairly, it is easy for him to justify stealing from the organization. He believes that he is only taking what he would be getting if the organization was fair.
- The organization must take appropriate disciplinary or administrative actions in cases of theft. There can be no acceptable level of internal theft. If employees believe that an activity considers a certain level of loss to be acceptable, then material theft will grow rapidly. Everyone will consider the material he takes to be within the 2 percent that the activity expects to lose.
- Policies on internal thefts must be enforced. Frequently, a thief is a long-term employee whose honesty has always been above question. In these cases, the temptation is very strong to let the offender go with a "slap on the hand." However, this is a clear signal to other employees that diversion of Army resources for personal use is acceptable to management. If an employee chooses to steal and is caught, he should be prosecuted.
- Publicity campaigns using posters and other media should be used to disseminate command policies on internal theft. Crime hot lines are also useful in increasing reports of employee thefts.

PILFERAGE

B-226. The protection of property, including the prevention of pilferage of government supplies and equipment, is one of the primary functions of the MP and civil service security forces. This function may include protecting supplies and equipment while in storage areas, during the issue process, or while they are in transit.

B-227. Pilferage is probably the most common and annoying hazard with which security personnel are concerned. It can become such a financial menace and detriment to operations that a large portion of the security force's efforts may have to be devoted to its control. Pilferage, particularly petty pilferage, is frequently difficult to detect, hard to prove, and dangerous to ignore.

B-228. Military property loss throughout the world would increase millions of dollars each year if subjected to uncontrolled pilferage. However, the risks incurred cannot be measured in terms of dollars alone. Loss of critical supplies for tactical units could result in loss of life or a danger to national defense. In some areas, losses could assume such proportions as to jeopardize an installation's mission. All installations and facilities can anticipate loss from pilferage. Actual losses will depend on such variable factors as the type and amount of materials, equipment, and supplies produced, processed, and stored at the facility; the number of persons employed; social and economic conditions in surrounding communities; command attitude; and physical-security measures used. Because these factors differ greatly in various types of installations and in different geographical locations, each must be considered separately.

B-229. To determine the severity of this hazard at any given installation or facility, it is necessary to determine the amount of loss that may be occurring.

Unfortunately, this is not always an easy task. Accounting methods may not be designed to pinpoint thefts; consequently, such losses remain undisclosed or they are lumped together with other shrinkages, thus effectively camouflaging them.

B-230. Inventory losses may be inaccurately labeled as pilferage for the following reasons:

- Failure to detect shortages in incoming shipments.
- Improper stock usage.
- Poor stock accounting.
- Poor warehousing.
- Improper handling and recording of defective and damaged stock.
- Inaccurate inventories.

B-231. In some cases, inventory losses may be impossible to detect because of the nature and quantities of materials involved. Stock inventory records may not be locally maintained, or there may be no method for spot checks or running inventories to discover the shortages.

PROFILE OF PILFERERS

B-232. Physical-security personnel must be able to recognize and counteract two types of pilferers—casual and systematic. A casual pilferer is one who steals primarily because he is unable to resist the temptation of an unexpected opportunity and has little fear of detection. There is usually little or no planning or premeditation involved in casual pilferage, and the pilferer normally acts alone. He may take items for which he has no immediate need or foreseeable use, or he may take small quantities of supplies for the use of family or friends or for use around his home. The degree of risk involved in casual pilferage is normally slight unless a very large number of persons are involved.

B-233. Casual pilferage occurs when the individual feels the need or desire for a certain article, and the opportunity to take it is provided by poor security measures. Though it involves unsystematic theft of small articles, casual pilferage is nevertheless very serious. It may have a great cumulative effect if permitted to become widespread, especially if the stolen items have a high cash or potential value.

B-234. There is always the possibility that casual pilferers, encouraged by successful theft, may turn to systematic pilferage. Casual pilferers are normally employees of the installation and usually are the most difficult to detect and apprehend.

B-235. A systematic pilferer is one who steals according to preconceived plans. He steals any and all types of supplies to sell for cash or to barter for other valuable or desirable commodities. He may work with another person or with a well-organized group of people, some of whom may be members of a cleaning team or even be in an advantageous position to locate or administratively control desired items or remove them from storage areas or transit facilities.

B-236. The act of pilferage may be a one-time occurrence, or such acts may extend over a period of months or even years. A large quantity of supplies with great value may be lost to groups of persons engaged in elaborately planned and carefully executed systematic-pilferage activities. Systematic pilferers may or may not be employees of the installation; if they are not, they frequently conspire with employees.

TARGETS FOR PILFERAGE

B-237. Both casual and systematic pilferers have certain problems to overcome to accomplish pilferage objectives. These problems include the following:

- A pilferer's first requirement is to locate the item or items to be stolen. For the casual pilferer, this may be accomplished through individual search or even accidental discovery. In systematic pilferage, more extensive means are generally used. These may consist of surveilling by members of the group or checking shopping and storage areas or documents by those who have access to them.
- The second requirement is to determine the manner in which he can gain access to and possession of the desired item. This may involve something as simple as breaking open a box, or it may be as complex as surveying security factors (such as physical safeguards or security procedures) for weaknesses. It may also involve attempting to bribe security forces, altering or forging shipping documents or passes, or creating disturbances to divert the attention of security personnel while the actual theft is taking place.
- The third requirement is to remove the stolen items to a place where the thief may benefit from his act. Articles of clothing may be worn to accomplish this. Small items may be concealed in many possible places on the thief's body or in vehicles. Through falsification of documents, whole truckloads of supplies may be removed from their proper place without immediate discovery.
- Finally, to derive any benefit from his act, the pilferer must use the item himself or dispose of it in some way. The casual pilferage of supplies is intended primarily to satisfy the need or desires of the thief. The systematic pilferer usually attempts to sell the material through "fences," pawnbrokers, or black-market operations.

B-238. Detection of use or disposal can help prevent similar pilferage through investigation and discovery of the means used to accomplish the original theft. Similarly, each of the problems faced by would-be pilferers offers an opportunity for constructive preventive measures. Careful study of the possible opportunities for the pilferer to solve his problem is essential in security work.

B-239. The primary concern of a systematic pilferer in selecting a target is its monetary value. Since he steals for personal profit, the systematic pilferer looks for items from which he can realize the greatest financial gain. This means he must also have or be able to find a ready market for items he may be able to steal. He pilfers small items of relatively high value (such as drugs, valuable metals, or electronic items). However, we cannot discount the possibility that a systematic pilferer may, if the profit is substantial, select a target of great size and weight. As a rule, bulk-storage areas contain most of the material that may be selected by systematic pilferers.

B-240. The casual pilferer is likely to take any item easily accessible to him. Since he normally will remove the item from the installation by concealing it on his person or in his POV, size is also an important consideration. Monetary value and available markets are not of any great concern to the casual pilferer, because he usually does not have any idea of selling the property he steals. Storage areas containing loose items are more likely to tempt casual pilferers than bulk-storage areas.

METHODS OF PILFERAGE

B-241. There are many ways that pilfered items may be removed from installations. Because the motives and targets likely to be selected by systematic and casual pilferers are very different, the methods of operation for each are very different.

B-242. As stated above, the casual pilferer steals whatever is available to him and generally removes it from the installation by concealing it on his person or in his automobile. The methods of the systematic pilferer are much more varied and complex. The means he may use are limited only by his ingenuity.

B-243. Shipping and receiving operations are extremely vulnerable to systematic pilferage. It is here that installation personnel and truck drivers have direct contact with each other and readily available means of conveyance. This offers a tempting opportunity for collusion.

B-244. One individual must not have control of all shipping and receiving transactions. Obviously, this procedure invites manipulation of government bills of lading and inaccurate storage and movement procedures through failure of one activity to compare bills and invoices with another activity. The opportunities for monetary kickbacks increase without a sound system of checks and balances.

B-245. Railway employees assigned to switching duties on the installation can operate in a similar manner. However, this operation is more difficult because a railway car normally cannot be directed to a location where stolen property can be easily and safely removed.

B-246. Tanker trucks used for shipping petroleum products may be altered to permit pilferage of the product. Trash- and salvage-disposal activities offer excellent opportunities to the systematic pilferer to gain access to valuable material. Property may be hidden in waste materials to be recovered by a accomplice who removes trash from the installation.

B-247. Other methods that may be used by systematic pilfers to remove property from military installations include—

- Throwing items over fences to be retrieved later by themselves or by accomplices.
- Packaging property and sending it to outside addresses through mail channels.
- Conspiring with security personnel.

- Wearing loose-fitting clothing to conceal small items.
- Removing items by using vehicles belonging to outside contractors and vendors.

CONTROL MEASURES FOR CASUAL PILFERAGE

B-248. Specific measures for preventing pilferage must be based on careful analyses of the conditions at each installation. The most practical and effective method for controlling casual pilferage is to establish psychological deterrents. This may be accomplished in a number of ways, such as—

- Searching individuals and vehicles leaving the installation at unannounced times and places.
- Conducting spot searches occasionally to detect attempts of theft.
- Making employees aware that they may be apprehended if they attempt to illegally remove property.

B-249. Care must be taken to ensure that personnel are not demoralized nor their legal rights violated by oppressive physical control or unethical security practices. An aggressive security-education program is an effective means of convincing employees that they have much more to lose than to gain by engaging in acts of theft. Case histories may be cited where employees were discharged or prosecuted for pilferage. Care must be taken in discussing these cases to preclude the identification of individuals because of possible civil suits for defamation of character. Also, it is generally poor policy to publicize derogatory information pertaining to specific individuals. It is important for all employees to realize that pilferage is morally wrong no matter how insignificant the value of the item taken.

B-250. It is particularly important for supervisory personnel to set a proper example and maintain a desirable moral climate for all employees. All employees must understand that they have a responsibility to report any loss to the proper authorities. Adequate inventory and control measures should be instituted to account for all material, supplies, and equipment. Poor accountability, if it is commonly known, provides one of the greatest sources of temptations to the casual pilferer.

B-251. Identifying all tools and equipment by some mark or code (where feasible) is necessary so that government property can be identified. Installation tools and equipment have counterparts on the civilian economy and cannot otherwise be identified as government property. Another control method requires individuals to sign for tools and equipment. The use of the signature control method reduces the temptation to pocket the item.

B-252. In establishing any deterrent to casual pilferage, physical-security officers must not lose sight of the fact that most employees are honest and disapprove of theft. Mutual respect between security personnel and other installation employees must be maintained if the facility is to be protected from other more dangerous forms of human hazards. Any security measure that infringes on the human rights or dignity of others will jeopardize rather than enhance the overall protection of the installation.

CONTROL MEASURES FOR SYSTEMATIC PILFERAGE

B-253. Unlike the casual pilferer, the systematic thief is not discouraged by psychological controls. Nothing short of active physical-security measures are effective in eliminating loss from this source. Some of these measures include—

- Establishing security surveillance at all exits from the installation.
- Establishing an effective package- and material-control system.
- Locating parking areas for POVs outside of the activity's perimeter fencing.
- Eliminating potential thieves during the hiring procedure by careful screening and observation.
- Investigating all losses quickly and efficiently.
- Establishing an effective key-control system.
- Establishing adequate security patrols to check buildings, grounds, perimeters, and likely locations for clandestine storage of property removed from its proper location.
- Installing mechanical and electrical intrusion-detection devices where applicable and practical.
- Coordinating with supply personnel to establish customer ID to authenticate supply-release documents at warehouses and exit gates.
- Establishing appropriate perimeter fencing, lighting, and parking facilities and effective pedestrian, railway, and vehicle gate-security control.

HOW TO STOP EMPLOYEE THEFT

B-254. No matter what it is called—internal theft, peculation, embezzlement, pilferage, inventory shrinkage, stealing, or defalcation—thefts committed by employees are behind at least 60 percent of crime-related losses. So many employees are stealing so much that employee theft is the most critical crime problem facing businesses today.

B-255. Although employee theft results in part from factors beyond control, the extent of employee theft in any business is a reflection of its management—the more mismanagement, the more theft. An effective stop-employee-theft policy must include at least the following:

- Pre-employment screening.
- Analysis of opportunities for theft.
- Analysis of how employees steal.
- Management-employee communication.
- Prosecution of employees caught stealing.

B-256. Each employer must reduce losses as much as possible. A police state need not be created. Large monetary expenditures need not be made.

B-257. The best way to stop employee theft is to simply not hire those employees inclined to steal. The best way is also impossible. The employer must set up a screening process that will weed out obvious security risks. Many experts believe that personnel screening is the most vital safeguard against internal theft. The following are some basic guidelines for the employer:

- Have the applicant fill out a written application. Ensure that the written application does not discriminate and that it conforms to any applicable laws.
- Solicit references, but keep in mind that those contacted will give favorable opinions. Ask primary references for secondary references. In contacting the latter, make it clear that the applicant did not refer them.
- Interview. During an interview, assess the applicant's maturity and values. Observe his gestures.
- Use psychological deterrents. Inform the applicant that your business routinely runs a background security check or that fingerprints will be taken. The hope is that the dishonest applicant will not be back.
- Obtain credit-bureau reports, but only after following guidelines set forth in the Fair Credit Reporting Act.

Opportunities, Methods, and Control

B-258. Cases of employee theft have been documented in almost every conceivable phase of business operations, from theft of petty cash to theft of railroad cars. An infinite variety of methods have been used. Some of the areas that are most vulnerable are—

- Shipping and receiving.
- Inventory.
- Accounting and record keeping.
- Cash, check, and credit transactions.
- Accounts payable.
- Payroll.
- Facility storage units.

B-259. Some of the methods used include—

- Pilfering (one item at a time).
- Stealing from cash registers or altering cash-register records.
- Issuing false refunds.
- Using the back door or trash containers.
- Taking advantage as a supervisor.
- Avoiding package controls.
- Embezzling.
- Forging checks.
- Stealing credit cards.
- Manipulating computers and stealing computer time.
- Conspiring with night cleaning crews.
- Duplicating keys or using a master key that is not properly controlled.
- Conspiring with outsiders (such as inflating insurance claims).

B-260. Too many opportunities exist for employees to exploit. Reduce these opportunities and the losses will be reduced. Reduce opportunities by using the following controls:

- Perform random spot checks on all phases of business in addition to regular, comprehensive audits.
- Check the payroll. Make sure that you are not paying a fictitious or dead employee.
- Take physical inventory seriously.
- Know what you own. Be able to identify it.
- Do not allow one employee to perform all functions. Separate receiving, purchasing, and accounts payable. Separate accountants from cash.
- Control payment authorizations.
- Keep blank checks secured. Do not presign or use uncoded, unnumbered checks.
- Reconcile cancelled checks with original invoices or vouchers.
- Secure exits. Restrict employees to one exit, preventing exit from the rear of buildings.
- Establish strict package control.
- Inspect cash-register receipts daily. Inspect the tape and ensure that the employee is identified on the slip. Deposit money daily.
- Issue ID badges to decrease employee presence in unauthorized areas.
- Simplify red tape; make it harder for the employee to disguise theft.
- Locate employee parking away from the business area.
- Establish a usage schedule of supplies to isolate irregularities.

Management-Employee Communication

B-261. Leadership must be firm, yet reasonable. Most employees pattern their values after their leaders, so a good example must be set. If you expect employees to remain honest, do not take home office supplies or goods.

B-262. Train new employees, advising them of the company's values and the standards by which they will be expected to perform. Explain all security procedures, stressing their importance. Emphasize that any deviations will be thoroughly investigated.

B-263. Establish grievance procedures; give your employees an outlet for disagreement and be receptive to all grievances submitted. Ensure that employees are aware of grievance procedures and that no reprisals are taken.

B-264. Regularly evaluate employee performance and encourage employees to evaluate management. Unrealistic performance standards can lead either to desperation and anger (resulting in dishonesty) or to get-even attitudes. Regularly review salaries, wages, and benefits—do not force employees to steal from you.

B-265. Delegate responsibility. Unless decision making exists among lower and middle levels, there is a tendency for development of an "it's-us-against-them" attitude. Delegate accountability as well; no decision is valid if it is lost in a "pass-the-buck" routine.

SECTION IV — ARMY PROPERTY AT THE LOCAL LEVEL

B-266. Proper accountability by commanders and subordinate personnel cannot be overemphasized. To ensure accountability of property, commanders must establish, implement, and supervise an installation's, activity's, or organization's security program.

B-267. Weaknesses in security procedures at the installation, activity, or organizational levels involving military property create vulnerability supported by criminal activity. Criminal activity includes—

- Theft.
- Fraud.
- Property diversion.
- Property manipulation.

B-268. Commanders and subordinate personnel must conduct a risk analysis; identify military property and; in the interest of monetary value and mission accomplishment, design mandatory security measures for specific property.

B-269. Security doctrine (as outlined in this manual) should be used to the maximum extent in securing Army property vulnerable to theft, destruction, or manipulation. Certain categories of property (such as in Table B-2, page B-62) must be assessed for security vulnerability and protective treatment. Security protective measures addressing this military property should be documented in the installation's physical-security plan. If the security measures recommended in Table B-2 are implemented using established doctrine, they should eliminate or reduce the property's vulnerability. This will reduce the incidents of theft, pilferage, and manipulation at the installation.

MOTOR VEHICLES

B-270. Security of tactical vehicles should be based on a uniform and cost-effective approach. For example, to ensure that a tactical vehicle without a locking device is properly secured, install a clamp, chain, and locking device as illustrated in Figure B-7, page B-63. To install the security device properly while maintaining safety, refer to Technical Bulletin (TB) 9-2300-422-20. Army motor-vehicle security should also incorporate the use of the following:

- Key/lock security and accountability.
- Protective lighting.
- Fencing.
- Walking patrols, as appropriate.
- Frequent observation and visits by mobile patrols or unit personnel (such as the charge of quarters [CQ], the staff duty officer [SDO], and the staff duty noncommissioned officer [SDNCO]).

Table B-2. Recommended Security Measures

Property	Inventoried by—					Inspected by—							
	Security Plan	Hand Receipt/Property Book	Secured by IDS	PS Plan	Regulation	Unit Leaders	PS Officer	SDO	Maint Officer	Supply Officer	Dining-Facility Officer	CQ	SDNCO
Arms/ammunition	x	x	x	x	x	x	x	x	x	x			x
Small arms	x	x	x	x	x	x	x	x	x	x			x
Explosives	x	x	x	x	x	x	x	x	x	x			x
Communications/electronic equip	x	x	x	x	x	x	x	x	x	x			x
Hand tools, tool sets/kits, and shop equip	x	x	x	x	x	x	x		x	x			x
Subsistence items	x	x		x	x	x				x	x		x
Controlled substances, precious metals, and tax-free items	x	x	x	x	x	x	x	x		x			x
Accounts	x	x	x	x	x	x							
POL products	x	x		x	x	x	x	x	x	x		x	
Repair parts	x	x	x	x	x	x	x		x				
Aircraft	x	x	*	x	x	x	x	x	x				
Vehicles	x	x		x	x	x	x	x	x	x		x	x
Towed weapon systems/components	x	x	x	x	x	x	x	x	x				
Carriage-mounted weapon systems	x	x	x	x	x	x	x	x	x				x
Construction material	x		*	x		x	x			x			x
Special-issue clothing (CTA)	x	x	x	x	x	x	x	x			x		
Individual clothing and equipment	x	x	x	x	x	x		x		x			
Organizational equip/components	x	x	*	x	x	x	x	x	x	x		x	x
Compasses, binoculars, and flashlights	x	x		x	x	x	x			x			x
Medical-unique items	x	x	x	x	x	x	x	x		x			x
Housekeeping supplies and equipment	x	x	x	x	x	x				x		x	
Housing furniture	x	x		x	x	x				x		x	
Mess equipment	x	x	x	x	x	x	x			x	x		x
Office machines	x	x	x	x	x	x	x			x			x
Expendable/consumable supplies	x		x	x	**	x		x		x	x		x

*** Depends on facility availability and cost effectiveness**
**** Depends on local policy**

Figure B-7. Typical Clamp and Chain Installation

CONSUMER OUTLETS

B-271. The lack of initiative at the management level within operational consumer outlets does little to prevent or reduce pilferage. Such shortcomings are identified as—

- Failure to present a professional image.
  - Lack of continuing interest, motivation, and direction.
  - No alertness to internal control of pilferage.
- Failure to institute and implement methods of operational effectiveness and efficiency. These methods include—
  - A clearly defined delegation of responsibility.
  - Insistence on stringent accountability
  - Orientation and training programs for subordinate supervisors and current and new employees.
- Failure to emphasize and enforce established criteria for continual employment.
  - Rules of conduct.
  - Standards of job performance. (Officially request appropriate action for employees guilty of criminal acts or infractions conducive to criminal acts.)
  - Inattentive job attitudes of subordinate supervisors.
  - Inadequate personal checks of established accounting and inventory procedures. **NOTE: Checks on both a regular and unannounced basis tend to control access to official stock records and to ensure careful and organized storage or stocking of merchandise.**
  - Infrequent observation of an employee's job performance.

  - Failure to report misconduct, criminal or otherwise, to superiors or responsible law-enforcement personnel in the activity.
  - Failure to implement recommendations made during physical-security inspections or crime-prevention surveys.

PILFERAGE

B-272. Pilferage may be accomplished by individual employees, by a team of employees, or by employees and patrons in collusion. These actions can be greatly reduced by tightening supervision and security in the following areas:

- **Merchandise display or dispensing areas.** The following measures can reduce merchandise pilferage in display areas:
  - Detecting unauthorized price reductions.
  - Preventing or making it difficult to alter price tags.
  - Checking procedures for declaring merchandise old, shopworn, damaged, or salvage.
  - Providing more unpackaged items for personal consumption.
  - Discouraging the careless waste of foods and other perishable items.
- **Cash registers.** The following are methods of pilferage from cash registers:
  - Stealing directly from an unattended register.
  - Rerunning register tapes at lower figures (this is preventable if the reset key is maintained by the supervisor).
  - Clearing the register at a lower total figure than actual receipts for the operational period.
  - Reporting overrings and refunds falsely.
- **Theft of merchandise.** The following methods enable merchandise to be pilfered:
  - Underringing.
  - Reusing cash-register tapes. This occurs when employees fail to provide patrons with tapes or when patrons allow employees to retain tapes. The tapes allow employees to package merchandise and remove it from the premises.
  - Removing items from bags or containers carried out by employees.

SHOPLIFTING

B-273. Shoplifting is usually confined to sales areas and is committed by casual and systematic pilferers. Items that are most frequently pilfered—

- Are relatively small in size.
- Have a high degree of consumer desirability.
- Are easily carried in pocketbooks or secreted on the person.

B-274. Amateur, adult shoplifters share the following characteristics:

- Theft is from a sudden temptation (impulse theft). There is success in the initial theft, which leads to more temptation, stronger impulses, and more thefts.
- They rarely have a genuine need for the item.

- They usually have enough money to pay for the item.
- They display symptoms of nervousness and uneasiness.

B-275. Juvenile shoplifters have the following traits:

- They act on a dare or "to belong."
- They may be coached or directed by an adult.

B-276. Professional shoplifters share these characteristics. They—

- May be talkative and are usually polite and deliberate.
- Look for opportunities continually.
- Do not take many chances.
- Are very capable of spotting security personnel.
- Steal for resale.
- Have "fences" (usually).
- Steal "to order" (often). They may have a list describing the items to be pilfered.
- Use innovative techniques.

B-277. Genuine cases of kleptomania are rare. Kleptomaniacs share the following characteristics. They—

- Take items without regard to their value or use.
- Steal compulsively and often openly.
- Are nervous and shy.

B-278. Narcotic addicts as shoplifters have the following characteristics. Only MP/security personnel should attempt apprehension. These types of shoplifters—

- Are desperate for money and they fear imprisonment.
- Take big chances.
- Take the merchandise and exit the premises quickly.
- Steal when they are at their lowest physical and/or psychological ebb.
- Are dangerous if you try to apprehend them (sometimes violent).

B-279. Alcoholics and vagrants as shoplifters share the following traits. They—

- Usually steal because of need.
- Are often under the influence of liquor at the time of theft.
- Take the merchandise quickly, then exit the premises.
- Are less likely to steal regularly at a single location.

B-280. The most pilferage occurs when employee coverage is low or when employees are untrained, inexperienced, or indifferent to the issue. The ineffective use of floor space aids shoplifters by creating congestion in the patron's traffic flow. Allowing an emphasis on small rooms or partitioned areas causes congestion that clusters, isolates, or partially hides displays. Shoplifting involves the use of one or more of the following means to obtain items:

- Palming or placing an open hand on a small article, squeezing the muscles of the hand over the article to grasp it, and lifting the still open and apparently empty hand.
- Using fitting rooms to put on tight or close-fitting garments under clothing worn into the store.
- Trying on hats, gloves, sweaters, and jackets, then exiting the store.
- Stepping around counters and removing items from unlocked showcases.
- Handling several items at once and replacing all except the item pilfered.
- Using accomplices to create a diversion of employee attention when secreting items on the person. Such items include:
  - Clothing.
  - Pocketbooks or handbags.
  - Umbrellas.
  - Various items placed in packages or paper sacks containing merchandise paid for at other departments.

ARSON

B-281. On an installation there is little incentive for a professional arsonist to operate since the government owns the buildings and insurance fraud in collusion with the property owner is not possible. However, arson can still be a problem. In the civilian community, most deliberate fires are not set for profit. They are set to "get even" with the property owner or just for the excitement of watching something burn. Military installations are susceptible to this type of crime. On many Army posts, there is a large number of empty, wooden structures that are ideal targets for revenge seekers or vandals.

B-282. Arson is an easy crime to perpetrate, and it is relatively difficult to collect the information needed to convict an arsonist. However, this does not mean we are helpless in combating this crime. Several major cities, including Seattle, Denver, Houston, and Philadelphia, have developed successful programs to reduce the number of arsons. Some of the successful, proactive measures that have been developed are—

- Securing or disposing of materials that could be used to start fires. Ensuring that the regulations on the storage of gasoline, paint, and solvents are enforced and that paint lockers are locked.
- Enforcing command policies on the police of the post. Removing piles of trash, scrap lumber, and other material that burns easily.
- Securing empty buildings (especially empty wooden structures) and posting them as off-limits areas.
- Patrolling areas susceptible to arson.
- Encouraging participation in neighborhood watches, taxi patrols, and other community programs that increase surveillance.
- Encouraging the reporting of suspicious activities through the establishment of crime hot lines.

- Establishing a close working relationship between fire fighters and law-enforcement personnel to ensure that fires of suspicious origin are reported and thoroughly investigated.
- Offering rewards for information leading to the apprehension of the arsonist (if there is an outbreak of arson).

SECTION V — COMMUNITY CRIME-PREVENTION PROGRAMS

B-283. There are a number of crime-prevention programs set up in communities to help deter and detect crime. These programs are supported and often headed by members of the community in conjunction with local law-enforcement agencies. The programs are as diverse in nature as they are in number, yet they complement one another.

NEIGHBORHOOD WATCH PROGRAM

B-284. A neighborhood watch program is an organized network of citizens interacting with other neighbors and the police in preventing and detecting crime in their neighborhood. Law-enforcement efforts to reduce crime cannot be accomplished effectively without the support and cooperation of all citizens. A strong community involvement with neighbors helping themselves and other neighbors in becoming more alert to activities in the neighborhood, protecting their property, and reporting suspicious activities is essential to an effective crime-prevention program.

B-285. The Army Neighborhood Watch Program is designed to encourage Army service members and their families to actively participate in protecting their own property and the property of their neighbors, joining community crime-prevention programs, and reporting suspicious activities to MP officers. The program is designed to develop the following:

- The awareness of community crime trends and prevention efforts.
- The knowledge of quarters' security procedures.
- A cooperative system of surveillance over each neighbor's property.
- Accurate observation and reporting of suspicious activities.
- The establishment of a reliable, two-way information link between the community and MP forces.

B-286. Most neighbors know the routines of the other families that live near them. They know what cars are normally parked in the neighborhood and when families are on vacation or out of the area. Neighbors are in a very good position to recognize burglars and other intruders. Also, residents are in a good position to recognize safety hazards and crime-conducive conditions near their homes.

B-287. To capitalize on these advantages, neighborhood watch programs organize blocks in family-housing areas or floors in troop billets to improve police and community interaction. They also disseminate information on crime problems and countermeasures.

BLOCK CLUBS

B-288. Block clubs are the basic components of an installation-wide neighborhood watch program. The geographical size of a block club may vary widely depending on the population's density and the nature of the terrain. The key factor is that the terrain organized into a single block club should promote a feeling of unity and mutual assistance. In a troop billet area, block clubs could be organized along company lines, by individual barracks, or by floor. In a family-housing area, a block club could cover one high-rise apartment building, one block, or one or more streets that are so situated that the residents identify themselves as a subcommunity within the housing area.

Organization

B-289. Individual residents, community-service organizations, or MP units can initiate block clubs. Existing organizations (such as the PTO) have frequently established contacts within the community and have sponsored the organization of block clubs. Regardless of the approach used to organize the installation, every family or resident in the area should be contacted and encouraged to attend a block club (see Figure B-8).

B-290. A block captain and a deputy block captain should be elected at the initial block meeting. These individuals serve as "spark plugs" to sustain interest in their geographic area. They also represent their block club at district meetings. The effectiveness of the presentation at the initial block-club meeting is critical. A PMO representative should explain the crime problems on the installation and clearly outline the functions of a block club and how it can affect the crime problem. The main objective should be to generate enthusiasm and build a foundation upon which an effective neighborhood watch can be built. The use of camcorders or other audiovisual programs can generate interest. Association with programs in the civilian community may also help to generate interest.

B-291. The functions of a block club should be clearly identified. Block clubs—

- Serve as the "eyes and ears" of the police.
- Encourage the implementation of individual countermeasures, such as marking personal property.
- Disseminate crime-prevention information.
- Improve police and community relations

Block-Club Districts

B-292. A block-club district should be organized as an intermediate link between individual block clubs and the installation's crime-prevention council. These districts should cover at least one housing area and should be composed of the block captains and deputies from each block club in the district's area. District leaders should be elected from the block captains and should serve as members of the installation's crime-prevention council. The primary functions of districts are to transmit information from the crime-prevention council to the block clubs and to develop incentive awards to recognize effective participation by the clubs.

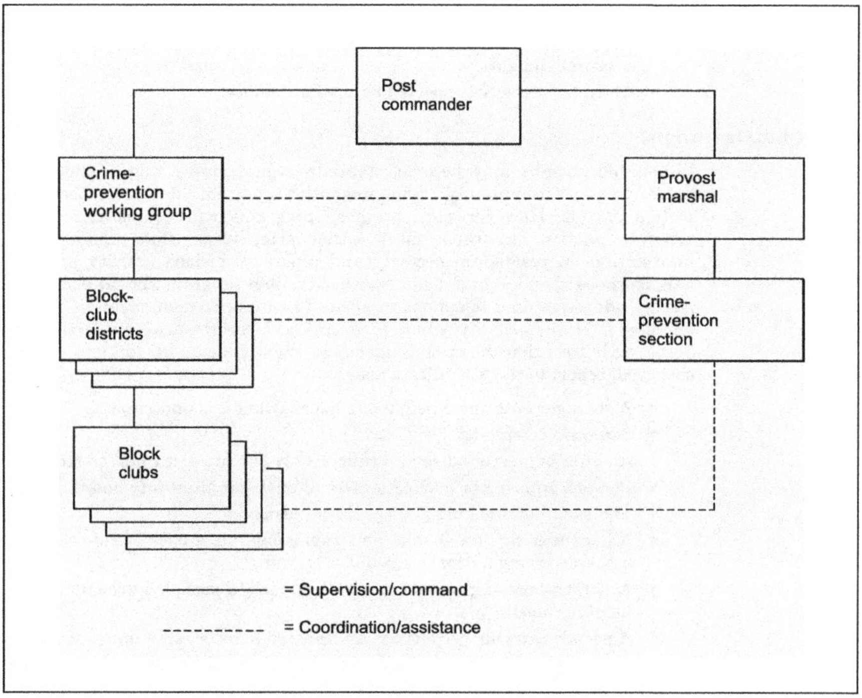

Figure B-8. Block-Club Organization

Maintaining Interest

B-293. Maintaining interest in neighborhood watch programs is a major problem. Typically, a particular incident generates the level of interest required to initially organize a system of block clubs and districts; but as the particular problem is overcome, interest wanes and the clubs gradually dissipate. A review of programs that have maintained their effectiveness for extended periods indicates that successful programs have the following characteristics:

- They are a formal organization with elected block captains and district leaders.
- They address a wide range of problems.
- They feature an incentive award program to recognize individuals and clubs that participate effectively.

- Their members attend periodic workshops to train leaders on various aspects of crime prevention.
- They have clear-cut, attainable objectives. The members need to see accomplishments.
- They are actively supported by the local police.

Block-Club Operations

B-294. Block clubs may become umbrella organizations that conduct the entire range of community crime-prevention programs, including foot and mobile patrols. However, each of these programs will be discussed in a separate section. In their most basic form, block clubs disseminate information on residential security and report suspicious activity in their neighborhood. Members of a neighborhood watch program should exchange names, addresses, and telephone numbers to enhance communication among neighbors. They must learn how to record and report suspicious activities. Some activities that block-club members should be alert for and, when observed, report to the MP officers are—

- A stranger entering a neighbor's house when it is unoccupied.
- Someone screaming.
- An offer of merchandise at a ridiculously low price (it could be stolen).
- Persons entering or leaving a place of business after duty hours.
- The sound of breaking glass or an explosion.
- A person going door to door and then going into a back or side yard, or a person trying a door to see if it is locked.
- A person loitering around a school, a park, a secluded area, or in the neighborhood.
- A person carrying property at an unusual hour or in an unusual place.
- A person exhibiting unusual mental or physical symptoms (he may be in need of medical help).
- A vehicle being loaded with valuables when parked by a closed business or untended residence.
- A business transaction conducted from a vehicle.

ARMY NEIGHBORHOOD WATCH MEETINGS

B-295. A PMO representative should attend the initial neighborhood block meeting to explain the relationships of crime prevention and the neighborhood watch programs. The meeting should be publicized with handout invitations announcing the time, location, and purpose of the meeting. If possible, the meeting should be held in the neighborhood (such as the training room in the troop billets, a local school, or a residence in family quarters). In addition to explaining the program concepts, the initial meeting could include the following:

- Installation and community criminal statistics concerning the nature and volume of housebreakings, larcenies, and other crimes.
- An exchange of names and telephone numbers (if applicable) of attendees. This information should also be placed on a neighborhood block sheet that is a geographical diagram of the block showing the

location of each room, apartment, and building address number in the block. The names and phone numbers of participants would be added to the address number of each residence drawn on the diagram. This block sheet should be distributed to block members at a subsequent meeting.

- Determination of the second meeting date and location.

B-296. Other agenda items and activities for subsequent meetings could include the following discussions:

- Residential security procedures and the conduct of quarters security inspections.
- Operation ID—marking and recording of personal property.
- The observation and report of suspicious activities.
- Other crime-prevention measures and procedures being implemented at the installation (such as rape prevention and citizen escort patrols).
- Fire prevention, personal safety, and other related activities.

OPERATION ID

B-297. Operation ID, which entails marking of property to make it identifiable and traceable to its owner if lost or stolen, was initiated by the Monterey Park, California, Police Department in 1963 and has been adopted by more than 80 percent of the police departments in the US. Operation ID is a low-cost, highly effective crime-prevention program. However, its success is contingent upon the willingness of individuals and communities to actively participate in marking and identifying their personal property.

B-298. Operation ID is designed to encourage Army service members and their families to mark their personal property with a standard Army-wide, owner-applied number. This numbering system permits the positive ID of the property and determines the location of the owner in case of theft or loss.

B-299. The principal advantages of Operation ID are theft deterrence and recovery of personal property. Marked stolen property is more difficult to dispose of, and illegal possession can result in the prosecution of a thief. Recovered lost or stolen property can only be returned if there is some means of identifying and locating the rightful owner.

METHODS OF IDENTIFYING PROPERTY

B-300. Various methods of establishing positive ID and ownership of property in case of loss are available for individuals. Each method has advantages and limitations, and a combination of these methods would be required to ensure the ID of all high-value personal property.

- Inscribing the owner's applied number with an etching or engraving tool would allow the recovering agency to visually identify the number inscribed on the property for notification and subsequent return of the property. Electrostatic markers are available for use at no cost to the individual. However, some personnel are reluctant to use this method since it can mar the property. The inscription should be made in a location that can be readily seen by the recovering agency but which would not deface the property's appearance or reduce its value. Some high-value personal items such as coins, jewelry, and silver cannot be inscribed with an owner's applied number. Another method of identifying these items would be required.

- Using invisible fluorescent ink, powder, or paste to mark the property can make it easier for the agency recovering the property to use an ultraviolet light and identify the owner. This marking will not mar the property. However, the fluorescent markings and ultraviolet light are an additional cost, and many agencies do not have ultraviolet lights to inspect recovered property.
- Using a laser photographic process to identify diamonds. Every diamond emits a unique reflection when penetrated by a low-level laser light. A laser photographic process has been developed to record a diamond's pattern of light reflection on film. Several jewelers throughout the country have the laser photographic equipment available. To register a diamond, two photographs are taken. One is provided to the owner and the other to a central registry. If a diamond is lost or stolen, the recovering agency can take the diamond to a jeweler that maintains the laser photographic equipment for print. This photograph would then be forwarded to the registry for owner ID.
- Photographing the personal item. Individuals can photograph personal high-value items that cannot be engraved. Although the agency recovering the lost or stolen items could not identify the owner, use of photographs could assist in verifying ownership if it is known that the item has been recovered. In addition, a photograph would assist in submitting claims against the government or private insurance companies, as appropriate.

RECORDING PERSONAL PROPERTY

B-301. Individuals should record identifying data (such as brand name, model, serial number, and value of the personal items), even if they use other methods of identifying property. This information would assist in determining what items may be lost, stolen, or damaged through fire, explosion, or other hazards. This information can also be used in claims against the government or private insurance companies, as appropriate.

IDENTIFICATION NUMBERING

B-302. There are various types of owner-applied numbering systems used to mark personal high-value items. Criteria that should be considered in determining which owner-applied numbering system to incorporate include—

- Uniqueness, where no two people have the same identifier.
- Permanence, so that the owner-applied number will not change.
- Ubiquity, so that an identifying number is available to any individual who desires one.
- Availability, where the identifying number can be easily obtained and remembered.
- Indispensable, so that there are incentives requiring an individual to have the number.

- Privacy, so that the number is not a means of infringing upon an individual's right to privacy.
- Uniformity, so that the owner-applied number would be readily recognized by law-enforcement agencies who handle or come into contact with the recovered property.
- Traceability, so that the property owner can be identified and located.

B-303. The most commonly used owner-applied numbering systems include—

- Driver's license number with the issuing state abbreviation prefix.
- Social security number.
- Personal numbers assigned to individuals by a local law-enforcement agency.
- Personal numbers with the marking agency's National Crime Information Center originating-agency ID number.
- A private numbering system maintained by a commercial organization.

STANDARD ARMY NUMBERING SYSTEM

B-304. To ensure that criteria for property ID numbering are met, the standard Army-wide, owner-applied numbering system is designated as the service member's social security number with a "USA" prefix. Upon recovery of lost or stolen property by other military or civilian law-enforcement agencies, the USA prefix, owner-applied number would alert the recovering agency that the property belongs to a member of the Army. The recovering agency can then contact the nearest Army installation's PMO/security office concerning the property and the owner-applied number inscribed. The service number can then be identified and located through the Army worldwide locator system. Since this locator system lists only social security numbers and locations of active Army service members, family members should also use the service member's social security number with the USA prefix when marking their personal items.

IDENTIFYING AND LOCATING OWNERS OF RECOVERED PROPERTY

B-305. The installation's PMO/security officer should be the initiator of tracer actions to identify and locate the owner of recovered property marked by the standard Army Operation ID numbering system. Continuous liaison should be maintained with local civilian law-enforcement agencies and other military installations to ensure that they are cognizant of the standard Army owner-applied numbering system and will contact the PMO/security office upon recovery of private property marked with this system. The PMO/security officer should readily accept custody of the private property if the recovering agency is willing to release the property and it is not required as evidence for criminal prosecution.

B-306. Upon notification of recovered property and the inscribed Army standard owner-applied number, the PMO/security officer should contact the servicing military personnel office (MILPO) for assistance in determining the name and location of the property owner by using the Army's worldwide locator microfiche.

B-307. If the owner-applied social security number is not listed on the Army's worldwide locator microfiche and the servicing MILPO is unable to provide the requested information, the PMO/security office should contact the following offices:

- For enlisted members: USA Enlisted Records and Evaluation Center, ATTN: PCRE-RF-L, Fort Benjamin Harrison, Indiana 46249.
- For officers: PERSCOM, ATTN: TAPC-MSR-S, 200 Stovall Street, Alexandria, Virginia 22332-0444.
- For warrant officers: PERSCOM, ATTN: TAPC-OPW, 200 Stovall Street, Alexandria, Virginia 22332-0444.

B-308. In case the owner-applied number cannot be identified by either the servicing MILPO or the USA Enlisted Records and Evaluation Center, the Army Reserve Personnel Command (AR-PERSCOM) should be contacted for assistance. The AR-PERSCOM retains files on separated, retired, and Reserve component Army members. Written requests should be forwarded to: Commander, AR-PERSCOM, ATTN: ARTC-PS, 1 Reserve Way, St. Louis, Missouri 63132.

B-309. When the name and location of the service member associated with the owner-applied number inscribed on the property is verified, the PMO/security office should notify the service member in writing that the property has been recovered. The notification should ascertain if the recovered property belongs to the service member, if the service member ever reported the property as lost or stolen, and if a claim was submitted to the SJA claims service for loss or theft of the property. Notification should also state where the property is located and a point of contact that the owner can deal with for the return of the property. If the property is to be retained as evidence for legal proceedings, the owner should be informed that the property will be returned upon completion of the proceeding. A copy of this letter should be provided to the US Army Claims Service, Fort Meade, Maryland 20755. The Army Claims Service will advise the PMO/security office if a claim was or was not submitted.

B-310. If the owner cannot be located, recovered property in the custody of the PM/security offices should be disposed of according to AR 190-22 and DOD 4160.21-M.

USE OF THE NATIONAL CRIME INFORMATION CENTER

B-311. Stolen articles may be entered into the National Crime Information Center (NCIC) file if a theft report has been made, if the item is valued at $500 or more, and if it has a unique manufacturer's assigned serial number and/or owner-applied number. Entering stolen personal property items meeting the above criteria into the NCIC or other police information systems as outlined in AR 190-27 is encouraged.

NEIGHBORHOOD WALKS

B-312. While most people are unwilling to participate more actively than as observers in a neighborhood watch, there are some individuals who want to become more actively involved in securing their neighborhoods. For this segment of the population, the organization of "neighborhood walks" provides a welcome opportunity to make a more active contribution. The basic idea is simple; residents patrol on foot through their own neighborhoods to observe and report crime. In practice, it is a little more complicated; however, neighborhood walks can have a dramatic impact on the crime rate, so the effort expended is worthwhile. The points that must be considered are—

- **Patrol composition.** Both adults and teenage children can volunteer to participate in neighborhood walks. However, there should be at least one adult in each party. There should be at least two people in each patrol group. Groups of four to six individuals are desirable since, by their numbers alone, they discourage attacks on the walkers. Larger groups are also more fun, and this is important when volunteers are providing the manpower.
- **Times/patrol duration.** As with other crime-prevention programs, maintaining a high level of interest can be a problem. A successful walk program in Philadelphia schedules groups for one 2-hour patrol per month. More frequent tours caused high drop-out rates among participants. Neighborhood walks should be conducted only during those times when the crime rate is the highest. Normally, there are not enough volunteers to conduct walks at times other than peak crime periods.
- **Functions.** Members of neighborhood-walk groups must understand that they are to observe only and not actively intervene in criminal acts. Participants and the government are legally liable for their actions during walks. When a crime or a suspicious activity is spotted, the neighborhood walkers should report it to the police. In Philadelphia, the walkers are equipped with horns. When a crime is spotted, they activate their horn and go to the nearest house to call the police. When residents hear the walker's horn, those in the immediate area turn on all of their lights and sound their own horns. The noise and increased lighting invariably causes the criminal to flee.
- **Neighborhood escorts.** In addition to observing and reporting criminal activity, neighborhood patrols can escort children and older persons between community-service facilities and residences. They can also request that owners secure property when they find it unsecured; for example, when there are unsecured bicycles parked on a front lawn.

VIGILANTISM

B-313. While active community participation is essential, vigilantism must be discouraged at all costs. Both formal law codes and US common law offer few protections for private citizens who take the law into their own hands. Block captains and installation crime-prevention officers must be alert for indications that neighborhood patrols are doing more than observing and take swift remedial action when required. Of course, all nonpolice participants must be prohibited from carrying weapons of any type while engaged in crime-prevention programs. Experience in neighborhoods having much higher violent crime rates than found on Army installations has demonstrated that passive devices like horns or whistles were adequate to discourage attacks.

These devices, plus the assignment of four to six individuals to a neighborhood patrol, provide sufficient protection.

MOBILE PATROLS

B-314. Some communities with high street-crime rates have been successful in organizing private citizens into mobile-patrol programs. Like neighborhood foot patrols, these mobile patrols serve as the "eyes and ears" of the police but do not actively intervene when they spot a crime in progress.

B-315. In a typical program, block captains or police crime-prevention officers assign specific patrol areas to each private mobile patrol. In addition, each patrol receives training on the patrol's functions, communications procedures, and emergency actions. Normally, patrols are instructed to blow their auto horns steadily when they observe a crime in progress. This is usually sufficient to drive off the criminal.

B-316. Most private citizen patrols use cellular telephones or citizen's band (CB) radios as direct communications links with the supporting police department. Installation CB radio clubs are often willing to sponsor anticrime patrols under police supervision. Commercial taxi companies that operate on military installations are also excellent candidates to organize into patrols. The cab drivers normally cover most of the high-crime-rate areas on the installation. Because of the frequency with which they cover them, they are familiar with the routine conditions in each area and are quick to spot suspicious activity.

B-317. The government cannot provide gasoline for POVs used for anticrime patrols; however, it may often provide magnetic signs to affix to the vehicle for identifying it as part of the police-sponsored, neighborhood-patrol program. As in the case of foot patrols, the installation crime-prevention officer must be alert for signs of vigilantism and must take positive action to discourage it if it appears.

PROJECT LOCK

B-318. Nearly one million automobiles are stolen in the US every year. The total value of cars stolen is around the billion dollar mark, making auto theft the nation's costliest crime involving property. Of even greater importance is the social impact of auto theft. For a growing number of young people each year, stealing cars represents the first step toward a life of crime.

B-319. Police agencies have been diligent in the apprehension of auto thieves and the recovery of stolen vehicles, but the auto-theft problem seems to be more amenable to improvement through prevention rather than punishment. Barring strict security, auto theft is one of the easier crimes to commit. All vehicles left unattended are vulnerable, and widespread prevention by police surveillance is a physical impossibility.

B-320. The problem has grown to serious proportions despite determined law enforcement because motorists continue to be negligent or unaware of their responsibility. As long as people invite theft by leaving their cars unlocked or leaving the key in the ignition, auto thefts will continue to climb. Almost half of all stolen cars each year had been left with keys in the ignition; nine out of ten of the stolen vehicles had been left unlocked.

B-321. If a significant reduction is to be made, the motorists themselves must make it. Widespread adoption of accepted and effective prevention practices by motorists presents the most logical and immediate improvement to this growing problem.

B-322. In 1963, the Boston police department conducted a broad information campaign with the assistance of the National Automobile Theft Bureau and the Insurance Information Institute. Since then, more than 525 "Lock Your Car" campaigns have been held in about 400 communities in 49 states.

B-323. In the months following these campaigns in such cities as Denver, Chicago, Atlanta, and San Francisco, significant reductions in the number of auto thefts (ranging from 9 to 54 percent) have been recorded. Undoubtedly, the "Lock Your Car" campaigns contributed to reducing auto-theft statistics.

B-324. Project Lock is designed to permit sponsoring groups to conduct one-day or one-week "Lock Your Car" campaigns. Its purpose is twofold; it—

- Alerts the public on the importance of locking cars and removing keys as a deterrent to auto theft.
- Contributes to the welfare of youths by preventing the commission of a first crime.

B-325. The following materials can be ordered to support Project Lock:

- Windshield flyers.
- Identifying insignia to be worn by inspectors.
- Tally cards for noting cars left with keys in ignitions or unlocked doors.

B-326. Groups considering sponsorship of this campaign should consider installation areas known to have car-theft problems. The PMO should assign patrolmen to accompany the teams on inspection day. Usually, the teams' routes can be arranged to fit the regular patrols of the MP officers. If uniformed police will not be available, the campaign should not be held.

COMMUNITY ORIENTATION

B-327. If the commander desires, a community orientation meeting might be held a month or so in advance of the campaign. This would be more desirable for a week-long rather than a day-long campaign. If such a meeting is planned, the PMO should issue invitations to representatives of the installation crime-prevention council, service clubs, women's clubs, PTOs, high schools, and churches.

B-328. The meeting should be opened with an introduction of the installation commander. After appropriate comments, the commander would read a proclamation setting the date for Lock-Your-Car Week. If possible, a representative of the National Automobile Theft Bureau should be asked to address the meeting. An alternative would be to have the PM review national and local trends in auto thefts stressing the importance of the forthcoming campaign. In conclusion, a representative of the sponsoring group might review the schedule of activities for the campaign.

PROGRAM ACTIVITIES

B-329. If the campaign is conducted in one day, a kick-off breakfast might be substituted for the orientation meeting. The program and the attendance might be similar if a weeklong campaign is planned and the orientation meeting were held. The kick-off breakfast might have a simple program with attendance limited to the sponsoring group and police representatives. An alternative to the breakfast would be a luncheon at which results to the moment are reported. Whether the campaign will last a week or a day, the general activities will be similar.

B-330. With the assistance of the MP officers, the installation should be zoned according to the established MP patrol areas where possible. The size and number of the zones will depend on the number of police and volunteer personnel that will be available. However, the zones should cover most of the installation's service areas, troop billet areas, and family-housing areas.

B-331. The inspection teams are each composed of a uniformed patrolman and three to five sponsoring group members, cover their assigned zones and place flyers under the windshield wipers of all cars found to have keys in the ignition or to be unlocked. Under no circumstances should the flyers be placed inside the cars, even through open windows.

B-332. It is suggested that flyers be ordered early enough to allow a local printer to inscribe an overleaf statement such as "This public service is provided as a courtesy of the MP force." The tally cards should be used to record the number of cars inspected, the number that were unlocked, and the number with keys in the ignition.

SCHOOL PROGRAM

B-333. The crime-prevention office might sponsor a poster contest for art students. It should be announced at least a month before the campaign to allow the entries to be placed in public in advance of the date. The PMO should present awards in an office ceremony. The ceremony should take place at noon on the campaign day or midweek if the campaign is longer.

PUBLIC ADDRESSES

B-334. Close to or during the campaign date, addresses by crime-prevention professionals should be scheduled for programs of service clubs, women's clubs, PTOs, and other civic groups. Well in advance of the campaign, the PAO should be visited by the PM and the other sponsors to develop comprehensive internal-information programs in support of the installation's Project Lock. Plans for the public-affairs program should include advance publicity for the campaign, coverage of events during the campaign, and wrap-up coverage following its completion. During the inspection days, findings on how many cars were unlocked and how many had keys in the ignition should be reported regularly to a headquarters (preferably to the MP unit).

CONCLUSION

B-335. Project Lock has been outlined to provide basic suggestions for a "Lock Your Car" day or week. No procedure can be designed to fit all needs or circumstances, and variations often will be desirable. However, if the suggestions and the materials contained in this section are used, Project Lock will not be difficult to organize and conduct. It provides better publicity opportunities than most public-service projects. It presents an opportunity for active participation by a number of crime-prevention groups. It has been field-tested and found to be an outstanding success.

SECTION VI — EVALUATION

B-336. Evaluations of installation crime-prevention programs (see Figure B-9, page B-80) do not rely on control groups, a tight control of variables, or elaborate statistical analyses to produce worthwhile results. Most often, the resources to conduct a formal evaluation that will stand up to rigorous academic scrutiny are not available.

CRIME-PREVENTION PROGRAMS

B-337. Several inherent difficulties in data collection on crime-prevention programs make it difficult to determine with 100 percent accuracy that a particular reduction in the crime rate was a result of a particular crime-prevention measure (unless a very elaborate analysis and control system are used). These difficulties include the control of variables, the displacement effect, and unreported crime.

CONTROL OF VARIABLES

B-338. In its simplest form, the type of evaluation most commonly used in academic or scientific settings seeks to determine the relationship between two variables. By varying the independent variable (for example, the dosage of a drug), the effect on the dependent variable (for example, a pulse rate) is determined while all other variables (such as food intake) are held constant.

B-339. Although more complex in form, the same model can be used to evaluate complex programs. Experimental and control groups can be selected. The "treatment" can be administered by researchers or those taught by researchers. The results can be analyzed for their statistical significance. However, since crime-prevention programs deal with human subjects, certain complications arise. The degree of success may have nothing to do with the efficacy of the program, but only with the way it was introduced or with the personal predilections of the groups involved. There is no standard population; human beings are not standardized as mice are for laboratory purposes. A program found successful at one installation may be a failure in another.

B-340. These considerations also apply in the evaluation of crime-control programs. This evaluation is further complicated by another problem—the people whose behavior is to be modified (the offenders) cannot be treated directly or separated into experimental and control groups; they will not stand up and be counted. Although public-health programs often encounter this problem, they often deal with physical cause-and-effect links between treatment and improvement. The same is not true for crime-control programs.

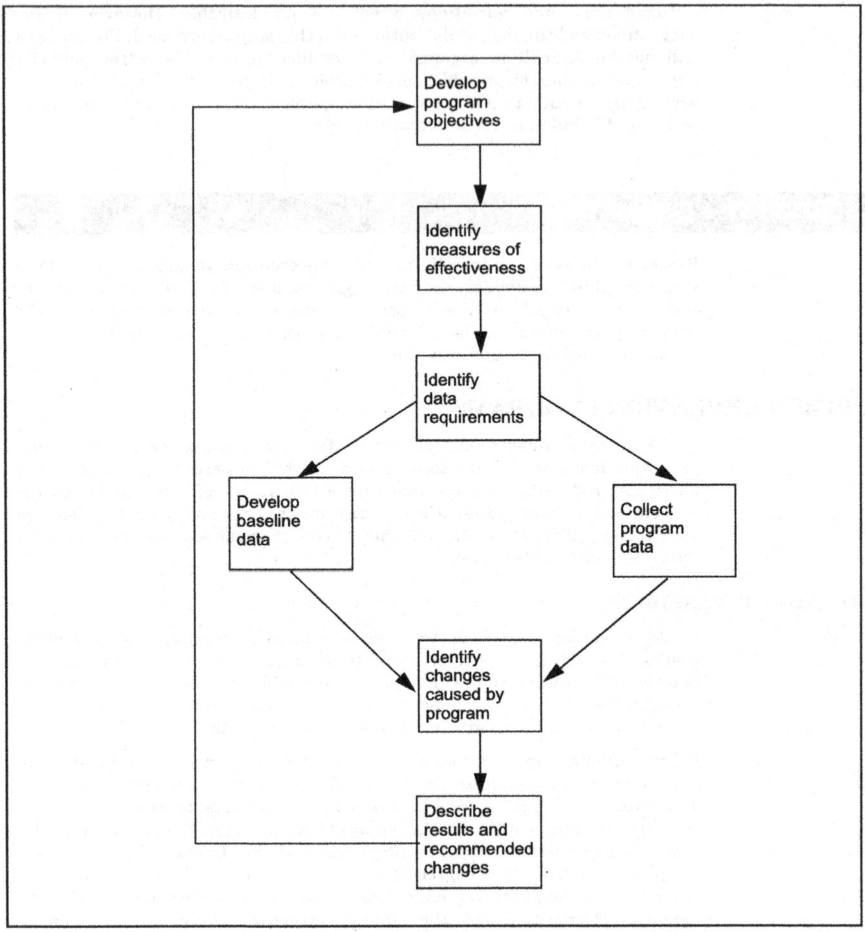

Figure B-9. Crime-Prevention Program Evaluation

The effectiveness of these programs is normally determined by looking at statistics of reported crime and arrests, which are more indirect indicators.

B-341. In a crime-control program, it may be impossible to classify variables as dependent and independent; they may all affect and be affected by each other. Furthermore, because of the difficulty in determining why people behave the way they do, a number of intervening and antecedent variables may go unnoticed. Police programs designed to reduce crime may have their most direct effect on the behavior of the general public toward the police, which in turn affects the crime rate.

B-342. Evaluations are not necessarily restricted to the analysis of objective crime data; they can also include subjective considerations and perceptions. These subjective evaluations can be of significant benefit in augmenting the statistical analyses of the program's results. They are especially helpful in assessing why and how a program worked and whether a statistical outcome is actually evidence that the program was successful. Interviews of participating agency personnel and residents of the program's area are usually used to supply this information. They can give the evaluator new insight into the actual program operation.

DISPLACEMENT EFFECT

B-343. In many cases where crime reductions have been measured and attributed to programs, it is unclear whether there has been an actual reduction in crime or whether the crime has been displaced. The amount of displacement depends to an extent on the offender's characteristics. An opportunistic offender can be pictured as having a relatively elastic demand—if the risk is too high, he will forgo the crime. An addict offender is typically pictured as having a relatively inelastic demand for the product because of his inelastic demand for drugs—despite the risks, he needs the product.

B-344. Deterrents may have little effect on perpetrators of expressive crimes. These are crimes in which the perpetrator is emotionally involved and is expressing these emotions. Most assaults and homicides fit this category. On the other hand, deterrents may have a strong effect on instrumental crimes, those that are seen by the offender only as a means to an end (usually money). If alternative avenues to the same end are made more attractive by comparison, the offender may well be deterred. Deterrence may produce a diversion to legal alternatives to crime; it also may cause displacement to illegal alternatives.

Displacement to Other Crimes

B-345. There is no immutable law that says a burglar cannot hold up a liquor store and a robber cannot burglarize a warehouse. If a specific crime or a set of crimes is the target of a crime-control program, offenders may decide to avoid the target crimes and ply their trade in other ways. Some offenders will be deterred from all crime if their crime specialty is the object of a crime-control program, but the extent of this deterrence should not be overestimated. The statutory categories of crime should not be confused with categories that serve to classify offenders.

B-346. In some cases, the result of displacing offenders to other crimes is beneficial. If the targeted crimes are more serious than the ones to which offenders are diverted, the net effect on the program may be the reduced danger to society. Of course, the converse may also be true; closing off the vulnerable and more easily protected targets of crime may cause an offender to commit more serious crimes with a net increase in the danger to society.

B-347. In some instances, the individual effect may be substantial but the overall effect may be negligible. Protecting a small fraction of premises against burglary will reduce the number of crimes committed against them, but the burglary rate against unprotected premises may go up.

Displacement to Other Tactics and Targets

B-348. Offenders can change their manner of committing a crime when a new program is established to counter their activity. One example of this took place in 1969 in a section of the Bronx that was having a rapid increase in outdoor crime. The crimes took place primarily in the evening hours when people were returning from work. The program instituted by the police consisted of intensive sweeps of randomly selected city blocks, coupled with plainclothes police officers patrolling the streets. It succeeded in reducing the number of offenses committed during the evening hours, but at the expense of increasing the number taking place in the late afternoon when patrolmen were taking their lunch hours or were occupied with school crossings or shift changes.

Displacement to Other Areas

B-349. The most frequently discussed type of crime displacement is from one area to another. For instance, it has been cynically suggested that the goal of the New York subway police is to chase crime into the streets where it belongs. More seriously, some recent police-helicopter-program evaluations have been questioned because they did not consider possible area displacements.

B-350. One type of boundary of interest to crime displacement is the jurisdictional boundary between the installation and surrounding cities. It has been conjectured that the crime reduction experienced in some central cities has been at the expense of the surrounding suburbs that have experienced increased crime rates.

B-351. An initial study of crime displacement was performed for the Washington, DC, area. It concluded that, although the decrease in Washington's crime rate was concurrent with an increase in the suburban crime rate, there is no evidence that the reduction in reported crime in Washington, DC, has resulted in a corresponding crime increase in the nearby suburbs.

B-352. The area-displacement effect can be measured with some degree of reliability. Three zones can be defined for the purposes of the measurement—the area containing the crime-control program (zone 1), a border around the area (zone 2), and the area chosen as the control area (zone 3). The width of the border may depend on the type of program implemented. If the program involves police helicopters, a quarter-mile-wide border may be necessary; for a patrol car, one or two blocks may suffice.

B-353. Crime rates before program initiation should be determined for all three zones. If zone 2 records a greater increase in crime than zone 3 while zone 1's crime rate decreases, then the increase in zone 2 can be attributed to two factors—

- The general increase in crime rate verified by any increase in zone 3.
- The increase caused by a displacement of crime from zone 1.

B-354. A displacement of this crime does not mean that the program is ineffective. It may suggest that the program should be expanded for all three zones.

UNREPORTED CRIME

B-355. Crime statistics are based on crimes reported to the police. It is well known that many crimes go unreported. Victimization studies can determine the extent of unreported crime and its change from year to year by area of the country or the reasons for failure to report them. These victimization studies are best suited to determining long-term effects. They are not well suited to most crime-control evaluations in which short-term changes must be assessed.

B-356. The amount of unreported crime is important but not for planning crime-control programs affecting police activities. The extent of unreported crimes is of little significance unless a program affects it. If a program encourages reported crime, the reported-crime rate may increase despite the program's effectiveness.

B-357. Ironically, a lowered reported-crime rate may be the direct result of an increase in the actual crime rate. Taking reports from victims of crimes occupies a substantial amount of a patrolman's time. Many of these crimes are minor and have no potential for solution. In an effort to increase the police department's time on patrol, some police chiefs have stopped the practice of sending a patrolman to get reports from the victim of a minor crime. This requires the victim to travel to the police station to report the crime. If the crime is minor or is seen by the victim to be unsolvable or the theft is not covered by insurance, the victim may decide not to inconvenience himself by going to the police station to report the crime. Therefore, the number of crimes reported to the police may drop. This may lead to a larger number of unreported crimes and prevent a complete picture to the crime-prevention counsel. Conversely, an actual decrease in crime due to the increased effectiveness of the police may produce an increase in the reported crime rate.

CRIME RATES

B-358. It has been pointed out that the crime rates, as presently calculated, do not reflect the true situation. For example, the rape rate should be calculated by dividing the annual number of rape cases by the number of women (since they are the population at risk). You would expect that the rate of commercial burglaries would be less in a residential area than a commercial area. When calculated on the basis of "per thousand people," this would be true; however, these rates should be obtained by dividing the number of cases by the number of commercial establishments (the population at risk) in each area.

B-359. The victim or the target is only one aspect of the crime. The offender can also be calculated into the rate. For example, the potential offenders in stranger-to-stranger crimes are usually considered to be males between 16 and 25 years of age. Therefore, one would expect fewer of these crimes in a city full of pensioners and retirees than in a city of the same population but with a higher proportion of young men. This fact is of minor importance in evaluating crime-control programs, since the age distribution of people in a city or a section of a city does not normally change greatly over the evaluation period. However, the former factor (the population at risk of becoming victimized) can be misleading if it is not taken into account. If possible, crime rates in experimental and control areas should be compared to the population that risks becoming victims of the target crimes.

MEASURES OF EFFECTIVENESS

B-360. The goals of the program determine the criteria that are used to measure its effectiveness. These goals and criteria should not be seen as confining; the evaluator should be amenable to broadening the criteria, especially if the program to be evaluated is a new one. For example, the program might be beneficial in some unforeseen way, wholly outside the original criteria. Conversely, the program may be an overall failure but a success according to the evaluation. It may be that the specified measures were the wrong ones to use for the program or should not have been used alone.

B-361. Programs aimed at controlling crime should not be evaluated solely for their effect on crime. Most programs cannot, by their very nature, focus on one specific objective alone. They normally are multifaceted in their effect and should be evaluated with respect to all of their facets. Similarly, the measures of effectiveness discussed in this section may not be adequate for every crime-control program, but they comprise some of the more useful measures that can be used.

B-362. This section concentrates on the two evaluation types—internal and external. Internal and external refer to whether the evaluation is conducted on the program's inner workings and logic or the external effect of the program (which depends on the program type). An internal evaluation of a crime-control program involving the use of new police patrol techniques would include the analysis of police response time and how it was effective in controlling crime or why it was successful in one area and not in another. The external evaluation would focus only on the effectiveness of the program in reducing crime rates or solving crimes, not on how or why or the conditions under which the results were achieved.

B-363. Evaluating how well a program achieved its goals is not the only purpose of an evaluation; how and why the results were achieved are of equal importance. External measures relate to the former evaluation; internal measures are concerned with the latter. The following examples will further serve to highlight the differences between these measures:

- Many crime-control programs depend on good community relations in order to achieve their goals. In these cases, a public-affairs campaign is often instituted concurrent with the crime-control program. The success of the public-affairs campaign should not be interpreted as program success. It may be a necessary part of the program, but it does not substitute for the results of the program in controlling crime. Testimonials from people involved in the program should also be considered only as a supplement to the evaluation based on external measures.

- A study undertaken for the President's Crime Commission showed that for certain types of incidents, the probability of arrest increased as the response time decreased. As a result of this finding, many police departments purchased new equipment or tried novel techniques to reduce response time without first determining whether their workloads included enough of the incidents for which quick response is useful. If this measure (response time) is to be used, it should be recognized as an internal measure and not substituted for the external evaluation.

INTERNAL MEASURES

B-364. Each program will have its own internal measures of effectiveness based on the logical elements of which it is constituted. This section covers only the internal measures of effectiveness that are common to most crime-control program evaluations. The measures covered include the crime rate, the clearance rate, the arrest rate, the crime-seriousness index, and the fear of crime.

CRIME RATE

B-365. The crime rate (the number of a specified type of crime committed per resident in a specified time period) is normally considered to be a measure of deterrence. If the crime rate decreases, it is presumed that potential offenders have modified their behavior to some extent and have committed fewer crimes. This is based on the assumption that the program has made the target crimes unattractive by increasing the actual or perceived risk of apprehension, by reducing the expected return from the crime, or by making alternative forms of behavior more attractive than the target group of offenses.

B-366. These deterrent effects use different means for their accomplishment. Most crime-control programs are police-oriented and concentrate on the risk-related aspects of deterrence. Victim-oriented programs focus on reducing the expected return. Many social and recreational programs deal with making alternatives more attractive. Regardless of the orientation of the programs, their deterrent effects are determined by measuring reported crime rates.

B-367. Reported crime rates can be changed by a number of factors, some of which are misleading. The public may feel that the police are becoming less effective in dealing with crimes and, therefore, report them less often. Conversely, if the public perceives that the police are becoming more effective, they may begin to report crimes that previously would have gone unreported. Another apparent crime-rate reduction may be due to the police not recording crimes that have been reported to them. Displacement effects that can produce misleading crime-rate reductions were discussed earlier.

B-368. There may also be an actual reduction in crime due to a program's deterrent effect. In some cases, the reduction in crime can be attributed to psychological deterrence. That is, the police department may have instituted some change (such as painting all police cars canary yellow) in preexisting patterns of operation that may cause a change in the behavior patterns of potential offenders. This type of deterrence is rarely long-lived.

B-369. On the other hand, there may have been a change instituted by the police that had the desired effect of increasing the actual risk of apprehension and, therefore, reducing the number of target offenses. An example of this is the police-operated burglar alarm of commercial establishments. In the experimental program, the number of alarms were increased almost tenfold compared to the (nonalarmed) control establishments. There was 1 capture in 36 control-group burglaries (2.8 percent), while there were 12 captures in 46 experimental-group burglaries (26 percent). Crime displacements to other crimes, tactics, targets, and areas reduced the actual effectiveness of the program, but this example shows that a significant change can be made in the actual risk of apprehension. Preliminary results indicate that the rate of increase of commercial burglaries has been decreased from about 15 percent per year to about 0 percent (at the expense of a greater increase in residential burglaries).

B-370. It is difficult but useful to distinguish between actual deterrence (due to an actual increase in risk) and deterrence that is purely psychological in nature (due to a perceived increase in risk). If it is suspected that part of the deterrent effect may be transient, a long-term study would be of benefit. In this way, the half life of the psychological deterrence can be gauged, which can give some indication of the extent to which resources should be committed to the program.

B-371. Some forms of psychological deterrence are almost entirely counterproductive. They may appear effective to those who would not commit a crime and ineffective to those who are "in the business" and study the presumed deterrent more closely. For example, a tear-gas pen may give a person a sense of security that is entirely without foundation. It may be dangerous to him if he actually attempts to use it when faced with an assailant.

B-372. One investigator has pointed out that for given criminal situations, nondelinquents perceive a higher risk of apprehension than do delinquents; in all probability, the delinquents have a more realistic assessment of the situation. A purely psychological deterrent may have the unfortunate effect of making only a cosmetic improvement. This gives the general population the impression that there has been a change for the better, while in reality the situation may not have changed or may have changed for the worse because of the division of resources to a nonexistent solution.

B-373. The crime rate can be used as a measure of effectiveness. However, the evaluator should delve into the determination of the crime rate to see if any change in the rate reflects a change in reporting procedures or the deterrent effect (with tangible evidence).

CLEARANCE RATE

B-374. The clearance rate is normally considered to be a measure of the ability of police to solve crimes. A cleared crime is one in which the police have identified the offender and have sufficient evidence to arrest him. The clearance rate is the percentage of total crimes that were cleared.

B-375. This measure of effectiveness should be used with care. A decreasing clearance rate may not mean that a police department is becoming less effective, and an increasing clearance rate may not mean that it is becoming more effective. This is due to a number of factors, primarily the public's conception of the role of the police with respect to crime and the present method of collecting crime data.

B-376. Often overlooked in discussion about crime is the role of the public in assisting the police. Police rely on community support to legitimize their authority as well as to help them carry out their work. If a segment of the community becomes alienated from the police (for whatever reason) and offers them little assistance in pursuing offenders, crime rates in these areas may rise. However, it is not only alienation of community groups that reduces the ability of the police to deal with crime; the profit motive is also to blame. Many store owners that have been robbed refuse to give their clerks time off (with pay) to help the police in their investigation. They absorb the loss of a robbery easily (it rarely comes close to the amount lost from shoplifting, employee theft, and damaged goods) and are unwilling to increase it by helping the police. They may feel that the chances of apprehending the offender are too slim, or they may be afraid of retribution if the offender discovers their assistance. They may also be afraid that their insurance will be cancelled.

B-377. If a police department begins a drive to increase its clearance rate, the increase may be forthcoming without any real change in police effectiveness. A survey of three police departments found that arrests for felonies were not made by the police in about 43 percent of the cases, in which there was probable cause, while the police were accompanied by witnesses. Making arrests in such instances would inflate the clearance rate quite easily. However, it should be noted that the police officer has a great deal of discretion in the exercise of his power of arrest. He may feel that the arrest charges will not hold up. One measure of the arrest quality is the percentage of arrests that leads to prosecutions.

B-378. In summary, clearance rate can be a measure for determining the effectiveness of crime-control programs. Its use can be increased by careful selection and specification of the crime categories that are studied, by determining the manner in which the crimes were cleared, and by determining if there has been a change in where the police draw the line in the exercise of their discretion.

ARREST RATE

B-379. Another measure of effectiveness that is often used as a determinant of crime-control effectiveness is the arrest rate, calculated either per police officer or per resident for a specified time period. Most of the considerations concerning the clearance rate (discussed above) also apply to the arrest rate. However, the arrest rate is distinguished from the clearance rate by an additional factor—it is not related to the total number of offenses. For example, the number of arrests for drug violations has risen considerably over the past few years. However, this increase is indicative of the extent of the problem, not of the effectiveness of the solution. It has been described how drug arrests may be traded off against arrests for other offenses and vice versa, especially when informal arrest quotas are established. Therefore, the use of the arrest rate by itself does not appear to be appropriate as a measure of the effectiveness for most crime-control programs.

CRIME-SERIOUSNESS INDEX

B-380. Among the many criticisms of crime statistics is the contention that, even if the data were reliable and complete, we would still have only a count of the number of incidents without an indication of their relative seriousness. The crime-seriousness index was proposed to include some of the major disutilities of crime typically committed by juveniles. Crimes are weighted according to the degree and nature of injury to the victims—whether they were intimidated and the nature of the intimidation or whether premises were forcibly entered, and the kind and value of property stolen. The weights were determined by requesting a sample of people to estimate the relative seriousness of various crimes.

B-381. All of the factors used to determine the weight are (or should be) included in offense reports. It would not be difficult to calculate an incident-seriousness score based on these reports, either for a specific evaluation or as a matter of course. Use of the seriousness index has also been proposed to measure the relative performance of law-enforcement agencies.

B-382. The crime-seriousness index is not the ultimate weighting scheme. The seriousness appears to be calculated more from the offender's viewpoint and the event than from the victim or society's viewpoint. For example, most people would consider the murder of a robbery victim by his assailant to be more serious than the murder of one spouse by the other. With regard to property loss, there is a difference between loss suffered by an individual who is insured and one who is not covered.

B-383. The loss relative to the individual's income is also an important factor; the theft of a $100 television from a low-income family has a much greater impact than the loss of $10,000 of jewels from a wealthy family. Perhaps a better index of the relative value of property loss to the victim would be the value of the loss in relation to the amount of the individual's discretionary income (that is, income not used for the basic necessities of life). Of course, such information is not available on police crime reports.

FEAR OF CRIME

B-384. It has been pointed out that the perceived risk of crime is greater than the actual risk of crime, and that this perceived risk does not seem to be correlated with the actual crime rate. Unless the public feels safer in proportion to its increased actual safety, the full potential of improvements will not have been reached. Therefore, the goal of a crime-control program can be broadened to include not only improved public safety (deterrence), effectiveness (clearance rate), and reduced crime impact (seriousness); but also improved, more accurate, public perceptions of safety as well.

B-385. Measurements of perceived safety can be both direct and indirect. Public-opinion surveys with regard to perceptions about crime and safety have been made frequently. It is also possible to gauge the effect of this fear using indirect measures by observing what people do rather than what they say. The number of patrons of movie theaters and restaurants at night (or other observations of this type of activity) could be used to gauge the fear of crime.

B-386. A reliable measure of the public's perception of public safety has not been developed. Additional research is being done and needs to be done before this type of measure of effectiveness can be used with confidence.

Appendix C

Intelligence, Counterintelligence, and Threat Analysis

Intelligence and counterintelligence make up the first line of defense in an antiterrorism program. A well-planned, systematic, all-source intelligence and counterintelligence program is essential. The role of intelligence and counterintelligence in antiterrorism is to identify the threat. Additionally, counterintelligence provides a warning of potential terrorist attacks and provides information for counterterrorism operations. This appendix provides the elements of the intelligence cycle that have particular importance in a viable antiterrorism program. Effective intelligence and counterintelligence support requires effort, planning and direction, collection and analysis, production, investigation, and dissemination. The entire process provides decision makers with information and timely warnings upon which to recommend antiterrorism actions.

INFORMATION SOURCES

C-1. The primary sources of intelligence and counterintelligence for the antiterrorism program are open-source information, criminal information, government intelligence and counterintelligence, and local information.

- **Open-source information.** This information is publicly available and can be collected, retained, and stored without special authorization. The news media is an excellent open source of information on terrorism. The media reports many major terrorist incidents and often includes in-depth reports on individuals, groups, or various government counterstrategies. Government sources include congressional hearings; publications by the Defense Intelligence Agency (DIA), the FBI, the Central Intelligence Agency (CIA), the Department of State (DOS), and the national criminal justice reference services. Additionally, there are private data services that offer timely information on terrorist activities worldwide. Terrorist groups and their affiliates may also have manuals, pamphlets, and newsletters that reveal their objectives, tactics, and possible targets.
- **Criminal information.** Both military and civil law-enforcement agencies collect criminal information. Because terrorist acts are criminal acts, criminal information is a major source for terrorist intelligence. Commanders must work through established law-enforcement liaison channels because the collection, retention, and dissemination of criminal information are regulated. Local military criminal investigative offices of the CID; the Naval Investigative Service Command (NISCOM); the Air Force Office of Special Investigations (AFOSI); and Headquarters, US Marine Corps,

Criminal Investigations Division, maintain current information that will assist in determining the local terrorist threat.

- **Government intelligence and counterintelligence.** The Community Counterterrorism Board (CCB) is responsible for coordinating with the national intelligence agencies concerning combating international terrorism. These agencies include the CIA (the lead agency), the DIA, the National Security Agency (NSA), the DOS, the Department of Justice (DOJ), the FBI, the Department of Energy (DOE), the Federal Aviation Administration (FAA), the Department of Transportation (DOT) (including the USCG), and the DOD. Service intelligence and counterintelligence production organizations include the US Army Counterintelligence Analysis Center; the Navy Antiterrorism Analysis Center (NAVATAC); Headquarters, US Marine Corps, Counterintelligence; and the US AFOSI Operations Center. These organizations compile comprehensive intelligence and counterintelligence for distribution on a need-to-know basis throughout the services. In combatant commands, the J2 is responsible for the intelligence fusion center. The Counterintelligence Support Officer (CISO) provides interface between the combatant command, the component commands, and the joint staff.
- **Local information.** Other valuable sources of information are the soldiers, civil servants, family members, and individuals with regional knowledge (such as college faculty or members of cultural organizations). Local crime or neighborhood watch programs can be valuable sources of information and can serve as a means to keep individuals informed in dispersed and remote areas. Intelligence exchanges with local government agencies through cooperative arrangements can also augment regional information.

RESPONSIBILITIES OF US GOVERNMENT LEAD AGENCIES

C-2. The FBI is responsible for collecting and processing domestic terrorist information. Overseas, terrorist intelligence is principally a CIA responsibility; but the DOS, the DIA, and the HN are also active players. The MI activities are conducted according to Presidential executive orders, federal law, status of forces agreements (SOFAs), memorandums of understanding (MOUs), and applicable service regulations. Responsibilities of intelligence activities include the following:

- The combatant commander (through the commander's J2 and the CISO in coordination with the DIA, the CIA, the embassy staff, the country team, and applicable HN authorities) obtains intelligence and counterintelligence specific to the AO. The commander issues intelligence and counterintelligence reports, advisories, and assessments to the units within his command or those operating within his command's AO. This network is the backbone for communicating intelligence and counterintelligence information and advisories and for warning of terrorist threats throughout the region.
- The Secretaries of the military departments were asked (in DOD Directive 2000.12) to ensure that a capability exists to receive and evaluate data from a service perspective and that the capability exists

to disseminate all relevant data on terrorist activities, trends, and indicators of imminent attack. To accomplish this task, each Secretary appoints an MI agency (INSCOM, NISCOM, or AFOSI) to conduct intelligence and counterintelligence activities directed against terrorists and to detect, neutralize, or deter terrorist acts. To accomplish this mission, the military department intelligence agency establishes counterintelligence offices on an area basis to collect and disseminate information to combatant commanders. Each military department's intelligence agency—

  - Coordinates with appropriate US and HN agencies.
  - Provides overall direction and coordination of the service counterintelligence effort.
  - Operates a 24-hour operations center to receive and disseminate worldwide terrorist threat information to and from the combatant command's J2, applicable service staff elements, subordinate commands, and national agencies.
  - Provides service commanders with information on terrorist threats concerning their personnel, facilities, and operations.
  - Investigates terrorist incidents for intelligence, counterintelligence, and force-protection aspects (with the FBI or HN authorities).
  - Provides terrorist threat information in threat briefings.
  - Conducts liaison with representatives from federal, state, and local agencies (as well as HN agencies) to exchange information on terrorists.
  - Provides international terrorism summaries and other threat information to supported commanders. On request, provides current intelligence and counterintelligence data on terrorist groups and disseminates time-sensitive and specific threat warnings to appropriate commands.

- Service criminal investigative services (such as the CID, the NISCOM, and the AFOSI) collect and evaluate criminal information and disseminate terrorist-related information to supported installation and activity commanders as well as to the service lead agency. As appropriate, criminal investigative elements also conduct liaison with local military or security police and civilian law-enforcement agencies.
- Intelligence staff elements of commanders at all echelons—
  - Report promptly all actual or suspected terrorist incidents, activities, and early warnings of terrorist attacks to supported and supporting activities, to the local counterintelligence office, and through the chain of command to the service lead agency.
  - Initiate and maintain liaison with the security police or PMO; local military criminal investigative offices; local counterintelligence offices; security offices; HN agencies and; as required, other organizations, elements, and individuals.
  - Develop and present terrorism threat-awareness briefings to all personnel within their commands (in cooperation with the local counterintelligence offices).
- Law-enforcement staff elements will—

    - Report all actual or suspected terrorist incidents or activities to their immediate commander, supported activities, and service lead agency through established reporting channels.
    - Initiate and maintain liaison with local counterintelligence offices and military criminal investigative offices.
    - Maintain liaison with federal, HN, and local law-enforcement agencies or other civil and military antiterrorism agencies as appropriate.
- Installation, base, unit, and port security officers—
    - Report all actual or suspected terrorist incidents or activities to their immediate commander, supporting military law-enforcement office, other supported activities, local counterintelligence office, and local military criminal investigation office.
    - Conduct regular liaison visits with the supporting military law-enforcement office, counterintelligence office, and local criminal investigation office.
    - Coordinate with the supporting military law-enforcement office and counterintelligence office on their preparation and continual updating of the threat assessments.
    - Assist in providing terrorism threat-awareness training and briefings to all personnel and family members as required by local situations.

INFORMATION REQUIREMENTS

C-3. To focus the threat analysis, intelligence and counterintelligence officers develop information requirements to identify targets using the following terrorist considerations—

- Organization, size, and composition of group.
- Motivation.
- Long- and short-range goals.
- Religious, political, and ethnic affiliations.
- International and national support (moral, physical, and financial).
- Recruiting methods, locations, and targets (students).
- Identity of group leaders, opportunists, and idealists.
- Group intelligence capabilities and connections with other terrorist groups.
- Sources of supply and support.
- Important dates, such as religious holidays.
- Planning ability.
- Internal discipline.
- Preferred tactics and operations.
- Willingness to kill.
- Willingness for self-sacrifice.
- Group skills, demonstrated or perceived (for example, sniping, demolitions, masquerade, industrial sabotage, airplane or boat

operations, tunneling, underwater, electronic surveillance, poisons, or contaminants).

- Equipment and weapons (on hand and required).
- Transportation (on hand and required).
- Medical-support availability.
- Means and methods of C^2.
- Means and method of communication.

THREAT ANALYSIS AND ASSESSMENT

C-4. The preparation of the terrorist threat analysis is a continual process of compiling and examining all available information to identify terrorist targeting of US interests. A vulnerability analysis is a continual process of compiling and examining information on a facility's security posture. The threat analysis is then paired with the facility's vulnerability analysis to create the threat and vulnerability assessment. Threat analysis is an essential step in identifying the probability of a terrorist attack. To enhance the capability to collect and analyze information from many sources, the DIA maintains a terrorism database. The combatant command's J2 and CISO (in coordination with the DIA) focus this database information and regional information toward the intelligence and counterintelligence needs specific to the security of the command. However, this terrorism database is limited to foreign terrorist groups because of limitations on US intelligence-collection operations. A country's threat assessments, information and biographies about terrorist organizations, and incidents in the database can be disseminated to commands. Commands at all echelons then augment or refine the DIA's threat analysis to focus on their area of interest. This process, operative across the full range of military operations, promotes coordination between all levels of the intelligence, counterintelligence, and law-enforcement communities; broadens acquisition channels; and enhances timely distribution of information to the supported commander.

C-5. Several factors complicate intelligence and counterintelligence collection and operations. The small size of terrorist groups, coupled with their mobility and cellular organization, make it difficult to identify the members. Unlike other criminals, terrorist cadres often receive training in counterintelligence and security measures from foreign intelligence agencies or other terrorists. Additionally, the traditional orientation of police organizations is toward individual criminals, while MI organizations focus on conventional forces. Terrorist activity, therefore, requires some degree of reorientation for police and MI and counterintelligence collection and operations.

C-6. An intelligence system's ability to provide critical and timely information to the user depends not only on efficient collection and processing, but also on the ability to organize, store, and retrieve this information rapidly. This capability, coupled with early warning, careful observation, and assessment of threat activity, enhances the probability of accurately predicting the types and timing of terrorist attacks.

C-7. Commanders must carefully exercise judgment in estimating both the existing terrorist threat and the need for changes in antiterrorism measures. Key questions are—

- What has changed (mission, political climate, installation and unit personnel or equipment, terrorist capabilities)?
- What affect will the changes have on the security posture?

C-8. Extraordinary security measures, unless part of a deliberate deception during critical or high-threat situations, draw attention and detract from mission accomplishment. Sound physical security, personnel who are aware, accurate threat and vulnerability assessments, and well-rehearsed response plans reduce the probability of a successful terrorist venture. The goal is to make an attack too difficult or the level of risk unacceptable to the terrorist.

DETERMINATION OF THE THREAT LEVEL

C-9. This threat-analysis methodology is used by the DIA, the joint staff, and the unified and specified commands for selecting the level of threat for an installation. It is applicable in an overseas setting, but fails to address issues unique to the sustaining base within CONUS. In CONUS there is a lack of intelligence and counterintelligence information from the CIA, the DIA, or the military services for the CONUS-based threat. That information must be extracted from law-enforcement channels at the local, regional, and national levels. In that context, the factors mentioned in this appendix are not as clear as they are within the intelligence process in place overseas. A modified version of this appendix should be considered for assessing threat levels in CONUS.

C-10. Threat levels within CONUS historically have been either low or negligible. This trend will most likely continue at least through the next decade. Domestic groups not covered by DOD intelligence reporting pose the greatest threat to the CONUS-based military. Due to the lack of reporting and information on these groups, the domestic terrorist groups are not currently factored into the current terrorist threat program. Therefore, the law-enforcement community must become a key player in establishing the threat levels in the context of the recommended model. A possible methodology would be the establishment of an additional threat level between low and medium—one that allows local commanders more flexibility in implementing additional security measures. Table C-1 shows the procedure for determining the threat level.

Table C-1. Threat Levels

Threat Level	
Critical	Factors 1, 2, and 5 are present. Factors 3 or 4 may be present.
High	Factors 1, 2, 3, and 4 are present.
Medium	Factors 1, 2, and 4 are present.
Low	Factors 1 and 2 are present. Factor 4 may be present.
Negligible	Factors 1 and/or 2 may be present.
Explanation of Factors	
Factor 1: Existence. A terrorist group is present, assessed to be present, or able to gain access to a given locale.	
Factor 2: Capability. The acquired, assessed, or demonstrated level of capability to conduct terrorist attacks.	
Factor 3: Intentions. Recent demonstrated anti-US terrorist activity or stated and/or assessed intent to conduct such activity.	
Factor 4: History. Demonstrated terrorist activity over time.	
Factor 5: Targeting. Current credible information on activity indicative of preparations for specific terrorist operations and/or specific intelligence that shows an attack is imminent.	

Appendix D

Crisis-Management Plan

The following pages highlight areas of concern in crisis-management planning. This plan is not meant to be all-inclusive or rigidly followed. Figure D-1 is a sample format only. It does not reflect a format developed and approved for use with OPLANs or contingency plans (CONPLANs) prepared by the CINCs to fulfill tasks assigned in the Joint Strategic Capabilities Plan (JSCP) or as otherwise directed by the Chairman of the Joint Chiefs of Staff. Figure D-2, page D-4, is a sample of the Crisis-Management-Plan Checklist, which is Annex A or Appendix H to the crisis-management plan. This checklist will help ensure that the plan is sound.

Copy No. ________
Issuing Headquarters
Place of Issue
Date of Issue

Crisis-Management Plan

Ref: Maps, charts, and other relevant documents.

Time Zone: X

Task Organization: List units organized to conduct antiterrorism operations. Include attachments, supporting roles, and the delegation of operational control as necessary.

1. Situation. Identify essential information to understand ongoing events.

a. Terrorist force. Identify the terrorist's composition, disposition, methods of operation, and estimated strengths and capabilities that could influence the crisis-management operation. Refer to an appropriate annex.

b. Response force. Explain the response force's abilities and responsibilities. The response force's abilities can influence the crisis-management mission.

c. Attachments and detachments. Address here or refer to an annex.

d. Assumptions. Provide assumptions used as a basis for this plan (for example, the strength of the response force to be supported and the support available from other agencies).

Figure D-1. Sample Crisis-Management Plan

(1) Tactical-situation possibilities. Obtained from the commander's planning guidance.

(2) Personnel situation. Provided by the personnel officer.

(3) Logistics situation. Provided by the logistics officer.

(4) Legal-situation possibilities. Provided by the SJA.

2. Mission. Identify the antiterrorism mission (for example, detect, deter, contain, and neutralize terrorist threats and actions aimed at the disruption of the installation).

3. Execution.

a. Concept of operations. State the commander's tactical plan. The purpose is to inform. It may address how the commander will conduct combat-terrorism operations. It provides enough detail to ensure proper action by subordinates in the absence of specific instructions. If the required details are extensive, address them in an annex. If an operation involves two or more distinct phases, designate each phase and use subparagraphs (for example, Phase I and Phase II).

b. Tasks. Identify specific tasks for each command element charged with executing a crisis-management mission. When giving multiple instructions, itemize and indicate the priority or sequence.

c. Coordinating instructions. Include coordination and control measures applicable to two or more command elements.

4. Service Support. Provide a statement of service-support instructions and arrangements supporting the crisis-management operation. Use the following subparagraphs as required:

a. General. Outline the general plan for service support.

b. Materiel and services. Address supply, transportation, labor, and services required.

c. Medical evacuation and hospitalization. Provide the plan for evacuating and hospitalizing sick, wounded, or injured personnel. Address evacuation responsibilities and the air-evacuation policy.

d. Personnel. Provide required information and instructions to supporting unit personnel.

(1) Maintenance of unit strength.

(a) Strength reports. Provide instructions for submitting status reports. Include requirements for routine and special reports.

(b) Replacements. Address validating existing personnel requisitions, instructions for submitting requisitions, and instructions for processing and removing replacements.

(2) Personnel management. Address military and civilian personnel and civilian detainee management procedures.

(3) Development and maintenance of morale.

(a) Morale and personnel services. Provide postal and financial services, religious activities, personal hygiene, and special services activity information.

(b) Graves registration. Include evacuation procedures and handling personal effects.

Figure D-1. Sample Crisis-Management Plan (continued)

(4) Maintenance of discipline, law, and order. Obtain this guidance from the PMO/security officer.

(5) Miscellaneous. Include personnel administrative matters not specifically assigned to another coordinating staff section or included in preceding subparagraphs.

e. Miscellaneous. Provide special instructions or special reports not covered in preceding paragraphs.

5. Command and Signal. Provide instructions for the command and operation of communications-electronics equipment. Communications-electronics instructions may refer to an annex but should list the index and issue number of the command, control, and communications (C^3) operation instructions in effect. If not already issued, give instructions for the control, coordination, and establishment of priorities in the use of electromagnetic emissions. Command instructions include subordinate and higher unit CP locations and designated alternate CPs.

6. Acknowledgement Instructions.

/s/

Commander

Annexes as applicable

Distribution:

Figure D-1. Sample Crisis-Management Plan (continued)

Annex A

Crisis -Management Plan Checklist

Yes	No	
		1. Intelligence and/or Counterintelligence.
____	____	Does the plan allow for the threat-analysis process (collection, analysis, production, and dissemination) to help identify the local threat?
____	____	Does the plan consider restrictions placed on the collection and storage of information?
____	____	Does the plan indicate an awareness of sources of information for the threat-analysis process (MI, counterintelligence, federal agencies, and state and local authorities)?
____	____	Does the plan allow for liaison and coordination of information (such as establishing a committee)?
		2. Threat Assessment.
____	____	Does the plan identify the local threat (immediate or long term)?
____	____	Does the plan identify other threats (such as national and international groups that have targeted or might target US installations)?
____	____	Does the installation incorporate factors for assessing the threat? Does it address —
____	____	Geography of the area concerned?
____	____	Law-enforcement resources?
____	____	Population cultural resources?
____	____	Communication capabilities?
____	____	Does the plan establish a priority of identified weaknesses and vulnerabilities?
____	____	Is the threat assessment periodically updated?
		3. Security Countermeasures.
____	____	Does the plan have specified THREATCONs and recommended actions?
____	____	Do security countermeasures include a combination of physical operations and sound-blanketing security measures?
____	____	Do the THREATCONs correspond to DOD 0-2000.12-H, Appendix BB?

Figure D-2. Sample Crisis-Management-Plan Checklist

Yes	No	
		4. OPSEC.
____	____	Have procedures been established that prevent terrorists from readily obtaining information about plans and operations (for example, not publishing the commanding general's itinerary and safeguarding classified material)?
____	____	Does the plan allow for in-depth coordination with the installation's OPSEC program?
____	____	Has an OPSEC annex been included in the CONPLAN?
		5. Personnel Security.
____	____	Has the threat analysis identified individuals vulnerable to terrorist attacks?
____	____	Has a training program been established to educate both military and civilian personnel in the proper techniques of personnel protection and security commensurate with the local threat and the type of position held?
		6. Physical Security.
____	____	Are special-threat plans and physical-security plans mutually supportive?
____	____	Do security measures establish obstacles to terrorist activity (such as guards, HN forces, lighting, and fencing)?
____	____	Does the special-threat plan include the threats identified in the threat statements of higher headquarters?
____	____	Does the physical-security officer assist in the threat analysis and corrective action?
____	____	Does the installation have and maintain detection systems and an appropriate assessment capability?
		7. Security Structure.
____	____	Does the plan indicate that the FBI has primary domestic investigative and operational responsibility in the US and US territories?
____	____	Has coordination with the SJA been established?
____	____	Does the plan allow for close cooperation between principal agents of the military, civilian, and HN communities and federal agencies?
____	____	Does the plan clearly indicate parameters for the use of force, including briefing any elements augmenting MP assets?
____	____	Is there a mutual understanding between all local agencies (military, local, FBI resident or senior agent-in-charge, HN forces, and local law enforcement) that might be involved in a terrorist incident on the installation regarding authority, jurisdiction, and possible interaction?

Figure D-2. Sample Crisis-Management-Plan Checklist (continued)

Yes	No	
___	___	Has the SJA considered the ramifications of closing the post (such as possible civilian union problems)?
___	___	Does the plan identify the DOS as having primary investigative and operational responsibilities overseas?
		8. Operations-Center Training.
___	___	Has the operational command and coordination center been established and exercised?
___	___	Is the operations center based on the needs of the installation while recognizing manpower limitations, resource availability, equipment, and command?
___	___	Does the plan include a location for the operations center?
___	___	Does the plan designate alternate locations for the operations center?
___	___	Does the plan allow for the use of visual aids (chalkboards, maps with overlays, and bulletin boards) to provide status reports and countermeasures?
___	___	Does the plan create and designate a location for a media center?
___	___	Have the operations and media centers been activated together within the last quarter?
___	___	Does the operations center have SOPs covering communications and reports to higher headquarters?
___	___	Does the operations center offer protection from a terrorist attack?
		9. Reaction-Force Training.
___	___	Has the force been trained and exercised under realistic conditions?
___	___	Has corrective action been applied to shortcomings and deficiencies?
___	___	Has the reaction force been formed and mission-specified trained (for example, building entry and search techniques, vehicle assault operations, countersniper techniques, and equipment)?
___	___	Has the reaction force been tested quarterly (alert procedures, response time, and overall preparedness)?
___	___	Has responsibility been fixed for the negotiation team? Has the negotiation team been trained and exercised under realistic conditions?
___	___	Does the negotiation team have the proper equipment?
		10. General Observations.
___	___	Was the plan developed as a coordinated staff effort?
___	___	Does the plan outline reporting requirements (logs, journals, and after-action reports)?

Figure D-2. Sample Crisis-Management-Plan Checklist (continued)

Yes	No	
____	____	Does the plan address the media's presence?
____	____	Does the plan include communication procedures and communication nets?
____	____	Does the plan consider the possible need for interpreters?
____	____	Does the plan consider the need for a list of personnel with various backgrounds to provide cultural profiles on foreign subjects and victims as well as to assist with any negotiation efforts?
____	____	Does the plan provide for and identify units that will augment MP assets?
____	____	Does the plan delineate specific taskings for each member of the operations center?
____	____	Does the plan provide for a response force for each phase of antiterrorism activity (initial response, negotiation, and assault)?
____	____	Does the plan designate service-support communications?
____	____	Does the plan make provisions for the notification of an accident-and-incident control officer?
____	____	Does the plan provide for EOD support?
____	____	Does the plan take into consideration the movement from various locations, including commercial airports, of civilian and military advisory personnel with military transportation assets?
____	____	Does the plan allow for the purchase or use of civilian vehicles, supplies, and food (if needed)? (This includes items used to satisfy a hostage demand.) Does the plan make provisions for paying civilian employees overtime if they are involved in a special-threat situation?
____	____	Does the plan take into consideration the messing, billeting, and transportation of civilian personnel?
____	____	Do appropriate personnel have the necessary language training?
____	____	Is WMD support available?

Figure D-2. Sample Crisis-Management-Plan Checklist (continued)

Appendix E

Office Security Measures

The office environment should afford executives the greatest degree of physical security. Executives usually work in facilities where attackers must pass by guards, security checkpoints, office workers, aides, or secretaries before reaching them. Unfortunately, the high media value of attacking executives in security strongholds where they are clearly associated with government activity increases the value of such attacks to terrorists. Hence, there may be a need to add security measures to offset the escalating capability of attack on more secure office areas by terrorist groups.

PHYSICAL-SECURITY SURVEY

E-1. A thorough physical-security survey of an office facility should be conducted. Offices of defense components attached to US embassies abroad should have these surveys performed by the DOS. Other DOD facilities should have surveys performed by the cognizant physical-security and facilities-engineering staffs. The best way to approach a physical-security site survey is to think like an intruder. Consider how approaches to the installation or facility could be made, how access to the building that houses executive offices could be gained, and how attacks on offices or other frequently used facilities could be mounted.

SECURITY-ENGINEERING ASSESSMENT

E-2. The next step in evaluating the need for supplemental physical-security measures is a thorough and detailed assessment of the weapons and tactics that terrorists might use to attack the structure in which DOD executives work. Security engineers and architects need technical threat data or assessments containing the following information—

- The mode of attack, such as—
  - Standoff weapons (man-portable AT/antiaircraft weapons, sniper rifles, rock grenades, and mortars).
  - Close combat weapons (submachine guns, pistols, knives, and garrotes).
  - Contact weapons (bombs, incendiary devices, and mines).
- Perimeter penetration aids (such as power tools, hand tools, or explosives), if used.
- The time of attack.
- The attacking force's size.
- The anticipated degree of outside support or autonomy.

E-3. Engineering design requirements are developed from the security engineering assessments. The data is used to—

- Assess the ability of building components to resist the effects of the threat.
- Identify appropriate security window-glazing materials and window treatments to determine what is required to achieve the desired penetration resistance times for anticipated threats.
- Calculate the total amount of delay time. This time is achieved by using camouflage, deception, barriers, and security devices to permit response forces to reach the scene of a terrorist attack in time to thwart the attack, capture or eliminate the terrorists, and rescue executives and their staffs or dependents.

TECHNICAL ASSESSMENT OF RESPONSES

E-4. After establishing a basic-design threat, engineers need data on the anticipated performance of response forces to be arrayed against the design threat and the expected or desired behavior of the protected executive. Some specific information needed includes—

- The response force's size, capability, supporting weapons, response time, and estimated effectiveness against the range of attacks.
- The desired options for the executive's protection—evacuate on warning, on detection, only if attacked, or only if forced to capitulate or do not evacuate.

E-5. Security planners need to know how long the structure that houses executives can withstand an attack before help arrives. Matching threat capabilities and anticipated operations by response forces establishes significant physical-security-system performance parameters. These can be quantified and used to develop detailed plans, drawings, and physical-security equipment-acquisition plans.

PHYSICAL-SECURITY ENHANCEMENT MEASURES

E-6. Several physical-security measures intended to provide additional protection for executives can be considered based on the requirements defined through the detailed analyses outlined above. The primary purpose of such measures should be to increase the time required by persons outside an installation to reach the executives housed at an installation. A secondary purpose of such measures should be to reduce or eliminate hazards to executives that might result from violence in the vicinity. Examples of physical-security measures to consider are—

- Increase the threat-detection time by installing sensors on perimeters and barriers. This includes—
  - Combining surveillance systems including seismic, acoustic, and IR sensors at or beyond the outer perimeter.
  - Supplementing surveillance systems with CCTV/imaging IR systems tied into the alert response-force staging area.

- Extending restricted areas or exclusion zones and relocating access-control points from the executive office area to a point closer to the installation's boundary.
- Enlarging and extending intrusion-detection sensors from within the installation to its perimeter, allowing the IDS to collect additional data necessary and sufficiently classify and identify an intrusion before the response force arrives.
- Enhancing both the number and the phenomenology of surveillance and detection systems within the executive office area as well as approaches leading to and from it in conjunction with measures listed below.

- Increase the threat's delay time between the perimeter and the executive office building. This includes—
  - Installing vehicle barriers and realigning roadways to eliminate straight, level stretches of road in excess of 50 meters in length.
  - Increasing concentric rings of fences, Jersey barricades, planters, bollards, and vehicle/personnel barriers.
  - Enhancing access-control areas supplemented by fire doors/security doors kept in a closed condition between the entrance to the building that houses executive offices and the executive office area.
- Confuse, camouflage, and deceive observers by hiding an executive's location. Accomplish this by—
  - Relocating executives to buildings not usually associated with office activities (barracks, motor pools, research and development [R&D] facilities, and so forth).
  - Constructing office areas in the barracks, motor pool, R&D facilities, and so forth.
  - Adding executive styles, decorative lighting, and window treatments to several different areas of office buildings to minimize the differences in external appearances between executive and nonexecutive offices.
- Increase the delay time between the entrance to the building that houses executives and the executive office area. Execute this by—
  - Adding fire doors, access-control points, dead-end corridors, and midcorridor physical barriers to complicate access to the executive office areas.
  - Adding security devices that, when activated, disrupt the intruder's ability to retain his thought processes (for example, flashing strobe lights, fog generators, noise generators, sirens, and fire-extinguishing systems).
- Increase the delay time by making access more difficult within the executive office structure. This may be accomplished by—
  - Substituting high-security doors and door frames for standard doors in areas leading to or from executive offices.
  - Installing high-security grating, wire mesh, or other materials to bar access to the executive office area through utility tunnels or conduits.

  - Strengthening walls, floors, and ceilings by substituting steel-plate, concrete-filled, steel-reinforced cinder blocks or other ballistic-resistant materials for plaster/lath or wallboard room dividers, thereby protecting against explosive devices that are used as tools to breach a barrier.
- Increase the protection for building occupants against weapons and explosives effects. This includes—
  - Substituting blast- or bullet-resistant panels for glass windows or adding a fragment-retention film at least 4 millimeters thick to the interior of glass windows.
  - Adding exterior screens/plates to cover window areas and protect against gunfire and grenade/bomb fragments.
  - Installing blast curtains, metal blinds, metal shutters, or other window treatments in executive offices to protect interior space from glass shards and other small projectiles.
  - Strengthening walls to resist weapons and explosives effects by adding steel plates, reinforced concrete, or other retrofitting measures.
  - Adding steel plates or other ballistic materials in crawl spaces above dropped ceilings or extending walls separating the executive office area from other portions of an office building from floor to floor, thereby preventing unobserved and undetected access to the space between dropped ceilings.
- Increase the hold time to contain penetrators by—
  - Adding positive-action controls to a facility's doors and gates so that gates default to a closed and locked condition unless manually released.
  - Adding positive-action controls to access-control areas so that persons inside an access-control area can neither advance nor withdraw without affirmative action by a security officer posted outside the access-control area.

E-7. These measures are used to facilitate the apprehension of terrorists. There may be some instances when defeating terrorist attempts to gain access to the executive enhances the security of the executive and the response force. This is accomplished by channeling the terrorists out of the facility and installation along one route, leaving alternative routes available to evacuate executives and other key personnel.

E-8. Install emergency executive-support facilities (including a safe haven and an emergency evacuation facility) by—

- Installing helicopter landing aids on a structure's roof or on an adjacent field far removed from parking areas.
- Installing a safe haven or other reinforced security structure adjacent to a helicopter landing facility to provide a secure waiting place for executives until a rescue helicopter with additional supporting air and ground units can extract the executives.

Appendix F

Physical-Security Plan

It is essential and in the best interest of security that each installation, unit, or activity maintains and uses a detailed physical-security plan. The plan should include at least special and general guard orders, access and material control, protective barriers/lighting systems, locks, and IDSs. All physical-security plans have the potential of being classified documents and must be treated accordingly. Figure F-1 depicts a sample physical-security plan.

Map Reference

Copy No. ________
Issuing Headquarters
Place of Issue
Date of Issue

Physical-Security Plan

1. Purpose. State the plan's purpose.

2. Area Security. Define the areas, buildings, and other structures considered critical and establish priorities for their protection.

3. Control Measures. Define and establish restrictions on access and movement into critical areas.

a. Categorize restrictions as to personnel, materials, and vehicles:

(1) Personnel access:

(a) Establishment of controls pertinent to each area or structure.

- Authority for access.
- Criteria for access.
  - Unit personnel.
  - Visitors.
  - Maintenance personnel.
  - Contractor personnel.
  - National guard.
  - Emergency response teams (police, fire, ambulance, and so forth).

Figure F-1. Sample Physical-Security Plan

(b) Identification and control.

- Description of the system to be used in each area. If a badge system is used, a complete description covering all aspects should be used in disseminating requirements for ID and control of personnel conducting business on the installation.
- Application of the system:
  - Unit personnel.
  - Visitors to restricted areas.
  - Visitors to administrative areas.
  - Vendors, tradesmen, and so forth.
  - Contractor personnel.
  - Maintenance or support personnel.
  - Fail-safe procedures during power outages.

(2) Material control.

(a) Incoming.

- Requirements for admission of material and supplies.
- Search and inspection of material for possible sabotage hazards.
- Special controls on delivery of supplies or personal shipments in restricted areas.

(b) Outgoing.

- Documentation required.
- Controls, as outlined in paragraph 3a(2a).
- Classified shipment not involving nuclear/chemical material.

(c) Nuclear/chemical material.

- Controls on movement of warheads/chemicals on the installation.
- Controls on shipments or movement of training warheads/chemicals.
- Controls on pickup or delivery of warheads/chemicals outside the installation.

(3) Vehicle control.

(a) Policy on search of military and privately owned vehicles.

(b) Parking regulations.

Figure F-1. Sample Physical-Security Plan (continued)

(c) Controls for entrance into restricted and administrative areas:

- Military vehicles.
- POVs.
- Emergency vehicles.
- Vehicle registration.

b. Indicate the manner in which the following security aids will be implemented on the installation:

(1) Protective barriers:

(a) Definition.

(b) Clear zones.

- Criteria.
- Maintenance.

(c) Signs.

- Types.
- Posting.

(d) Gates.

- Hours of operation.
- Security requirements.
- Lock security.
- Barrier plan.

(2) Protective lighting system:

(a) Use and control.

(b) Inspection.

(c) Action taken in case of commercial power failure.

(d) Action taken in case of failure of alternate power source.

(3) Emergency lighting system:

(a) Stationary.

(b) Portable.

Figure F-1. Sample Physical-Security Plan (continued)

(4) IDSs:

(a) Security classification.

(b) Inspection.

(c) Use and monitoring.

(d) Action taken in case of alarm conditions.

(e) Maintenance.

(f) Alarm logs or registers.

(g) Tamper-proof provisions.

(h) Monitor-panel locations.

(5) Communications:

(a) Locations.

(b) Use.

(c) Tests.

(d) Authentication.

(6) Security forces: General instructions that would apply to all security-force personnel (fixed and mobile). Detailed instructions such as special orders and SOP information should be attached as annexes. Security-force facets include—

(a) Composition and organization.

(b) Tour of duty.

(c) Essential posts and routes.

(d) Weapons and equipment.

(e) Training.

(f) Use of MWD teams.

(g) Method of challenging with signs and countersigns.

(h) Alert forces:

- Composition.
- Mission.
- Weapons and equipment.
- Location.
- Deployment concept.

Figure F-1. Sample Physical-Security Plan (continued)

(7) Contingency plans: Required actions in response to various emergency situations. Detailed plans for situations (counterterrorism, bomb threats, hostage negotiations, disaster, fire, and so forth) should be attached as annexes.

(a) Individual actions.

(b) Alert-force actions.

(c) Security-force actions.

(8) Use of air surveillance.

(9) Coordinating instructions. Matters that require coordination with other military and civil agencies such as—

(a) Adjacent installations or units.

(b) State and local agencies.

(c) Similar host-country agencies.

(d) Federal agencies.

The coordination/interaction allows for an exchange of intelligence information on security measures being used, contingency plans, and any other information to enhance local security.

On an installation, the host activity shall assume responsibility for coordinating physical-security efforts of all tenants, regardless of the components represented, as outlined in the support agreements and the host-activity security plan. Applicable provisions shall be included in, or be an appendix to, the support agreement. A formal agreement will contain definite assignment of physical-security responsibility for the items stored. The agreement should address—

- Maximum qualities to be stored.
- Physical safeguards to be used.
- Frequency of, and responsibility for, physical inventories or reconciliations.
- Reporting of losses for investigation.
- Lock and key control.
- The unit that has overall responsibility.

Procedures for authorization and ID of individuals to receipt for and physically take custody of Army property. The purpose of such coordination is protection in depth. Authority, jurisdiction, and responsibility must be set forth in a manner that ensures protection and avoids duplication of effort.

/s/ Commander

Figure F-1. Sample Physical-Security Plan (continued)

ANNEXES

F-1. Annexes to the plan should include, but are not limited to, the following. More information can be found in AR 190-13.

- **Annex A.** The installation threat statement (intelligence). This annex should contain the Terrorism Counteraction Plan (refer to AR 190-13).
- **Annex B.** A bomb-threat plan. As a minimum, the bomb-threat plan should provide guidance for—
  - Control of the operation.
  - Evacuation.
  - Search.
  - Finding the bomb or suspected bomb.
  - Disposal.
  - Detonation and damage control.
  - Control of publicity.
  - After-action report.
- **Annex C.** An installation closure plan.
- **Annex D.** A natural-disaster plan. This plan will be coordinated with natural-disaster plans of local jurisdictions. At a minimum, the natural-disaster plan should provide guidance for—
  - Control of the operation.
  - Evacuation.
  - Communication.
  - Control of publicity.
  - After-action report.
- **Annex E.** A civil-disturbance plan. It is the commander's responsibility to formulate a civil-disturbance plan based on local threats. (For example, commanders of chemical facilities should anticipate the need to develop crowd-control procedures to handle antichemical demonstrations.)
- **Annex F.** A resource plan to meet the minimum-essential physical-security needs for the installation or activity.
- **Annex G.** A communication plan. This plan is required to establish communications with other federal agencies and local law-enforcement agencies to share information about possible threats. The communications plan should address all communication needs for annexes B through F above.
- **Annex H.** A list of designated restricted areas.
- **Annex I.** A list of installation MEVAs.
- **Annex J.** A contingency plan. In most instances, it will be necessary to increase security for AA&E and other sensitive property, assets, and facilities during periods of natural disasters, natural emergencies, or increased threat from terrorists or criminal elements. Therefore, CONPLANs should include provisions for increasing the physical-security measures and procedures based on the local commander's assessment of the situation. Such contingencies may include hostage negotiations, protective services, and special-reaction teams. These

provisions should be designed for early detection of an attempted intrusion, theft, or interruption of normal security conditions.

- **Annex K.** Work-stoppage plan. This is a requirement for conducting a physical-security survey.

TACTICAL-ENVIRONMENT CONSIDERATIONS

F-2. In a tactical environment, the development of a physical-security plan is based on METT-TC (using the OPORD format and the higher headquarters' order). The order may be specific about the tasks the unit will perform. Time available may be limited and the scheme of maneuver may be dictated, but the leader must still evaluate the mission in terms of METT-TC to determine how MP elements can best carry out the commander's order.

F-3. Consider each of the following factors and compare courses of action to form a base for the physical-security plan. When the plan is firm, issue it as an order.

- Concepts for reconnaissance, coordination with adjacent and/or supporting units, and troop movement.
- Physical-security installation configurations and facilities. Areas to consider may include drop zones, landing zones, ranges, and training areas.

MISSION

F-4. The mission is usually the emplacement of defensive security rings to protect the populace against insurgents. The number of defensive security rings depends on the particular site and situation. The following questions must be evaluated:

- What is the mission?
- What specific and implied tasks are there to accomplish the mission?
- What is the commander's intent?

ENEMY

F-5. The commander identifies insurgent units operating in the area and tries to determine the type and size of the unit; the enemy's tactics, weapons, equipment, and probable collaborators; and the inhabitants' attitudes toward the insurgents. The following questions must be evaluated:

- What is known about the enemy?
- Where is the enemy and how strong is he?
- What weapons does the enemy have?
- What is the enemy doing?
- What can the enemy do in response to MP actions?
- How can we exploit the enemy's weaknesses?

TERRAIN AND WEATHER

F-6. The commander can use observation and fields of fire, cover and concealment, obstacles, key terrain, and avenues of approach (OCOKA) to plan for the physical-security defensive sites. The following questions must be evaluated:

- How will the terrain and weather affect the operation?
- How fast can movement be accomplished, and how much space does the terrain and unit formations take up?
- Will the weather affect the terrain or personnel?
- Has the weather already affected the terrain?

TROOPS

F-7. The commander must consider available equipment, the reaction time, reaction forces, communication assets, organization of troops, and medical support (if available). The following questions must be evaluated:

- What are the present conditions of vehicles and personnel?
- What is the status of ammunition and supplies?
- Who is best able to do a specific task?
- How much sleep have the soldiers had in the past 24 hours?
- What other assets are available to support the mission?
- How many teams/squads are available?
- What supplies and equipment are needed?
- What fire support is available and how can it be obtained?

TIME AVAILABLE

F-8. This factor is critical since the inhabitants must be ready to respond to an insurgent attack with little or no warning. The following questions must be evaluated:

- How much time is available to conduct planning?
- How long will it take to reach the objective?
- How long will it take to prepare the position?
- How much time do subordinates need?
- How long will it take the enemy to reposition forces?

CIVILIAN CONSIDERATIONS

F-9. The commander also must consider nonbelligerent third parties (such as dislocated civilians, personnel of international businesses and relief organizations, and the media). Every commander must prepare a site overlay that shows, as a minimum, the following:

- The attitude of the HN toward US forces.
- The population density near the objective.
- The condition of the local civilians.
- The possible effect of refugees and dislocated civilians on the mission.

Appendix G

Personal-Protection Measures

Terrorists frequently emulate military organizations as they develop, plan, train, and carry out terrorist attacks against DOD assets. Terrorists have a critical need for information regarding the whereabouts, habits, working environments, home environments, and other potential points of leverage against their targets. The three intelligence-collection methods used by terrorists against potential targets are human intelligence (HUMINT), photographic intelligence (PHOTINT), and signal intelligence (SIGINT).

PERSONAL PROTECTION

G-1. The measures that follow are useful in providing personal protection for US government employees and DOD civilian contractors in CONUS or OCONUS facilities.

OVERCOME ROUTINES

G-2. The reduced probability of success in kidnapping or killing a target makes the target far less desirable. Perform the following measures to prevent daily routines from being observed:

- Vary your route to and from work and your arrival and departure times.
- Vary your exercise schedule, using different routes and distances. It is best not to exercise alone.
- Do not divulge family or personal information to strangers.
- Enter and exit buildings through different doors, if possible.
- Avoid other routines.

MAINTAIN A LOW PROFILE

G-3. Americans are easy to identify in an overseas area. Perform the following measures to reduce easy ID:

- Dress and behave in public in a manner consistent with local customs. Items that are distinctively American should not be worn or displayed outside American compounds.
- Reduce visibility in the local community.
- Avoid flashing large sums of money, expensive jewelry, or luxury items.
- Avoid public disputes or confrontations, and report any trouble to the proper authorities.
- Ensure that personal information (home address, phone number, or family information) is not divulged.

PREPARE FOR UNEXPECTED EVENTS

G-4. All DOD personnel, contractors, and their family members should implement the following general measures:

- Get into the habit of checking in with friends and family.
- Know how to use the local phone system.
- Know the locations of civilian police, military police, government agencies, and the US embassy.
- Know certain key phrases in the local language.
- Set up simple signal systems that can alert family members or associates that danger is present.
- Carry ID showing your blood type and any special medical conditions.
- Keep personal affairs in good order.
- Avoid carrying sensitive or potentially embarrassing items.

WORKING ENVIRONMENT

G-5. The working environment is not immune from attempted acts by criminals or terrorists. DOD installations in CONUS and OCONUS usually provide a level of basic security comparable or superior to the basic level of security provided in the surrounding community. The following are general practices that can help reduce the likelihood of a terrorist attack:

- Establish and support an effective security program.
- Discourage the use of office facilities to store objects of significant intrinsic value unless it is mission essential.
- Train personnel to be alert for suspicious activities, persons, or objects.
- Arrange office interiors so that strange or foreign objects left in the room will be recognized immediately.
- Provide for security systems on exterior doors and windows.
- Ensure that access-control procedures are rigorously observed at all times for access to—
  - The installation.
  - Buildings within an installation.
  - Restricted or exclusion areas within buildings.
- Use an ID badge system containing a photograph.
- Identify offices by room number, color, or object name and not by rank, title, or the name of the incumbent.
- Avoid using nameplates on offices and parking places.

OFFICE PROCEDURES

G-6. In an office, the following steps can be taken to make intelligence collection and targeting more difficult for terrorists:

- Telephone and mail procedures:
  - When answering the telephone, avoid using ranks or titles.
  - When taking telephone messages, do not reveal the whereabouts or activities of the person being sought.

- When leaving telephone messages, place them in unmarked folders; do not leave them exposed for observers to identify caller names and phone numbers, persons called, and messages left.
- When opening mail, use a checklist to help identify letter bombs or packaged IEDs.

- Visitor-control procedures:
  - Place strict limitations on access to the executive office area.
  - Lock doors (from the inside) from the visitor-access area to executive offices or other restricted areas of a facility.
  - Ensure that receptionists clear all visitors before they enter inner offices.
  - Permit workmen or visitors access to restricted areas or exclusion areas under escort and only with proper ID. Confirm the work to be done before admitting workmen to restricted areas of the facility.
  - Limit publicity in public waiting areas to information that does not identify personnel by name, position, or office location.
  - Avoid posting unit rosters, manning boards, or photo boards where visitors or local contractors can view them.
  - Restrict the use of message boards, sign-in/-out boards, and other visual communications to general statements of availability.
- General working procedures:
  - Avoid carrying attaché cases, briefcases, or other courier bags unless necessary.
  - Avoid carrying items with markings that identify the owner by rank or title, even within the office environment.
  - Avoid working alone late at night and on days when the remainder of the staff is absent.
  - Ensure that office doors are locked when the office is vacant for any lengthy period, at night, and on weekends. If late-night work is necessary, work in conference rooms or internal offices where outside observation is not possible.
  - Ensure that the security office retains the office keys.
  - Ensure that papers, correspondence, communications materials, and other documents are not left unattended overnight.
  - Ensure that maintenance activity and janitorial services in key offices, production offices, or maintenance facilities are performed under the supervision of security personnel.
  - Prohibit the removal of property, material, or information stored on any media from the facility without proper written authorization.
  - Consider prohibiting the importation of property, material, or information stored on any media into the facility unless such items have been properly inspected.
  - Lock offices not in use to prohibit unauthorized access of stored material that could be used to hide IEDs or intelligence-collection devices.
  - Minimize the use of vehicles or vehicle markings that make it possible to readily identify the vehicle and its occupants as US-government or DOD-contractor personnel.

- Ensure that all personnel have access to some sort of duress alarm to annunciate and warn of a terrorist attack.
- Ensure that secretaries and guard posts are equipped with covert duress alarms that can be used to alert backup forces.
- Avoid placing office furnishings directly in front of exterior windows.

SPECIAL PROCEDURES FOR EXECUTIVE ASSISTANTS

G-7. The following suggestions are intended to be a guide for secretaries and executive assistants who may find themselves performing personnel-security duties as collateral duty. Executive assistants and security personnel should regularly train and exercise procedures used in case they must evacuate mission-critical personnel to safe havens.

- Request the installation of physical barriers (such as electromagnetically operated doors) to separate offices of senior executives from other offices.
- Request the installation of a silent trouble-alarm button with a signal terminating in the security department.
- Admit visitors into the executive area when they are positively screened in advance or are personally recognized.
- Do not inform unknown callers of an executive's whereabouts, home address, or telephone number.
- Store a fire extinguisher, a first aid kit, and an oxygen bottle in the office area.
- Remain calm and listen carefully when receiving a threatening call.
- Do not accept packages from strangers until satisfied with the individual's identity and the nature of the parcel.
- Keep travel itineraries for all personnel confidential.
- Distribute daily schedules for senior officers and civilian officials on a limited basis.

HOME ENVIRONMENT

G-8. The following discussion is intended to assist personnel in formulating plans to obtain housing outside US government compounds or DOD facilities. Personnel assigned to government housing may also find the antiterrorism and security tips presented below helpful in reducing the threat of violence and loss of property.

G-9. For general residential-security routines, discuss with family members the importance of—

- Varying routines in their daily activities.
- Blending in with the local environment.
- Avoiding unnecessary publicity and photographs that identify individual family members.
- Being alert to individuals, parked or abandoned vehicles, unusual utility work, or gatherings of people inconsistent with the residential environment.

SECURITY PRACTICES AT HOME

G-10. The following measures are specifically recommended for residential implementation. These measures are an extension of office antiterrorism-security practices.

- Do not use nameplates or uniquely American symbols on the exterior of residences occupied by DOD personnel overseas.
- Do not use nameplates on parking places, and avoid parking private or government vehicles in the same location day after day.
- Ensure that all family members answer the telephone politely but that they provide no information as to the name of the occupants until the caller's identity has been established.
- Treat all telephone conversations as though anyone who wanted to listen in was doing so.
- Examine carefully all mail delivered to the residence.

SOCIAL AND RECREATIONAL ACTIVITIES

G-11. DOD personnel are encouraged to participate in many social and recreational activities. The following precautions are recommended:

- Respond to formal social invitations in person (where possible) or by direct telephone contact.
- Be attentive to the security environment of social gatherings.
- Avoid the development of patterns with respect to time of arrival or departure at social events.
- Avoid prolonged presence at social functions where there is a high concentration of persons thought to be terrorist targets.
- Refrain from excessive use of alcohol at social functions; remain clearheaded and unimpaired.
- Vary routes to and from social events held at a central facility.
- Minimize appearances in uniform or formal attire.
- Decline invitations to appear in publicity photos.
- Participate in recreational activities within the American compound or at a DOD installation whenever possible.

NOTE: Refer to DOD 0-2000.12-H, Graphic Training Aid (GTA) 19-4-3, and Joint Services (JS) Guide 5260 for further guidance and explanation regarding protective measures.

Appendix H

Bombs

Terrorists have frequently used homemade devices or IEDs to carry out their attacks against DOD personnel, facilities, and assets. The IEDs are ideal terrorist weapons. They are relatively inexpensive to make, and the components of many IEDs are common items that can be obtained from many sources and are difficult to trace. The IEDs can be large or small and be designed so that they are transported to the attack site in components for last-minute assembly. Such design concepts make detection more difficult and provide an additional increment of personal safety to the terrorists.

GENERAL

H-1. The use of IEDs can enhance the violence that gives terrorist groups their ability to intimidate or coerce a target population. The detonation itself creates a highly visual, newsworthy scene, even hours after the detonation occurs. Bombs can detonate anywhere, without apparent reason and without warning. The use of bombs in a terror campaign emphasizes the authorities' inability to safeguard the public and maintain law and order. Bombs are ideal weapons because they can be designed to give terrorists opportunities to escape from the scene of their crimes.

CONCEALING BOMBS

H-2. Given the question, "Where have terrorists placed bombs in the past, and where should we look for them?" results in no easy answer. Table H-1, page H-2, lists a few obvious locations that should be examined. Terrorists who use bombs as their weapons of choice can be very creative in designing and placing their weapons.

H-3. Bombs can be found anywhere people can place them. Without becoming paranoid and seeing a bomb under every rock and behind every tree, the practical answer to the above questions is: "Where they can be easily placed without the bomber being caught."

DAMAGE AND CASUALTY MECHANISMS

H-4. The IEDs and other explosive devices inflict casualties in a variety of ways, including the following:

- Blast over pressure (a crushing action on vital components of the body; eardrums are the most vulnerable).
- Falling structural material.
- Flying debris (especially glass).
- Asphyxiation (lack of oxygen).

Table H-1. Potential IED Hiding Places

Outside Areas	
• Trash cans • Dumpsters • Mailboxes • Bushes	• Street drainage systems • Storage areas • Parked cars
Inside Buildings	
• Mail parcels or letters • Inside desks/storage containers • Ceilings with removable panels • Areas hidden by drapes or curtains • Recent repaired/patched segments of walls, floors, or ceilings	• Restrooms • Trash receptacles • Utility closets • Boiler rooms • Under stairwells
In Plain Sight	

- Sudden body translation against rigid barriers or objects (being picked up and thrown by a pressure wave).
- Bomb fragments.
- Burns from incendiary devices or fires resulting from blast damage.
- Inhalation of toxic fumes resulting from fires.

H-5. It is impossible to calculate a single minimum safe distance from an IED or other explosive device. The safe distance varies with each device and its placement. As a rule, the farther away from a bomb, the safer the intended or collateral targets are. Blast effects, fragmentation injuries, and injuries resulting from flying debris diminish greatly as the distance between a bomb and possible targets increase. The amount of material in the device, the type of explosive material, the manner in which the device is constructed, and the location or the container in which it is placed all have a bearing on the specific destructive potential for each IED.

H-6. The following are four general rules to follow to avoid injury from an IED:

- Move as far from a suspicious object as possible without being in further danger from other hazards such as traffic or secondary sources of explosion (such as POL storage).
- Stay out of the object's LOS, thereby reducing the hazard of injury because of direct fragmentation.
- Keep away from glass windows or other materials that could become flying debris.
- Remain alert for additional or secondary explosive devices in the immediate area, especially if the existence of a bomb-threat evacuation assembly area has been highly publicized.

H-7. Some terrorists have used two especially devious tactics in the past to intensify the magnitude of casualties inflicted by bombing attacks. In some instances, they have detonated a small device to lure media attention and curiosity seekers to the site; a larger, more deadly device has detonated some time after the first device, thereby inflicting a large number of casualties.

H-8. Other terrorists have used a real or simulated device to force the evacuation of a facility only to detonate a much more substantial device in identified bomb-threat evacuation assembly areas. These attacks are especially harmful because the evacuation assembly areas often concentrate government or commercial office workers more densely than they are when dispersed throughout their usual workplaces.

TELEPHONIC THREATS

H-9. When receiving a telephonic threat, treat the call seriously. Often, an anonymous telephone call is made regarding a bomb or an IED. See Figure H-1, page H-4, for information to record/obtain when receiving these calls

H-10. When an anonymous warning or threat is received, initiate the bomb-threat data card and notify the PMO, security police, security forces, or other law-enforcement/security offices immediately. Local SOPs will determine subsequent actions. Immediate action may include a search without evacuation, the movement of personnel within the establishment, a partial evacuation, or a total evacuation. The following criteria helps determine what immediate action to take:

- Factors favoring a search before the movement of personnel:—
  - There is a high incidence of hoax telephone threats.
  - Effective security arrangements have been established.
  - Information in the warning is imprecise or incorrect.
  - The caller sounded intoxicated, amused, or very young.
  - The prevailing threat of terrorist activity is low.
- Factors favoring movement of personnel before searching:
  - The area (post or base) is comparatively open.
  - Information in the warning is precise as to the matters of location, a description of the device, the timing, and the motive for the attack.
  - A prevailing threat of terrorist activity is high.

EVACUATION DRILLS

H-11. Evacuation and search drills should be performed periodically under the supervision of the installation's or unit's senior officer. The drills should be held in cooperation with local police if possible. Personnel in adjacent buildings should be informed of drills to avoid causing unnecessary alarm.

H-12. Evacuation procedures depend on the circumstances. Prepare, publicize, and rehearse evacuation plans in advance. Address alarm systems, assembly areas, routes to assembly areas, personnel-evacuation responses, building and area clearances, and evacuation drills.

PERSONNEL-EVACUATION RESPONSE

H-13. The bomb-threat alarm system should be easily distinguished from the fire alarm. When the alarm sounds, personnel should—

- Lock up or secure all classified materials.
- Conduct a quick visual search of their immediate working area.

Instructions: Be calm. Be courteous. Listen, do not interrupt the caller. Notify supervisor/security officer by prearranged signal while caller is on line.

Name of Operator____________________Time__________Date_________

Caller's Identity

Sex: ❑ Male ❑ Female ❑ Adult ❑ Juvenile Approximate age: Years__________

Origin of Call

❑ Local ❑ Booth ❑ Internal (From within bldg)
❑ Long Distance If internal, leave line open for tracing the call.

Voice Characteristics		Speech		Language	
❑ Loud	❑ Soft	❑ Fast	❑ Slow	❑ Excellent	❑ Good
❑ High Pitch	❑ Deep	❑ Distinct	❑ Distorted	❑ Fair	❑ Poor
❑ Raspy	❑ Pleasant	❑ Stutter	❑ Nasal	❑ Foul	❑ Other ____
❑ Intoxicated	❑ Other____	❑ Slurred	❑ Lisp		
			❑ Other ____		

Accent	Manner		Background Noises	
❑ Local	❑ Calm	❑ Angry	❑ Factory Machines	❑ Trains
❑ Not Local	❑ Rational	❑ Irrational	❑ Bedlam	❑ Animals
Region____	❑ Coherent	❑ Incoherent	❑ Music	❑ Quiet
❑ Foreign	❑ Deliberate	❑ Emotional	❑ Office Machines	❑ Voices
Race_____	❑ Righteous	❑ Laughing	❑ Mixed	❑ Airplanes
			❑ Street Traffic	❑ Party Atmosphere

Bomb Facts

Pretend difficulty with your hearing. Keep caller talking.

If caller seems agreeable to further conversation, ask questions like —
When will it go off? Certain Hour - Time Remaining - What kind of bomb? - Where are you now?
How do you know so much about the bomb? - What is your name and address?

If building is occupied, inform caller that detonation could cause injury or death.
Did caller appear familiar with plant or building by his description of the bomb location?

Write out the message in its entirety and any other comments on a separate sheet of paper and attach to this checklist.

Action To Take Immediately After Call

Notify your supervisor/security officer as instructed. Talk to no one other than as instructed by your supervisor/security officer.

Figure H-1. Sample Bomb-Threat Data Card

- Open windows (wherever possible).
- Leave the building, taking only valuable personal belongings.
- Leave doors open and immediately proceed to the assembly area.

H-14. Opening the building will reduce internal damage due to blast effects. It will also somewhat mitigate the extent of debris flying out of or falling from the building should a detonation occur.

ASSEMBLY AREAS

H-15. Choose the routes to the assembly area so that personnel do not approach the IED at any time. Preselect the routes to the assembly area, but devise a system to inform personnel of the location of the suspected IED and alternate routes. Routes prevent confusion and bunching and avoid potential hazards (such as plate glass, windows, and likely locations of additional IEDs).

H-16. Assembly areas should be preselected and well known to personnel. Establish a clearly defined procedure for controlling, marshaling, and checking personnel within the assembly area. If buildings or establishments are in a public area, coordinate the assembly areas with local police. Assembly areas are selected using the following criteria:

- Locate assembly areas at least 100 meters from the likely target or building (if possible).
- Locate assembly areas in areas where there is little chance of an IED being hidden. Open spaces are best. Avoid parking areas because IEDs can be easily hidden in vehicles.
- Select alternate assembly areas to reduce the likelihood of ambush with a second device or small-arms fire. If possible, search the assembly area before personnel occupy the space.
- Avoid locating assembly areas near expanses of plate glass or windows. Blast effects can cause windows to be sucked outward rather than blown inward.
- Select multiple assembly areas (if possible) to reduce the concentration of key personnel. Drill and exercise personnel to go to different assembly areas to avoid developing an evacuation and emergency pattern that can be used by terrorists to attack identifiable key personnel.

BUILDING AND AREA CLEARANCE

H-17. Establish procedures to ensure that threatened buildings and areas are cleared. Prevent personnel from reentering the building. Establish a cordon to prevent personnel from entering the danger area. Establish an initial control point (ICP) as the focal point for the PMO and for MP control.

H-18. Cordon suspicious objects to a distance of at least 100 meters, and cordon suspicious vehicles to a distance of at least 200 meters. Ensure that nobody enters the cordoned area. Establish an ICP on the cordon to control access; relinquish ICP responsibility to the PMO or local police upon their arrival. Maintain the cordon until the PMO, security police, security forces, or local police have completed their examination or stated that the cordon may stand down.

SEARCHING FOR A SUSPECTED IED

H-19. Searches are conducted in response to a telephonic threat or a report of an unidentified object on or near premises occupied by DOD personnel. The following types of searches may be used when searching for a suspected bomb or IED:

- An occupant search is used when the threat's credibility is low. Occupants search their own areas. The search is completed quickly because occupants know their area and are most likely to notice anything unusual.
- A team search is used when the threat's credibility is high. The search is very thorough and places the minimum number of personnel at risk. Evacuate the area completely, and ensure that it remains evacuated until the search is complete. Search teams will make a slow, thorough, systematic search of the area.

H-20. The following procedures should be followed if a search for explosive devices must be conducted before qualified EOD teams arrive:

- Make an audio check, listening for unusual sounds.
- Sweep the area visually up to the waist, then sweep up to the ceiling. Do not forget the tops of cabinets and cupboards.
- Perform a thorough and systematic search in and around containers and fixtures.
- Pass search results as quickly as possible to the leader responsible for controlling the search area. Do not use a radio; it may detonate the explosive.

H-21. Circumstances might arise in the case of a very short warning period. In other instances, a threat of a bomb against some facilities (if true) might necessitate the evacuation of a very large area. In these circumstances, searching for the presence of an explosive device to identify its location, appearance, and possible operating characteristics may be warranted.

H-22. Personnel who have not been trained in IED search and ID techniques should not search for explosive devices. Two types of errors are very common—the false ID of objects as IEDs and the incorrect ID of IEDs as benign objects. Depending on the devices used to arm and trigger an IED, the search process could actually result in an explosion.

SEARCH ORGANIZATION

H-23. The person controlling the search should have a method of tracking and recording the search results (such as a diagram of the area). Delegate areas of responsibility to the search-team leader, who should report to the person controlling the search when each area has been cleared. Pay particular attention to entrances, toilets, corridors, stairs, unlocked closets, storage spaces, rooms and areas not checked by usual occupants, external building areas, window ledges, ventilators, courtyards, and spaces shielded from normal view.

DISCOVERY OF A SUSPECTED IED

H-24. When a suspicious object has been found, report its location and general description immediately to the nearest law-enforcement or supervisory person. Do not touch or move a suspicious object. Instead, perform the following steps:

- If an object appears in an area associated with a specific individual or a clearly identified area—
  - Ask the individual/occupant to describe objects they have brought to work in the past few days.
  - Ask for an accounting of objects.
  - Ask for a verbal description/ID of objects.
- If an object's presence remains inexplicable—
  - Evacuate buildings and surrounding areas, including the search team.
  - Ensure that evacuated areas are at least 100 meters from the suspicious object.
  - Establish a cordon and an ICP.
  - Inform personnel at the ICP that an object has been found.
  - Keep the person who located the object at the ICP until questioned.
  - Avoid reentering the facility to identify an object that may or may not be an IED.

REACTING TO AN EXPLODED IED

H-25. The following procedures should be taken when an explosive/IED detonates at a DOD facility:

- For explosions without casualties—
  - Maintain the cordon. Allow only authorized personnel into the explosion area.
  - Fight any fires threatening undamaged buildings without risking personnel.
  - Report the explosion to the PMO, security police, security forces, or local police if they are not on the scene.
  - Report the explosion to the installation operations center even if an EOD team is on its way. Provide as much detail as possible, such as the time of the explosion, the number of explosions, the color of smoke, and the speed and spread of fire.
  - Ensure that a clear passage for emergency vehicles (fire trucks, ambulances, and so forth) and corresponding personnel is maintained.
  - Refer media inquiries to the PAO.
  - Establish a separate information center to handle inquiries from concerned friends and relatives.
- For explosions with casualties—
  - Select a small number of personnel to help search for casualties.
  - Assign additional personnel the responsibility for maintaining the cordon to keep additional volunteers searching for casualties.

Maintain the cordon until the EOD team verifies no further presence of bombs/IEDs at the site and the fire marshal determines that risk of additional injury to searchers from falling debris is acceptable.

- Prepare a casualty list for notification of next of kin; delay publication of the list until its accuracy is determined.
- Arrange for unaffected personnel to contact their next of kin immediately.

H-26. Civilian management officials and subordinate military commanders continue to have important personal roles to fulfill during a bomb/IED attack on DOD personnel, facilities, and assets. Perform the following procedures when reporting an attack:

- Pass available information to the operations center.
- Avoid delaying reports due to lack of information; report what is known. Do not take risks to obtain information.
- Include the following information in the report:
  - Any warning received and if so, how it was received.
  - The identity of the person who discovered the device.
  - How the device was discovered (casual discovery or organized search).
  - The location of the device (give as much detail as possible).
  - The time of discovery.
  - The estimated length of time the device has been in its location.
  - A description of the device (give as much detail as possible).
  - Safety measures taken.
  - Suggested routes to the scene.
  - Any other pertinent information.

H-27. Perform the following procedures when providing emergency assistance to authorities:

- Ensure that the PMO, security police, security forces, and other emergency-response units from local police, fire and rescue, and EOD teams are not impeded from reaching the ICP. Help maintain crowd control and emergency services' access to the site.
- Evacuate through the doors and windows of buildings.
- Assist the on-scene commander by obtaining a building diagram showing detailed plans of the public-service conduits (gas, electricity, central heating, and so forth), if possible. If unavailable, a sketch can be drawn by someone with detailed knowledge of the building.
- Locate, identify, and make witnesses available to investigative agency representatives when they arrive on the scene. Witnesses include the person who discovered the device, witnessed the explosion, or possesses detailed knowledge of the building or area.

H-28. Performing the above steps will provide substantial assistance to the crisis-management team and give other personnel constructive, supportive actions to take in resolving the crisis. Care must be exercised, however, that additional explosive devices are not concealed for detonation during the midst of rescue operations. These attacks add to the physical damage and emotional devastation of bomb/IED attacks.

H-29. The use of bombs and IEDs during terrorist attacks against DOD personnel, facilities, and assets is a common occurrence. The procedures outlined in this appendix are intended to help a DOD facility respond to an attack before an explosive device detonates. The procedures are also intended to help mitigate the consequences of an attack in case efforts to find an explosive device and render it inoperable are not successful. Incurring the costs to DOD facilities and installations of detecting an explosive device and terminating a terrorist incident before the device can detonate are almost always preferable rather than exercising plans and options to respond to a detonation. Several of the security measures discussed will help reduce the likelihood of a successful bomb/IED attack against DOD assets.

Appendix I

Executive Protection

DOD Directive 2000.12 recognizes a need to provide protection to military officers and DOD civilians who are assigned to high-risk billets, who are (by the nature of their work) high-risk personnel, or who are assigned to facilities identified as high-risk targets. The directive defines these terms as follows:

- High-risk billet. Authorized personnel billet (identified and recommended by appropriate authority) that because of grade, assignment, travel itinerary, or symbolic value may make personnel filling them an especially attractive or accessible terrorist target.
- High-risk personnel. US personnel and their family members whose grade, assignment, travel itinerary, or symbolic value may make them an especially attractive or accessible terrorist target.
- High-risk target. US facilities and material resources that, because of mission sensitivity, ease of access, isolation, or symbolic value may be an especially attractive or accessible terrorist target.

NOTE: For purposes of this appendix, the term executive will be applied to all persons requiring additional security protection who are assigned to high-risk billets, designated as high-risk personnel, or identified as high-risk targets.

SUPPLEMENTAL SECURITY MEASURES

I-1. The specific supplemental security measures that may be furnished to executives are subject to a wide range of legal and policy constraints. US law establishes stringent requirements that must be met before certain security measures may be implemented. DOD regulations, instructions, and legal opinions may further constrain the implementation of some protective measures described in this chapter. The SOFAs and MOUs between the US and a foreign government will also limit the use of some supplemental security measures. Leases and other conditions imposed by contract for purchase of land or buildings by the US for DOD use may also limit the application of certain security techniques. All of these constraints should be carefully considered when conducting security surveys, developing plans, and implementing additional security measures to protect high-risk personnel.

EXECUTIVE PROTECTION GOALS

I-2. In the discussion that follows, several measures are outlined that can afford senior military officers and DOD personnel additional protection against terrorist acts. The purpose underlying these measures is to—

- Increase the interval of time between detecting a threat and the onset of hostile action against executives and their dependents.
- Increase the amount of time required by terrorists to gain physical access to executives from the onset of hostile actions, whether the executives are at home, at the office, or in transit.

I-3. The implementation of supplemental security measures should strive to achieve the following prioritized goals:

- Enhancements should hold the terrorist threat at bay until a response force arrives (delay at a distance).
- Enhancements in physical security should enable executives to flee to safety (delay to permit flight).
- Enhancements should permit the executive to retreat into a safe haven of sufficient strength and survivability. This should enable a response force to wage an effective counterattack to liberate executives and others accompanying them to a safe haven, including family members at home and colleagues and visitors at work (delay, hold, and counterattack).

I-4. The following supplemental measures should be applied with care. There is a clear trade-off between increasing the level of physical security at the office and at home and preserving the anonymity of executives, thereby avoiding telltale signs of activity that point to prominence or criticality. These measures can be expensive. Expense can be measured not just in terms of dollars, but also in terms of changes to organizational routine. Therefore, three questions must be resolved before implementing bold, disruptive, and expensive supplemental security enhancements:

- What are the most cost-effective means of enhancing the security of executives at risk?
- How many changes in organizational routines and personal behaviors will have to be made for security measures to be effective in reducing the risk of terrorist attacks and the vulnerability of executives to such attacks?
- What are the anticipated costs of additional security measures in terms of dollars, organizational functionality, and mission capability?

I-5. Security enhancements can be made to improve the security of executives and can be even more effective if executives and their families take full advantage of and reinforce those measures. If executives do not change their behavior to accommodate additional security and protective measures, then the behaviors can effectively defeat the purpose of additional protection. Additional increments of security can be obtained to defeat virtually any threat. However, there is a point at which it is no longer cost-effective to add layer upon layer of protective measures to defeat a threat.

RESIDENTIAL SECURITY MEASURES

I-6. While terrorist groups conduct intelligence operations to identify targets, mistakes have been made in the past. DOD personnel should avoid leasing residences previously used by representatives of governments or organizations known to be targets of various terrorist groups. DOD personnel leasing residences formerly used by representatives of such governments may be placing themselves unnecessarily at risk of being attacked as a result of mistaken identity.

I-7. An executive's entire lifestyle should be included in security surveys used to assess the need for supplemental physical-security measures at the office. The executive's home and transportation from home to office and back should also be examined for risk and vulnerability. The same principles used to identify supplemental security improvements in an office environment apply to an executive's home environment as well. The purposes of physical-security enhancements are to—

- Increase the amount of time terrorists need to initiate and complete an attack on executives while at home, thereby giving response forces more time to rescue executives and their dependents.
- Reduce potential harm to executives and their families because of a terrorist assault mounted against the residence.

I-8. The goals of enhanced residential physical-security measures are to—

- Increase the amount of time between detection of a threat and the onset of hostile actions.
- Delay the terrorists as long as possible. Prevent terrorist access to executives and their family members and make it difficult to leave the scene to escape prosecution. These measures should not further jeopardize the lives of executives and their family members.
- Provide a safe haven where executives and their family members may flee for security pending the arrival of a response force on the scene.

I-9. The following measures can be implemented selectively to help security personnel achieve these objectives:

- Increase the time interval between threat detection and the onset of hostile terrorist acts by—
  - Ensuring that all door locks and window clasps are working.
  - Ensuring that all doors and windows are properly secured to their frames and that the frames are properly anchored to the residential structure.
  - Locking driveway gates with a security lock to prevent entry.
  - Installing a through-door viewing device or visitor intercom.
  - Installing security lights to aid in viewing entrances.
- Increase the number of physical barriers between the outer perimeter of the residence and the interior of the residence by—
  - Adding heavy, remotely operated gates to all fences, walls, and perimeter barriers consistent with the penetration resistance of the barrier between the residence, the street, and adjacent neighbors.
  - Creating a vestibule or air lock between living quarters and the exterior of a residence, ensuring that no one can enter the residence directly from the outside.
  - Adding fire doors or security doors or gates between the residence's bedrooms and living areas.

- Increase the time required to penetrate exterior structural walls by explosives, hand-held power tools, and hand tools by—
  - Adding additional armor covered by aesthetically pleasing materials to exterior walls.
  - Adding a separate reinforced masonry wall around the residence.
- Increase the surveillance of the residence and decrease response time by—
  - Installing CCTV systems to permit remote viewing of all doors and windows accessible from the ground, nearby structures, trees, or easily acquired platforms (such as a van parked next to a wall).
  - Installing area IDSs between the residence's perimeter and the residence itself, varying the number and types of sensors, and adding backup communication channels between the IDS and a surveillance assessment/response dispatch center.
- Increase the residence's durability and survivability to a terrorist attack by—
  - Fitting windows with either venetian blinds or thick curtains to reduce the observability of activities within the residence and to reduce hazards of flying glass in case of nearby explosions or gunfire.
  - Installing backup power systems for security devices (surveillance systems, communication systems, and access-control systems).
  - Ensuring that backup communication is available with the installation or embassy's security department via a secure landline or two-way radio.
  - Fitting a panic-alarm bell to the outside of the house with switches on all floor levels. Such an alarm should also annunciate at the local police and cognizant DOD or DOS security office.
  - Installing a safe haven in the home.

TRANSPORTATION MEASURES

I-10. High-risk personnel are most accessible to terrorists while in transit in official or privately owned vehicles. Specific steps can be taken to reduce the vulnerability of executives in transit.

SPECIAL TRANSPORTATION IN TRANSIT FROM DOMICILE TO DUTY

I-11. As a rule, Congress has strongly opposed the provision of domicile-to-duty transportation by the federal government to its officers and employees. Only 16 officials are entitled by statute to such assistance. Congress did, however, grant authority to the President and the heads of executive agencies and departments to provide domicile-to-duty transportation under certain circumstances. According to the statute, "a passenger carrier may be used to transport between residence and place of employment an officer or employee with regard to whom the head of a Federal agency makes a determination, [provided] that highly unusual circumstances present a clear and present danger, that an emergency exists, or that compelling operational considerations make such transportation essential to the conduct of official business."

I-12. The phrase "highly unusual circumstances which present a clear and present danger" is understood to mean that—

- The perceived danger is real, not imaginary.
- The perceived danger is immediate or imminent, not merely potential.
- Proof is provided that the use of a government vehicle would provide protection not otherwise available.

I-13. Such a danger would exist where there is an explicit threat of terrorist attacks or riot conditions and such transportation would be the only means of providing safe passage to and from work.

I-14. The phrase "emergency exists" means that there is an immediate, unforeseeable, temporary need to provide home-to-work transportation for an agency's essential employees. The phrase "similarly compelling operational considerations" means that there is an element of gravity or importance for the need of government-furnished transportation comparable to the gravity or importance associated with a clear and present danger or an emergency. Congress suggested further, "in such instances, [it is expected] that home-to-work transportation would be provided only for those employees who are essential to the operation of the government."

I-15. The Secretary of Defense has the statutory authority to allow a CINC to use government-owned or -leased vehicles to provide transportation in an area outside of the US for members of the uniformed services and other DOD personnel under certain circumstances. These circumstances include and are limited to a determination by the CINC that public or private transportation in the area is unsafe or is not available. Under these circumstances, DOD may provide transportation (usually in government buses or passenger vans) to personnel and their family members if it will help the CINC and his subordinate commanders maintain the capability to perform or undertake assigned missions. This transportation is not intended for transporting personnel from their residences to their places of work. The Secretary of Defense and the Service Secretaries also have the statutory authority to provide transportation from home to duty stations and back on a limited basis. This authority is usually implemented by providing a nontactical armored vehicle (NTAV) to protect personnel.

I-16. It is a DOD policy to make NTAVs available where necessary to enhance the security of DOD personnel consistent with the requirements and limitations found in the statute. DOD issuances, service regulations, and CINC guidance stipulate detailed procedures by which DOD manages NTAV programs. The statute also establishes a procedure for Presidential waiver of the "buy American" requirement; DOD and service regulations provide for the delegation of Presidential authority from the President to the Secretary of Defense; to the Director, Defense Security Assistance Agency; and to the Director, DIA. DOD Instruction 5210.84 authorizes DOS acquisition and installation of light vehicle armoring to DOS specifications in local defense-component vehicles on a reimbursable basis. The level of protection provided to the Defense Component Office will comply with approved overseas security policy group armored-vehicle standards.

I-17. The DOD recognizes two classes of NTAVs—heavy and light. Heavy NTAVs are fully armored vehicles intended to protect occupants from attack by bombs; IEDs; grenades; and high-velocity, small-arms projectiles. Light NTAVs are less than fully armored vehicles and are intended to protect occupants from attack by medium-velocity, small-arms projectiles and at least some types of IEDs.

I-18. The dividing lines between heavy and light NTAVs have become less distinct over time as armoring techniques and materials have given greater capability to NTAVs that are not classified as heavy. As a practical matter, add-on vehicle-armoring kits are now in production which (when properly installed in an appropriately powered and suspended vehicle) will provide a level of protection approaching that of the heavy NTAVs.

Heavy NTAVs

I-19. Heavy NTAVs may be assigned to US personnel upon certification by a Service Secretary only under the following conditions:

- Highly unusual circumstances present a clear and present danger to the health and safety of a nominated protectee.
- Compelling operational considerations make such transportation essential to conducting official business.

I-20. If the physical-security survey concludes that a heavy NTAV is warranted, the nominated protectee's Service Secretary shall, on the advice and recommendation of a combatant commander, determine whether the use of a heavy NTAV is warranted. If so, the Service Secretary shall authorize the use of a vehicle for a renewable 90- to 360-day period. At the end of the period, the requirement will be reexamined and a recertification for the protection shall be issued by the Service Secretary.

I-21. Each of the services manages a portion of the DOD's NTAV program. Each service has issued supplementary mandatory guidance for processing requests for, as well as allocation and use of, these scarce assets.

I-22. Heavy NTAVs are complex systems requiring specialized maintenance and operation. Normally, they will be assigned to DOD personnel with a driver who has been properly trained in the operation and maintenance of the vehicle. The operator is not a chauffeur; he is an integral part of a supplemental security package provided by DOD to meet its obligations in protecting key assets.

Light NTAVs

I-23. Light NTAVs may also be provided to US employees and officers where highly unusual circumstances present a clear and present danger to the health and safety of a nominated protectee or compelling operational considerations warrant their use. This category of NTAV features add-on armoring. While they are a less-complex armoring system than those used in heavy NTAVs, light NTAVs afford substantial protection to occupants against a variety of threats. New developments in after-manufacture armoring kits for vehicles are occurring at a rapid pace, increasing the number of vehicle manufacturers and models for which other NTAV modifications are suitable. Each service and the DIA have instructions for implementing DOD policy that authorizes the use of other NTAVs to enhance personnel protection of high-risk persons.

PRIVATELY OWNED VEHICLES

I-24. High-risk personnel may wish to forego the use of POVs during periods of extreme risk. Considerations include selecting measures that—

- Deter secret entry, making undetected placement of IEDs in or under the vehicle difficult for terrorists to accomplish.
- Enhance the vehicle's ability to increase distance between it and pursuers.
- Assist response forces in case of an incident.
- Make the vehicle appear little different than its standard models.

INDIVIDUAL PROTECTIVE MEASURES

I-25. Executives can enhance their personal security in the office environment by—

- Discouraging staff members who are taking telephone messages from disclosing their whereabouts.
- Ensuring that caution is used when opening mail and being especially careful with letters or packages that might contain IEDs.
- Ensuring that access is strictly limited to their office area.
- Limiting publicity and keeping official biographies short. This includes using outdated photographs if a publicity photograph is essential.
- Ensuring that they are not working alone late at night and on days when the remainder of the staff is absent.
- Working in conference rooms or internal offices where outside observation is not possible if late-night work is necessary. Security officers should be notified of the work so that they can periodically look in.
- Ensuring that office furnishings are not placed directly in front of exterior windows.

OFFICIAL BUSINESS AWAY FROM THE OFFICE

I-26. The following suggestions reinforce efforts by executives to maintain the high level of security provided in the home or office environment while on official business outside of these locations:

- Discuss security requirements with the person planning the function.
- Travel to and from the function with escorts.
- Choose the route carefully.
- Avoid publicizing planned attendance at official functions (unless required).
- Attempt to sit away from both public areas and windows.
- Encourage the function's sponsor to close the curtains to minimize the likelihood that anyone outside will be able to see inside and determine who is attending the function. This is extremely important for an evening function, when a well-lit interior can be easily viewed from a darkened exterior.
- Request that external floodlights be used to illuminate the area around the building where an evening function will occur.

LOCAL OFFICIAL AND UNOFFICIAL TRAVEL

I-27. Executives can greatly enhance their personal security when conducting official and unofficial travel by following these general practices:

- Vary daily patterns, such as leaving and returning at different times.
- Consider escorts to and from work or travel with a neighbor.
- Establish a simple oral or visual duress procedure between executives and drivers (for example, a phrase or movement used by the executive or driver only if something is amiss).
- Vary taxi companies. Ensure that the ID photo on the license matches the driver. If uneasy for any reason, take another taxi.
- Attend social functions with others, if possible.
- Examine the car before entering to see if there has been any interference. A small mirror on a rod is a cheap and effective method to inspect underneath cars. Do not touch the vehicle until it has been thoroughly checked (inside, around, and under).
- Avoid leaving personal items exposed in the car (uniform items, service-issued maps, official briefcases, and so forth).

SECURITY PRACTICES WHILE DRIVING

I-28. Executives can take the following measures to enhance security while driving:

- Keep car doors locked. Do not open windows more than a few inches.
- Avoid overloading a vehicle, and wear seat belts.
- Park vehicles in parking areas that are either locked or monitored. Never park overnight on the street. Before entering vehicles, check for signs of tampering.
- Keep the trunk locked.
- Drive in the inner lanes to keep from being forced to the curb.
- Use defensive and evasive driving techniques. Drill with your driver by watching for suspicious cars and taking evasive action.
- Avoid driving close behind other vehicles (especially service trucks), and be aware of activities and road conditions two to three blocks ahead.
- Beware of minor accidents that could block traffic in suspect areas such as crossroads. Crossroads are preferred areas for terrorist or criminal activities because they offer escape advantages.

I-29. If a terrorist roadblock is encountered, use the shoulder or curb (hit at a 30- to 45-degree angle) of the road to go around it or ram the terrorist's blocking vehicle. Blocking vehicles should be rammed in a nonengine area, at a 45-degree angle, in low gear, and at a constant moderate speed. The goal is to knock the blocking vehicle out of the way. In all cases, do not stop and never allow the executive's vehicle to be boxed in with a loss of maneuverability. Whenever a target vehicle veers away from the terrorist vehicle, it gives adverse maneuvering room and presents a better target to gunfire.

INTERURBAN, NATIONAL, AND INTERNATIONAL TRAVEL SECURITY PRACTICES AND PROCEDURES

I-30. To enhance security in interurban, national, and international circumstances, executives should—

- Book airline seats at the last moment. Consider using an alias.
- Restrict the use of ranks or titles.
- Avoid allowing unknown visitors in the hotel room or suite.
- Keep staff and family members advised of the itinerary and subsequent changes. Clearly and emphatically restrict this information to those having a need to know.

HOME SECURITY PRACTICES AND PROCEDURES

I-31. To enhance security at home, executives should—

- Check the ID of persons entering the premises (electricians, plumbers, telephone-maintenance personnel, and so forth). When in doubt, call their office to verify their identity before allowing them in your home.
- Avoid opening the door to a caller at night until he is visually identified through a window or a door viewer.
- Close curtains in a room before turning on lights.
- Consider placing the telephone where you will not be seen from doors or windows when answering.
- Investigate the household staff (especially temporary staff members).
- Stay alert and be on the lookout for the unusual. Ensure that the home is locked and secure whenever the residence is unattended. Be cautious upon return and look for the movement of furniture or the placement of unusual wires.
- Note and report suspicious persons.
- Control house keys strictly.
- Park the car in a locked garage.
- Consider installing a panic-alarm bell to the outside of the house with switches located on all floor levels.
- Clear the area around the house of dense foliage or shrubbery.
- Test duress alarms (if available). Make certain that family members understand how they work as well as the importance of their use.
- Cooperate with law-enforcement personnel, and abide by their security recommendations.

SECURITY AT SOCIAL AND RECREATIONAL ACTIVITIES

I-32. The risk of terrorist incidents is always present for high-risk personnel or personnel assigned to high-risk billets. The following measures are intended to permit executives to live a close-to-normal life while still remaining mindful of the risks to their security.

- Ensure that the host is aware of and takes appropriate measures for your security.
- Have your personal staff assist a civilian host, if required.
- Arrange for visitors to be subject to adequate security control.

- Screen the invitation list, if possible.
- Vary the times of sporting activities (golfing, jogging, and so forth).

COMBATING-TERRORISM TRAINING FOR EXECUTIVES

I-33. Combatant commanders annually compile a list of high-risk billets in their AO. These lists are forwarded through the appropriate service personnel channels, enabling each service to identify, plan, and provide resources to meet training requirements. All personnel and adult family members en route to high-risk billets must attend the Individual Terrorism Awareness Course (INTAC) conducted at the US Army John F. Kennedy Special Warfare Center at Fort Bragg, North Carolina. During this one-week course, personnel will receive instruction in defensive-driving techniques and survival shooting as well as individual protective measures and hostage survival. These individuals should also attend the appropriate regional orientation course (Middle East, Asia/Pacific, Latin America, or Africa) offered at the US Air Force Special Operations School at Hurlburt Air Force Base (AFB), Florida. The service member whose duties will require frequent vehicle operation should attend an appropriate evasive-driving course. Information on current offerings may be obtained by contacting the service representative to the DOD Antiterrorism Coordinating Committee or the Combating Terrorism Branch in the Office of the Assistant Secretary of Defense (OASD) Special Operations/ Low-Intensity Conflict (SO/LIC).

TRAVEL TO POTENTIAL PHYSICAL-THREAT RISK AREAS

I-34. Personnel en route to potential physical-threat risk areas (as identified by the OASDSO/LIC) should attend one of the following courses:

- The Dynamics of International Terrorism Course conducted at the US Air Force Special Operations School at Hurlburt AFB, Florida. During this one-week course, personnel will receive lectures on threats by region (Europe, Middle East, Latin America, Asia/Pacific, and Africa), the history and psychology of terrorism, personnel combating-terrorism measures (vehicle, personal, airline, and physical security), and hostage survival.
- A Regional Orientation Course (Middle East, Latin America, Africa, Asia/Pacific) at the US Air Force Special Operations School at Hurlburt AFB, Florida. This one-week course offers personnel instruction in cultural, political/military, and individual security factors associated with the region.

I-35. Installation security personnel may also receive the above training if they have completed the Antiterrorism Instructor Qualification Course (AIQC) at Fort Bragg, North Carolina.

PROTECTIVE SECURITY DETAILS

I-36. Each service can provide bodyguards for key senior military officers, DOD civilians, other US officials, or foreign dignitaries requiring personal protection. Each Service Secretary is responsible for assigning protective security details (PSDs) to service members based on the recommendation of their counterintelligence and/or law-enforcement investigation staffs. The PSDs are assigned to DOD personnel who meet requirements established by service regulations. In general, PSDs may be assigned only to those personnel whose position or assignment places them at risk and whose continued availability to the National Command Authorities and the CINCs is vital to DOD's mission execution.

I-37. A PSD provides high levels of security to an executive by establishing a series of protective cordons around him. The establishment of defense in depth often means that the innermost protective layer is in close contact with the executive at all hours of the day and night.

I-38. A PSD is trained to maintain a low profile. It is concerned about the executive's visibility and its ability to blend into his surroundings. There is nothing more damaging to the security of an executive than the obvious, detectable presence of a PSD when all other measures to have him blend into the local environment have been successful. A PSD will strive to keep travel routes and means of transportation from being publicized. If this cannot be accomplished, the PSD may suggest editorial changes to the itinerary scheduled for release to limit details of planned travel from public disclosure. For example, routes to and from announced appointments usually do not need to be revealed.

I-39. During the course of a PSD's mission, its members may be asked to perform several different security functions. They may, for example, perform direct or indirect protection or escort duty. Direct protection is open and obvious; indirect is generally a surveillance measure. The security-guard unit may operate as an interior guard and may consist of one or more men stationed at fixed posts. A PSD's members should know the identity of each individual in the party of a protected official; executives can help by introducing them to each member of the official party.

I-40. The protected person's attitude is critical to the success of the PSD's mission. Executives have a right and a responsibility to make their wishes known with respect to their personal security. They also have an obligation to listen carefully to the head of the PSD who is trained and highly qualified to help make reasonable judgments about manageable risks. A PSD's members understand that their function is inherently intrusive and that executives can easily resent the loss of privacy that accompanies the protection offered. On the other hand, PSDs have jobs to do, not merely to protect executives, but to help safeguard mission-critical assets—senior military and civilian leaders.

I-41. One of the most demanding functions placed on a PSD is to limit the ability of individuals to circulate and approach the executive. This is often very frustrating to executives who wish to shake hands, engage in close conversations with visitors, and move freely and without impediment in a social situation. The PSDs are trained to strictly enforce limitations on the circulation of individuals, carefully checking each person for ID and ascertaining that he is authorized to be present at the occasion.

I-42. DOD personnel who are provided with PSDs and must conduct official business or hold social engagements in large rooms can take steps to minimize the disruptions to such functions. These steps include—

- Providing advance attendee lists to the head of the PSD.
- Having one or more members of the staff who know the attendees stand with PSD members and identify the attendees as they arrive.
- Informing attendees that they will be admitted only at specified entrances.

I-43. The PSD's members are highly trained security specialists. While in the company of executives, they will be accommodating and helpful. Executives should remember, however, that the primary function of the PSD's members is to protect them, not perform errands or carry out personal services. A PSD's members who are performing valet or other chores cannot effectively protect the senior officers or civilian officials to whom they have been assigned.

EXECUTIVE-PROTECTION SYSTEM INTEGRATION

I-44. This appendix has focused on supplemental security measures used to address terrorist threats to senior high-risk personnel within the DOD. Various methods and measures have been discussed that provide increments of security over and above the base level of security provided to all DOD personnel assigned to an installation, facility, activity, or unit. In making decisions to allocate protective resources to enhance the security of senior officers and senior DOD officials, it is essential to remember that measures must be applied systematically. Additional security measures implemented to protect high-risk personnel in the office environment must be carried over to official functions conducted outside the office. The security measures must also be extended to protected persons' private lives and, depending on the nature of the threat, the lives of their family members.

I-45. The converse is equally true. It makes no sense to provide domicile-to-duty transportation for a high-risk person and make no provision for additional protection at home, at the office, and at official business and social functions. In view of the total costs of security measured in dollars, time, and inconvenience to protected persons, their staffs, colleagues, and families, it may be more prudent to radically alter living and working arrangements than to try to augment security in a piecemeal manner. For example, it might be prudent to house high-risk personnel within a DOD installation rather than to try to secure a detached, private residence at a substantial distance from the operations base of a response force. The key to successful executive protection is to ensure that the level of protection afforded by physical-security measures, operational procedures in the office and at home, and PSDs is constant. The level of protection must be matched to the threat and must be sustainable. Executives have a special responsibility to set a personal example of combating-terrorism awareness; of attention to personal, family, office, information, and OPSEC concerns; and of combating-terrorism security measures implementation. By doing so, they make their colleagues and subordinates more aware and more conscious of their security environment and less likely to be victimized by terrorist attacks.

Appendix J

Resource Management

Programs need annual funding to operate. This funding is obtained by devising and documenting a resource program that looks seven years out. Looking ahead one or two years is generally not a problem. Commanders must be able to describe future requirements in the out years so that money will be available when the program arrives in future years.

FUNDING PROGRAMS

J-1. Physical security is dependent on integrated systems with budgetary constraints. This appendix serves to inform commanders of three basic funding programs contributing to physical security—

- The RJC6, which resources physical-security equipment purchase and maintenance.
- The QLPR, which resources law-enforcement operations (to include security guards and special-reaction teams).
- The VTER, which provides resources for projects and temporary programs that enhance any type of security due to an increase in terrorist threat.

PROJECTED REQUIREMENTS

J-2. Recurring VTER requirements are usually shifted to QLPR as standard requirements after a period of a few years. Installations send seven-year projected requirements to the MACOMs for submission into the Army's Program Objective Memorandum (POM), which is an annual significant event for resource managers at all levels. Installation requirements should be included in the POM build; otherwise, the installation's program loses visibility from the start of the funding process. Input format is determined locally, but a generic example of the format and the type of information requested is provided at Figure J-1, page J-2.

OBLIGATION PLAN

J-3. Budget execution deals with the current fiscal year (October through September), with the exception of some types of dollars (appropriations) that are multiyear (such as procurement dollars that are executable for three years). Once the resource manager notifies an installation of its available annual funding, an obligation plan by month or quarter is developed to display how and when the funds will be spent. This obligation plan is also used to forecast when the program will run out of money, which in turn will justify the submission of an unfinanced requirement (UFR) to obtain the proper level of resources. Examples of an annual obligation plan and UFR are located at Figures J-2 and J-3, page J-3.

Program Objective Memorandum (POM)

Installation: Fort McClellan, AL
Activity: Military Police School
Appropriation: Operation & Maintenance, Army (OMA)
Army Management Structure (AMS): 321731
Management Decision Package (MDEP): VTER
Element of Resource (EOR): 2500
Issue: Force Protection/Antiterrorism Courses

Resources	FY00	FY01	FY02	FY03	FY04	FY05	FY06
($000)	500	510	520	530	540	550	560

Justification/Impact:
Funding is required to continue the eight functional contract training courses conducted by the US Army Military Police School in Force Protection/Antiterrorism for the Army and selected DOD services and agencies. Courses focus on the five pillars of force protection and provide students with a variety of force-protection expertise to prevent, react to, and contain a terrorist event. If funds are not provided, the projected annual student load of 2,500 cannot be trained.

Figure J-1. Sample POM

TYPES OF APPROPRIATIONS

J-4. The most common types of appropriations that managers will be involved with are day-to-day operations or mission (OMA) funds and procurement (OPA) dollars that must be used for major projects or equipment buys over the $100,000 threshold. Army Materiel Command (AMC) uses research, development, and engineering (RD&E) dollars for operations at US Army Test and Evaluation Command (TECOM) facilities and for testing new equipment. Construction projects over $500,000 use military construction (MILCON) funds.

J-5. Resources needed for physical security are the result of planning. Commanders include physical-security programs and improvements as a part of all annual budgets. When the situation changes based on METT-TC, physical-security programs are reviewed, updated and, when necessary, approved. Commanders allocate resources consistent with the threat. Force-protection requirements provide the fundamental reasons for resourcing physical security. Security managers, force-protection officers, and PMs help identify security requirements and prioritize expenditures. When physical-security improvements are not properly planned, integrated, and budgeted for, commanders accept risks for physical-security failures.

Annual Obligation Plan
FY 99

Installation: Fort McClellan, AL
Activity: Military Police School
Appropriation: Operation & Maintenance, Army (OMA)
Army Management Structure (AMS): 321731
Management Decision Package (MDEP): VTER
Element of Resource (EOR): 2500
Issue: Force Protection/Antiterrorism

Phased Monthly Obligation Plan (Cumulative):

Month	$000
Oct	40
Nov	80
Dec	120
Jan	160
Feb	200
Mar	240
Apr	280
May	320
Jun	360
Jul	360
Aug	360
Sep	360

Figure J-2. Sample Annual Obligation Plan

Unfinanced Requirement (UFR)
FY 99

Installation: Fort McClellan, AL
Activity: Military Police School
Appropriation: Operation & Maintenance, Army (OMA)
Army Management Structure (AMS): 321731
Management Decision Package (MDEP): VTER
Element of Resource (EOR): 2500
Issue: Force Protection/Antiterrorism

Justification/Impact:
Funds in the amount of $360K have been provided to conduct the eight Force-Protection/Antiterrorism contract courses in the US Army Military Police School. This amount will only support training through June 99. An additional $120K is required to conduct training during 4th quarter FY99. Unless additional funds are provided, 625 students will not be trained.

Figure J-3. Sample UFR

Appendix K

Vulnerability Assessment

After a commander has obtained a threat analysis, he proceeds to complete the analysis by conducting the vulnerability and criticality assessments. (This appendix will discuss only the VA.) This process considers a mission review and analysis of the installation, base, unit, or port in relation to the terrorist threat. The review should assess the cost of antiterrorism measures in terms of lost or reduced mission effectiveness. It should then assess the level of acceptable risk to facilities and personnel given the estimated erosion of mission effectiveness. Often the best operational method and routine may be the worst to counter potential terrorist activities. This review and analysis is performed routinely and particularly for deployment.

ASSESSMENT CONSIDERATIONS

K-1. The installation, base, unit, or port assessment is derived from the results of the vulnerability and criticality assessments. The assessment provides the staff with the overall vulnerability to terrorist attack. The staff then develops the crisis-management plan, which addresses all terrorist threat levels regardless of the present level. The THREATCONs are then applied according to the local threat. The considerations are—

- **Vulnerability.** The VA is a self-assessment tool used to evaluate its vulnerability to terrorist attack. The more vulnerable an installation, base, unit, or port is, the more attractive it becomes to terrorist attack.
- **Criticality.** The criticality assessment identifies key assets and infrastructures located on and adjacent to the installation, base, unit, or port. These assets are normally symbolic targets that traditionally appeal to a specific terrorist group (such as headquarters buildings and monuments). It addresses the impact of the temporary or permanent loss of key assets or infrastructures to the ability of the installation, base, unit, or port to perform its mission. The staff determines and prioritizes critical assets. The commander approves the prioritized list. The assessment—
  - Selects key assets.
  - Determines whether critical functions can be duplicated under various attack scenarios.
  - Determines the time required to duplicate key assets or infrastructure efforts if temporarily or permanently lost.
  - Determines the vulnerability of key assets or infrastructures to bombs, vehicle crashes, armed assault, and sabotage.
  - Determines the priority of response to key assets and infrastructures in case of fire, multiple bombings, or other terrorist acts.

- **Damage.** The damage assessment determines the ability of the installation, base, unit, or port to plan for and respond to a terrorist attack against key assets and infrastructures.
- **Recovery procedures.** The recovery-procedures assessment determines the capability to recover from the temporary or permanent loss of key assets and infrastructures. Based on this assessment, the staff establishes recovery procedures to ensure the continued ability to perform the mission.

THREATCON LEVELS

K-2. Specific security measures should be directly linked with THREATCON levels. These considerations are—

- **THREATCON Normal.** This THREATCON level exists when a general threat of possible terrorist activity exists but warrants only a routine security posture.
- **THREATCON Alpha.** This THREATCON applies when there is a general threat of possible terrorist activity against personnel and facilities (the nature and extent of which are unpredictable) and when circumstances do not justify full implementation of THREATCON Bravo measures. It may be necessary to implement measures from higher THREATCONs either resulting from intelligence or as a deterrent. The measures in this THREATCON must be capable of being maintained indefinitely.
- **THREATCON Bravo.** This THREATCON applies when an increased and more predictable threat of terrorist activity exists. The measures in this THREATCON must be capable of being maintained for weeks without causing undue hardship, affecting operational capability, or aggravating relations with local authorities. While in Bravo, the installation should bring manning levels and physical-protection levels to the point where the installation can instantly transition to THREATCON Charlie or Delta.
- **THREATCON Charlie.** The transition to THREATCON Charlie must be done on short notice. It is a result of an incident occurring or the receipt of intelligence indicating that some form of terrorist action against personnel and facilities is imminent. Charlie measures should primarily focus on manning adjustments and procedural changes. Security forces will usually enhance their security presence by acquiring additional manning or by adjusting the work-rest ratio (such as moving from a 3:1 to a 6:1 ratio). At Charlie, off-installation travel should be minimized.
- **THREATCON Delta.** Since the transition to THREATCON Delta is immediate, Delta measures should primarily focus on manning adjustments and procedural changes. THREATCON Delta applies in the immediate area where a terrorist attack has occurred or when intelligence has been received that terrorist action against a specific location or person is likely. The security force's manning level is usually peaked in Charlie; therefore, Delta's additional manning will usually come from an augmentation force. Once in Delta, nonessential operations will cease in order to enhance the security and response

posture. Normally, this THREATCON is declared as a localized condition.

K-3. With exception of THREATCON Normal, all THREATCON levels have certain measures associated with them. These measures are listed in JP 3-07.2, Appendix J, and AR 525-13.

ASSESSING VULNERABILITY

K-4. A VA addresses the consequences of terrorist attacks in terms of the ability of units, installations, commands, or activities to accomplish their assignments successfully, even if terrorists have inflicted casualties or destroyed or damaged DOD assets. The VA focuses on two broad areas—

- Preventing and, failing that, substantially mitigating the effects of a terrorist act.
- Maintaining emergency preparedness and crisis response.

K-5. The VA provides the commander with a tool to assess the potential vulnerability of an installation, base, unit, or port activity; but it is not a substitute for sound judgment. The VA must stand on its own and be supported by valid considerations. Typically, a small group of knowledgeable individuals develop the VA. The VA team consists of personnel with required areas of expertise. Some of these team members are the—

- Assessment-team chief.
- Physical-security specialist.
- Structural engineer.
- Infrastructure engineer.
- Operations-readiness specialist.
- Intelligence and/or counterintelligence specialist.

K-6. The functions and responsibilities of each team member are outlined in JP 3-07.2. The following paragraphs provide basic information regarding these areas:

- The assessment-team chief's key responsibilities include overseeing the management, training, and performance of the vulnerability-team members; finalizing the assessment-team out briefing; and preparing the population dynamics and risk assessment.
- The physical-security specialist is responsible for the security and safety of the installation, facility, and personnel.
- The structural engineer examines a variety of potential terrorist weapon effects and structural responses. This function serves to better protect personnel from shocks and blasts by reducing damage through the technically appropriate use of standoff measures, hardening, blast shielding, and shatter-resistant window film (such as Mylar®) as described in Chapter 3. The structural engineer's main responsibility is threat and damage assessment from terrorist weapons estimates and suggestions for threat protection or damage-mitigation measures.
- The infrastructure engineer examines protection against the effects of WMD, protection against terrorist-incident induced fires, and utility systems that can be used to minimize terrorist-incident casualties

(including elements of power, environmental control, and life-support systems). The primary responsibilities include infrastructure security and fire, safety, and damage control.

- The operations-readiness specialist examines plans, procedures, and capabilities for crisis response, consequence management, and recovery operations should a terrorist incident occur. The main responsibilities of this position include emergency-medical and individual-readiness assessments.
- The intelligence and/or counterintelligence specialist has the primary responsibility of performing logical analyses and preparing possible conclusions regarding terrorist targets and target vulnerabilities. These are based on processed intelligence information and knowledge of terrorist capabilities and methods in view of US installation, facility, and personnel safety and security practices.

CONDUCTING THE ASSESSMENT

K-7. Upon its arrival, the assessment team provides an in briefing for the commander, staff, and designated technical point of contact. Site personnel should conduct a site-familiarization briefing and tour. Administrative activities may include establishing the team support area, setting up equipment, scheduling team and technical points of contact meetings and discussions, ensuring classified-material control, establishing a personnel locator, and organizing materials for the out briefing and site folder. Each assessment-team member conducts the assessment based on the specific responsibilities for each function within his area.

POST-ASSESSMENT ACTIVITIES

K-8. Within 30 days of the visit, a summary narrative report and an annotated briefing should be delivered to the installation commander. Follow-on assistance for the commander may be applicable in areas of technical characteristics of improvement options, cost estimates, and generic sources of materials and equipment.

DRILLS AND EXERCISES

K-9. Multiechelon war gaming of possible terrorist attacks is the best test, short of an actual incident, to analyze the response of an installation, base, unit, or port. Drills and exercises test suspected vulnerabilities and antiterrorist measures. These exercises and drills also train the staff as well as reaction-force leadership and help maintain a valid threat assessment by identifying and adjusting to changing threat capabilities.

Glossary

AA&E	arms, ammunition, and explosives
ABCS	Army Battle Command System
AC	alternating current
admin	administration
ADP	automated data processing
AF	Air Force
AFB	Air Force base
AFM	Air Force manual
AFMAN	Air Force manual
AFOSI	Air Force Office of Special Investigations
AFR	Air Force regulation
AIQC	Antiterrorism Instructor Qualification Course
AIS	automated information system
AL	Alabama
AM	amplitude-modulated
AMC	Army Materiel Command
AMS	Army management structure
AO	area of operations
AP	armor piercing

Apr April

AR Army regulation

AR-PERSCOM Army Reserve Personnel Command

ASP ammunition supply point

AT antitank

AT/FP antiterrorism/force protection

attn attention

Aug August

AWG American wire gauge

bldg building

BMS balanced magnetic switch

BTO barbed-tape obstacle

BUPERS Bureau of Naval Personnel

C^2 command and control

C^3 command, control, and communications

cav cavalry

CB citizen's band

CCB Community Counterterrorism Board

CCIR commander's critical information requirements

CCTV closed-circuit television

CD-ROM compact-disk, read-only memory

cdr	commander
CG	command guidance
chap	chapter
CIA	Central Intelligence Agency
CID	Criminal Investigation Division
CINC	commander in chief
CISO	counterintelligence support officer
CMU	concrete-masonry unit
CONEX	container express
CONPLAN	contingency plan
CONUS	continental United States
CP	command post
CPWG	crime-prevention working group
CQ	charge of quarters
CRIMP	Crime Reduction Involving Many People
CTA	common table of allowance
DA	Department of the Army
DARE	Drug Abuse Resistance and Education
DC	direct current
DC	District of Columbia
Dec	December

DIA	Defense Intelligence Agency
DOD	Department of Defense
DOE	Department of Energy
DOJ	Department of Justice
DOS	Department of State
DOT	Department of Transportation
DS	direct support
DTM	data-transmission media
DTOC	division tactical operations center
EDM	emergency-destruct measures
EECS	electronic entry-control system
EOD	explosive-ordnance disposal
EOR	element of resource
EPW	enemy prisoner of war
equip	equipment
ESS	electronic security system
FAA	Federal Aviation Administration
FBI	Federal Bureau of Investigation
FCC	Federal Communications Commission
Feb	February
FIS	foreign-intelligence services

FM field manual

FM frequency-modulated

ft foot, feet

FY fiscal year

G2 Assistant Chief of Staff, G2 (Intelligence)

GHz gigahertz

GTA graphic training aid

HN host nation

HQ headquarters

HUD Housing and Urban Development Administration

HUMINT human intelligence

Hz hertz

IAW in accordance with

ICP initial control point

ID identification

IDS intrusion-detection system

IED improvised explosive device

IG inspector general

IID improvised incendiary device

in inch(es)

INSCOM US Army Intelligence and Security Command

INTAC	Individual Terrorism Awareness Course
IPB	intelligence preparation of the battlefield
IR	infrared
ISS	information systems security
J2	Intelligence Directorate (Joint Command)
Jan	January
JS	Joint Service
JSAT	Joint Security Assistance Training
JSCP	Joint Strategic Capabilities Plan
JSIIDS	Joint-Service Interior Intrusion-Detection System
Jul	July
Jun	June
K	one thousand
kHz	kilohertz
LED	light-emitting diode
liq	liquid
LOS	line of sight
LOTS	logistics over the shore
LP	listening post
LRA	local reproduction authorized
MACOM	major Army command

maint	maintenance
Mar	March
MCO	Marine Corps order
MDEP	management decision package
MDMP	military decision-making process
METT-TC	mission, enemy, terrain, troops, time available, and civilian considerations
MEVA	mission-essential or vulnerable area
MI	military intelligence
MILCON	military construction
MILPO	military personnel office
MILVAN	military van
min	minimum
mm	millimeter(s)
MO	modus operandi
MOU	memorandum of understanding
MP	military police
MPACS	Military Police Automated Control System
MPMIS	Military Police Management Information System
MPR	military-police report
MS-DOS	Microsoft®-disk operating system
MWD	military working dog

N/A not applicable

naut nautical

NAVATAC Navy Antiterrorism Analysis Center

NBC nuclear, biological, and chemical

NCIC National Crime Information Center

NCO noncommissioned officer

NISCOM Naval Investigative Service Command

No. number

Nov November

NSA National Security Agency

NTAV nontactical armored vehicle

NVD night-vision device

O official

OASD Office of the Assistant Secretary of Defense

OCOKA observation and fields of fire, cover and concealment, obstacles, key terrain, and avenues of approach

OCONUS outside the continental United States

Oct October

OMA operations and maintenance, Army

OPA operations and procurement, Army

OPLAN operation plan

OPORD operations order

OPSEC operations security

Pam pamphlet

PAO public affairs office(r)

PD probability of detection

PERSCOM Personnel Command

PHOTINT photographic intelligence

PI police intelligence

PIR passive infrared

PM provost marshal

PMO provost marshal office

POL petroleum, oil, and lubricants

POM Program Objective Memorandum

POV privately owned vehicle

PS physical security

PSD protective security detail

PSI physical-security inspector

PTO Parent-Teacher Organization

PX post exchange

R&D research and development

RD&E research, development, and engineering

ref reference(s)

RF radio frequency

RII relevant information and intelligence

ROI report of investigation

RORO roll on/roll off

RPG rocket-propelled grenade

/s/ signed

S2 Intelligence Officer (US Army)

SAW squad automatic weapon

SDNCO staff duty noncommissioned officer

SDO staff duty officer

Sep September

SIGINT signal intelligence

SJA staff judge advocate

SO/LIC Special Operations/Low-Intensity Conflict

SOFA status of forces agreement

SOP standing operating procedure

St. Saint

STANO surveillance, target acquisition, and night observation

stat statute

STC sound-transmission coefficient

STD standard

STU secure telephone unit

TAACOM	Theater Army Area Command
TB	technical bulletin
TDY	temporary duty
TECOM	US Army Test and Evaluation Command
TEMPEST	Terminal Electromagnetic-Pulse Emanation Standard
THREATCON	threat conditions
TM	technical manual
TMDE	test, measurement, and diagnostic equipment
TNT	trinitrotoluene
TRADOC	US Army Training and Doctrine Command
TSC	triple-standard concertina
UCMJ	Uniform Code of Military Justice
UFR	unfinanced requirement
US	United States
USACE	US Army Corps of Engineers
USACIDC	US Army Criminal Investigation Command
USAMPS	US Army Military Police School
USCG	US Coast Guard
VA	vulnerability assessment
VIP	very important person
vol	volume
WMD	weapons of mass destruction

Bibliography

AR 12-15. *Joint Security Assistance Training (JSAT) Regulation (*Navy Instructions 4950.4; AFR 50-29). 28 February 1990.

AR 50-5. *Nuclear and Chemical Weapons and Material—Nuclear Surety.* 3 October 1986.

AR 50-6. *Nuclear and Chemical Weapons and Materiel, Chemical Surety.* 1 February 1995.

AR 190-11. *Physical Security of Arms, Ammunitions, and Explosives.* 30 September 1993.

AR 190-12. *Military Police Working Dogs.* 30 September 1993.

AR 190-13. *The Army Physical Security Program.* 30 September 1993.

AR 190-14. *Carrying of Fire Arms and Use of Force for Law Enforcement and Security Duties.* 12 March 1993.

AR 190-22. *Searches, Seizures and Disposition of Property.* 1 January 1983.

AR 190-27. *Army Participation in National Crime Information Center.* 28 May 1993.

AR 190-51. *Security of Unclassified Army Property (Sensitive and Nonsensitive).* 30 September 1993.

AR 190-56. *The Army Civilian Police and Security Guard Program.* 21 June 1995.

AR 190-59. *Chemical Agent Security Program.* 27 June 1994.

AR 380-5. *Department of the Army Information Security Program.* 25 February 1988.

AR 380-10. *Technology Transfer, Disclosure of Information and Contacts with Foreign Representatives.* 30 December 1994.

AR 380-19. *Information Systems Security.* 27 February 1998.

AR 380-67. *The Department of the Army Personnel Security Program.* 9 September 1988.

AR 381-20. *The Army Counterintelligence Program.* 15 November 1993.

AR 405-20. *Federal Legislative Jurisdiction.* 1 August 1973.

AR 525-13. *Antiterrorism Force Protection (AT/FP): Security of Personnel, Information, and Critical Resources.* 10 September 1998.

AR 600-8-14. *Identification Cards for Members of the Uniformed Services, Their Family Members, and Other Eligible Personnel* (AF 36-3026[I]; BUPERS I 1750.10A; MCO P5512.1B; CG M5512.1; Manual 29.2, Instructions 1 and 2). 1 March 1998.

AR 604-5. *Personnel Security Clearance, Department of the Army Personnel Security Program Regulation* 1 February 1984.

DA Form 2028. *Recommended Changes to Publications and Blank Forms.* 1 February 1974.

DA Form 2806-R. *Physical Security Survey Report.* April 1985.

DA Form 2806-1-R. *Physical Security Inspection Report (LRA).* April 1985.

DA Form 2819. *Law Enforcement and Discipline Report.* May 1988.

DA Form 3975. *Military Police Report.* May 1988.

DA Pam 190-12. *Military Working Dog Program.* 30 September 1993.

DA Pam 190-51. *Risk Analysis for Army Property.* 30 September 1993.

DOD 0-2000.12-H. *Protection of DOD Personnel and Activities Against Acts of Terrorism and Political Turbulence.* February 1993.

DOD 4160.21-M. *Defense Reutilization and Marketing Manual.* August 1997.

DOD Directive 2000.12. *DOD Antiterrorism / Force Protection (AT / FP) Program.* 13 April 1999.

DOD Instruction 2000.16. *DOD Combating Terrorism Program Standards.* 10 May 1999.

DOD Instruction 5210.84. *Security of DOD Personnel at US Missions Abroad.* 22 January 1992.

Federal Specification FF-L-2740. *Locks, Combination.* 12 October 1989.

FM 5-34. *Engineer Field Data.* 30 August 1999.

FM 19-4. *Military Police Battlefield Circulation Control, Area Security, and Enemy Prisoner of War Operations.* 7 May 1993.

FM 55-20. *Army Rail Transport Units and Operations.* 1 June 00.

GTA 19-4-3. *Individual Protective Measures.* July 1997.

JP 3-07.2. *Joint Tactics, Techniques, and Procedures for Antiterrorism.* 17 March 1998.

JS Guide 5260. *Service Member's Personal Protection Guide: A Self-Help Handbook to Combating Terrorism.* July 1996.

TB 9-2300-422-20. *Security of Tactical Wheeled Vehicles.* 27 August 1988.

TM 5-805-8. *Builders' Hardware.* 20 January 1992.

TM 5-820-4. *Drainage for Areas Other Than Airfields* (AFM 88-5, Chap 4). 14 October 1983.

TM 5-853-1. *(O) Security Engineering Project Development* (AFMAN 32-1071, Vol 1). 12 May 1994.

TM 5-853-2. *(O) Security Engineering Concept Design* (AFMAN 32-1071, Vol 2). 12 May 1994.

TM 5-853-3. *(O) Security Engineering Final Design* (AFMAN 32-1071, Vol 3). 12 May 1994.

TM 5-853-4. *Security Engineering Electronic Security Systems.* 12 May 1994.

USACE STD 872-50-01. *Entry Points for US Army Installations.* To be published within the next six months.

USACE STD 872-90-02. *FE5 Chain-Link Security Fence Details.* May 1992.

USACE STD 872-90-03. *FE6 Chain-Link Security Fence Details.* May 1992.

USACE STD 872-90-04. *FE7 Chain-Link Security Fence Details for Non-Sensored Fence.* May 1992.

USACE STD 872-90-05. *FE7 Chain-Link Security Fence Details for Sensored Fence.* May 1992.

USACE STD 872-90-06. *FE8 Chain-Link Security Fence Details for Sensored Fence.* May 1992.

USACE STD 872-90-07. *FE5 Chain-Link Fence Gate Details.* May 1992.

USACE STD 872-90-08. *FE6 Chain-Link Fence Gate Details.* May 1992.

USACE STD 872-90-09. *FE8 Chain-Link Fence Gate Details.* May 1992.

Index

Change Number 1

Headquarters
Department of the Army
Washington, DC, 17 May 2004

Intelligence

Contents

*This publication supersedes FM 34-1, 27 September 1994.

Preface

FM 2-0 is the Army's keystone manual for military intelligence (MI) doctrine. It describes—

- The fundamentals of intelligence operations.
- The operational environment (OE).
- Intelligence in unified action.
- The Intelligence Battlefield Operating System (BOS).
- Intelligence considerations in strategic readiness.
- The intelligence process.
- MI roles and functions within the context of Army operations.

This manual conforms to the overarching doctrinal precepts presented in FM 3-0.

This manual provides doctrinal guidance for the Intelligence BOS actions in support of commanders and staffs. It also serves as a reference for personnel who are developing doctrine; tactics, techniques, and procedures (TTP); materiel and force structure; and institutional and unit training for intelligence operations.

This manual provides MI guidance for all commanders, staffs, trainers, and MI personnel at all echelons. It forms the foundation for MI and the Intelligence BOS doctrine development, and applies equally to the Active Component (AC), United States Army Reserve (USAR), and Army National Guard (ARNG). It is also intended for commanders and staffs of joint and combined commands, US Naval and Marine Forces, units of the US Air Force, and the military forces of multinational partners.

Headquarters, US Army Training and Doctrine Command is the proponent for this publication. The preparing agency is the US Army Intelligence Center and School. Send written comments and recommendations on DA Form 2028 (Recommended Changes to Publications and Blank Forms) directly to: Commander, ATZS-FDT-D (FM 2-0), 550 Cibeque Street, Fort Huachuca, AZ 85613-7017. Send comments and recommendations by e-mail to *ATZS-FDC-D@hua.army.mil*. Follow the DA Form 2028 format or submit an electronic DA Form 2028.

Unless otherwise stated, masculine nouns and pronouns do not refer exclusively to men.

This manual contains Army tactical task (ART) description taken verbatim from the *Army Universal Task List* (AUTL). ART task descriptions are followed by a reference to FM 7-15 in parentheses and the ART number; for example (See FM 7-15, ART 5.1.1.).

PART ONE

Intelligence in the Operational Environment

Part One discusses MI's role in peace, conflict, and war. Supporting the warfighter with effective intelligence is the primary focus of Military Intelligence. Intelligence provides commanders and decisionmakers with the requisite information facilitating their situational understanding so that they may successfully accomplish their missions in full spectrum operations.

Chapter 1 describes the operational environment and the roles of MI within the operational environment. It introduces the Intelligence warfighting function, the intelligence tasks, and the intelligence process, which are the mechanisms through which MI supports the warfighter. This chapter also introduces the intelligence disciplines, which are explained in detail in Part Three of this manual.

Chapter 2 describes the interaction of MI within the nation's intelligence community structure, providing an overview of the intelligence community at the national level and the unified action level—joint, multinational, and interagency aspects of full spectrum operations. This chapter also discusses the concepts and components of intelligence reach.

Chapter 1

Intelligence and the Operational Environment

ROLE OF INTELLIGENCE

1-1. The commander requires intelligence about the enemy and the area of operations (AO) prior to engaging in operations in order to effectively execute battles, engagements, and other missions across the full spectrum of operations. Intelligence assists the commander in visualizing the AO, organizing the forces, and controlling operations to achieve the desired tactical objectives or end-state. Intelligence supports force protection (FP) by alerting the commander to emerging threats and assisting in security operations.

1-2. The unit may need to deal with multiple threats. The commander must understand how current and potential enemies organize, equip, train, employ, and control their forces. Intelligence provides an understanding of the enemy, which assists in planning, preparing, and executing military operations. The commander must also understand the operational environment and its effects on both friendly and enemy operations. The commander receives mission-oriented intelligence on enemy forces and the

AO from the G2/S2. The G2/S2 depends upon the intelligence, surveillance, and reconnaissance (ISR) effort to collect and provide information on the enemy and AO.

1-3. One of the most significant contributions that intelligence personnel can accomplish is to accurately predict future enemy events. Although this is an extremely difficult task, predictive intelligence enables the commander and staff to anticipate key enemy events or reactions and develop corresponding plans or counteractions. Intelligence analysis conclusions, assessments, or products that attempt to define, describe, present, or portray the future situation or conditions of the enemy, terrain, weather, and civil considerations.

1-4. The most important purpose of intelligence is to influence decisionmaking. Commanders must receive the intelligence, understand it (because it is tailored to the commander's requirements), believe it, and act on it. Through this doctrinal concept, intelligence drives operations.

THE INTELLIGENCE WARFIGHTING FUNCTION

1-5. *The intelligence warfighting function is the related tasks and systems that facilitate understanding of the operational environment.* (FM 3-0) It includes tasks associated with intelligence, surveillance, and reconnaissance (ISR) operations, and is driven by the commander. Intelligence is more than just collection. It is a continuous process that involves analyzing information from all sources and conducting operations to develop the situation. The intelligence warfighting function includes the following tasks:

- Support to force generation.
- Support to situational understanding.
- Perform intelligence, surveillance, and reconnaissance.
- Support to targeting and information operations.

1-6. The intelligence warfighting function is one of six warfighting functions—movement and maneuver, intelligence, fires, sustainment, command and control, and protection. *A warfighting function is a group of tasks and systems (people, organizations, information, and processes) united by a common purpose that commanders use to accomplish missions and training objectives* (FM 3-0). (See FM 3-0, chapter 4, for a detailed discussion of the warfighting functions.) The intelligence warfighting function is a flexible force of personnel, organizations, and equipment that, individually or collectively, provide commanders with the timely, relevant, and accurate intelligence required to visualize the battlefield, assess the situation, and direct military actions. Additionally, the intelligence warfighting function is—

- A complex system that operates worldwide, from "mud-to-space," in support of an operation, to include the ability to leverage theater and national capabilities.
- Requires cooperation and division of labor internally, higher, lower, adjacent, and across components and the coalition.

1-7. The intelligence warfighting function not only includes assets within the MI branch but also includes the assets of all branches or warfighting functions that conduct intelligence warfighting function tasks. Every Soldier, as a part of a small unit, is a potential information collector and an essential component to help reach situational understanding. Each Soldier develops a special level of awareness simply due to exposure to events occurring in the AO and has the opportunity to collect and report information by observation and interaction with the population (see paragraph 3-3).

1-8. Planning and executing military operations will require intelligence regarding the threat (traditional, irregular, catastrophic, and disruptive) and the AO. The intelligence warfighting function generates intelligence and intelligence products that portray the enemy and aspects of the environment. These intelligence products enable the commander to identify potential courses of action (COAs), plan operations, employ forces effectively, employ effective tactics and techniques, and implement protection.

1-9. The intelligence warfighting function is always engaged in supporting the commander in offensive, defensive, stability, and, when directed, civil support operations. Intelligence provides products that are timely, relevant, accurate, and predictive. Hard training, thorough planning, meticulous preparation, and aggressive execution posture the Army for success. In the current environment we must maintain intelligence readiness to support operations on "no notice." This support is comprehensive and reaches across full spectrum operations and levels of war to produce the intelligence required to successfully accomplish the mission. A combination of space, aerial, seaborne, and ground-based systems provide the most comprehensive intelligence possible. During force projection operations, the intelligence warfighting function supports the commander with accurate and responsive intelligence from predeployment through redeployment.

1-10. The intelligence warfighting function architecture provides specific intelligence and communications structures at each echelon from the national level through the tactical level. These structures include intelligence organizations, systems, and procedures for collecting, processing, analyzing, and delivering intelligence and other critical information in a useable form to those who need it, when they need it. Effective communications connectivity and automation are essential components of this architecture.

INTELLIGENCE TASKS (METL)

1-11. The personnel and organizations within the Intelligence warfighting function conduct four primary intelligence tasks that facilitate the commander's visualization and understanding of the threat and the AO. These tasks are interactive and often take place simultaneously. (Refer to FM 7-15 for the complete subordinate task listing.) Figure 1-1 shows these tasks tailored to the commander's needs.

INTELLIGENCE TASKS	COMMANDER'S FOCUS	COMMANDER'S DECISIONS
Support to Situational Understanding - Perform IPB - Perform Situation Development - Provide Intelligence Support to Force Protection - Conduct Police Intelligence Operations	**Plan a mission.** **Execute the operation.** **Secure the force.**	**Which COA will I implement?** **Which enemy actions are expected?**
Support to Strategic Responsiveness - Perform I&W - Ensure Intelligence Readiness - Conduct Area Studies of Foreign Countries - Support Sensitive Site Exploitation	**Orient on contingencies.**	**Should I increase the unit's level of readiness?** **Should I implement the OPLAN?**
Conduct Intelligence, Surveillance, and Reconnaissance - Perform Intelligence Synchronization - Perform ISR Integration - Conduct Tactical Reconnaissance - Conduct Surveillance	**Plan the mission.** **Prepare.** **Execute.** **Assess.**	**Which DPs, HPTs, etc., are linked to the enemy actions?** **Are the assets available and in position to collect on the DPs, HPTs, etc.?** **Have the assets been repositioned for contingency mission?**
Provide Intelligence Support to Effects - Provide Intelligence Support to Targeting - Provide Intelligence Support to IO - Provide Intelligence Support to Combat Assessment	**Destroy/suppress/ neutralize targets.** **Reposition intelligence or attack assets.**	**Is my fire (lethal or non-lethal) and maneuver effective?** **Should I refire the same targets?**

Figure 1-1. Intelligence Tailored to the Commander's Needs.

SUPPORT TO SITUATIONAL UNDERSTANDING

1-12. This task centers on providing information and intelligence to the commander, which facilitates the commander's understanding of the enemy and the environment. It supports the command's ability to make sound decisions. Support to situational understanding comprises four subtasks: perform intelligence preparation of the battlefield (IPB), perform situation development, provide intelligence support to FP, and conduct police intelligence operations.

Perform Intelligence Preparation of the Battlefield

1-13. The G2/S2 is the staff proponent for IPB. IPB is the staff planning activity undertaken by the entire staff to define and understand the battlefield and the options it presents to friendly and threat forces. IPB includes input from the whole staff. There is only one IPB in each headquarters with inputs from all affected staff cells; they are not separate warfighting function or staff section IPBs throughout the headquarters. It is a systematic process of analyzing and visualizing the operational environment in a specific geographic area for a specific mission or in anticipation of a specific mission. By applying IPB, the commander and staff gain the information necessary to selectively apply and maximize combat power at critical points in time and space. IPB is most effective when it integrates each staff element's expertise into the final products. To conduct effective IPB, the G2/S2 must—

- Produce IPB products that support the staff's preparation of estimates and the military decision-making process (MDMP).

- Identify characteristics of the AO, including the information environment, that will influence friendly and threat operations.
- Establish the area of interest (AOI) in accordance with the commander's guidance.
- Identify gaps in current intelligence holdings.
- Determine multiple enemy COAs (ECOAs) by employing predictive analysis techniques to anticipate future enemy actions, capabilities, or situations.
- Establish a database that encompasses all relevant data sets within and related to the operational environment.
- Determine the enemy order of battle (OB), doctrine, and TTP. Identify any patterns in enemy behavior or activities.
- Accurately identify and report hazards within the AO, including the medical threat and toxic industrial material (TIM).
- Accurately identify threat capabilities, high-value targets (HVTs), and threat models.
- Integrate IPB information into COA analysis and the MDMP.
- Update IPB products as information becomes available.

Perform Situation Development

1-14. Situation development is a process for analyzing information and producing current intelligence about the enemy and environment during operations. The process helps the intelligence officer recognize and interpret the indicators of enemy intentions, objectives, combat effectiveness, and potential ECOAs. Situation development—

- Confirms or denies threat COAs.
- Provides threat locations.
- Explains what the threat is doing in relation to the friendly force operations.
- Provides an estimate of threat combat effectiveness.

1-15. Through situation development, the intelligence officer is able to quickly identify information gaps, and explain enemy activities in relation to the unit's own operations, thereby assisting the commander in gaining situational understanding. Situation development helps the commander make decisions and execute branches and sequels. This reduces risk and uncertainty in the execution of the plan. The intelligence officer maintains, presents, and disseminates the results of situation development through intelligence input to the common operational picture (COP) and other intelligence products.

Provide Intelligence Support to Force Protection

1-16. Provide intelligence in support of protecting the tactical force's fighting potential so that it can be applied at the appropriate time and place. This task includes the measures the force takes to remain viable and functional by protecting itself from the effects of or recovery from enemy activities. FP consists of those actions taken to prevent or mitigate hostile actions against Department of Defense (DOD) personnel (to include Department of Army

[DA] civilians, contractors, uniformed personnel, and family members), resources, facilities, and critical information. These actions—

- Conserve the force's fighting potential for application at the decisive time and place.
- Incorporate coordinated and synchronized offensive and defensive measures.
- Facilitate the effective employment of the joint force while degrading the capabilities of and opportunities for the threat.

1-17. Intelligence to FP consists of monitoring and reporting the activities, intentions, and capabilities of adversarial groups and determining their possible COAs. Detecting the adversary's methods in today's OE requires a higher level of situational understanding, informed by current and precise intelligence. This type of threat drives the need for predictive intelligence based on analysis of focused information from intelligence, law enforcement, and security activities.

1-18. Intelligence analysis in support of FP employs analytical methodologies and tools to provide situational understanding and to predict the adversary's actions. Modified or standard time-event charts, association matrixes, activity matrixes, link diagrams, and overlays are beneficial in monitoring the actions of the adversary. Overlays may include (but are not limited to) threat training camps, organizations, finances, personalities, industrial sites, information systems, decisionmaking infrastructures, specific activities, and locations of previous attacks.

Conduct Police Intelligence Operations

1-19. Police intelligence operations (PIO) are a military police (MP) function that supports, enhances, and contributes to the commander's force protection program, COP, and situational understanding. The PIO function ensures that information collected during the conduct of other MP functions—maneuver and mobility support, area security, law and order, and internment and resettlement—is provided as input to the intelligence collection effort and turned into action or reports. PIO has three components: (See FM 7-15, ART 1.1.4)

- Collect police information.
- Conduct Police Information Assessment Program (PIAP).
- Develop police intelligence products.

Notes:

US Code, Executive Orders, DOD Directives, and Army Regulations contain specific guidance regarding prohibition on the collection of intelligence information on US citizens, US corporations, and resident aliens. These laws and regulations include criminal penalties for their violation. Any PIO directed against US citizens should undergo competent legal review prior to their initiation.

The inclusion of the PIO task branch in the Intelligence warfighting function does not change the intelligence process described in this manual.

SUPPORT TO STRATEGIC RESPONSIVENESS

1-20. Intelligence support to strategic responsiveness supports staff planning and preparation by defining the full spectrum of threats, forecasting future threats, and forewarning the commander of enemy actions and intentions. Support to strategic responsiveness consists of four subtasks: Perform I&W, ensure intelligence readiness, conduct area studies of foreign countries, and support sensitive site exploitation. (See FM 7-15, ART 1.2)

Perform Indications and Warnings

1-21. This activity provides the commander with forewarning of enemy actions or intentions; the imminence of threat actions. The intelligence officer develops I&W in order to rapidly alert the commander of events or activities that would change the basic nature of the operations. It enables the commander to quickly reorient the force to unexpected contingencies and shape the battlefield. (See FM 7-15, ART 1.2.1)

1-22. The G2/S2 at the operational and strategic levels develops I&W in order to rapidly alert the commander of events or activities that would change the basic nature of the operations so the commander can initiate the appropriate action in a timely manner. I&W reduce the risk of enemy actions that are counter to planning assumptions. I&W enable the commander to quickly reorient the force to unexpected events and to shape the battlefield by manipulating enemy activities.

Ensure Intelligence Readiness

1-23. Intelligence readiness operations support contingency planning and preparation by developing baseline knowledge of multiple potential threats and operational environments. These operations and related intelligence training activities engage the Intelligence warfighting function to respond effectively to the commander's contingency planning intelligence requirements. (See FM 7-15, ART 1.2.2) While still in garrison, intelligence defines the full spectrum of threats and forecasts future threats and dangers. Intelligence readiness operations accomplish the following:

- Provide intelligence to support contingency-based training and staff planning.
- Identify, consider, and evaluate all potential threats to the entire unit.
- Provide a broad understanding of the operational environment of the contingency area, which is developed through continuous exchange of information and intelligence with higher echelon and joint intelligence organizations.

Conduct Area Studies of Foreign Countries

1-24. Study and understand the cultural, social, political, religious, and moral beliefs and attitudes of allied, host nation (HN), or indigenous forces to assist in accomplishing goals and objectives. (See FM 7-15, ART 1.2.3)

Note: The inclusion of this task does not change the support to strategic responsiveness provided by MI organizations described in this manual.

Support Sensitive Site Exploitation

1-25. Sensitive site exploitation consists of a related series of activities inside a sensitive site captured from an adversary. A sensitive site is a designated, geographically limited area with special military, diplomatic, economic, or information sensitivity for the United States. This includes factories with technical data on enemy weapon systems, war crimes sites, critical hostile government facilities, areas suspected of containing persons of high rank in a hostile government or organization, terrorist money laundering, and document storage areas for secret police forces. These activities exploit personnel, documents, electronic data, and material captured at the site, while neutralizing any threat posed by the site or its contents. While the physical process of exploiting the sensitive site begins at the site itself, full exploitation may involve teams of experts located around the world. (See FM 7-15, ART 1.2.4)

Note: The inclusion of this task does not change the support to strategic responsiveness provided by MI organizations described in this manual.

CONDUCT INTELLIGENCE, SURVEILLANCE, AND RECONNAISSANCE

1-26. With staff participation, the intelligence officer synchronizes intelligence support to the ISR effort by focusing the collection, processing, analysis, and intelligence products on the critical needs of the commander. The operations officer, in coordination with the intelligence officer, tasks and directs the available ISR assets to answer the commander's critical information requirements (CCIRs). Through various detection methods and systematic observation, reconnaissance and surveillance obtains the required information. A continuous process, this task has four subtasks: perform intelligence synchronization, perform ISR integration, conduct tactical reconnaissance, and conduct surveillance. (See FM 7-15, ART 1.3, and FM 3-90).

Perform Intelligence Synchronization

1-27. The intelligence officer, with staff participation, synchronizes the entire collection effort to include all assets the commander controls, assets of lateral units and higher echelon units and organizations, and intelligence reach to answer the commander's priority intelligence requirements (PIRs) and information requirements (IRs). (See FM 7-15, ART 1.3.1)

1-28. The intelligence officer, with staff participation, supports the G3/S3 in orchestrating the entire ISR effort to include all assets the commander controls, assets of lateral units and higher echelon units and organizations, and intelligence reach to answer the CCIRs (PIRs and friendly force information requirements [FFIRs]) and other intelligence requirements. Intelligence synchronization activities include the following:

- Conducting requirements management (RM): anticipate, develop, analyze, validate, and prioritize intelligence requirements. Recommend PIRs to the commander. Manage the commander's intelligence requirements, requests for information (RFIs) from subordinate and lateral organizations, and tasks from higher headquarters. Eliminate satisfied requirements and add new requirements as necessary.

- Developing indicators.
- Developing specific IRs (specific information requirements [SIRs]).
- Converting the SIRs into intelligence tasks or ISR tasks. (See Figure 1-2 for the ISR task development process.) The S2/G2 assigns intelligence production and reach tasks to subordinate intelligence elements or personnel, submits RFIs to higher and lateral echelons, and coordinates with (or assists) the G3/S3 to develop and assign ISR tasks.
- Comparing the ISR tasks to the capabilities and limitations of the available ISR assets (in coordination with the operations officer).
- Forwarding SIRs that cannot be answered by available assets to higher or lateral organizations as RFIs.
- Assessing collection asset reporting and intelligence production to evaluate the effectiveness of the ISR effort.
- Maintaining situational understanding to identify gaps in coverage and to identify the need to cue or redirect ISR assets.
- Updating the intelligence synchronization plan. The G2/S2 manages and updates the intelligence synchronization plan as PIRs are answered and new requirements arise.

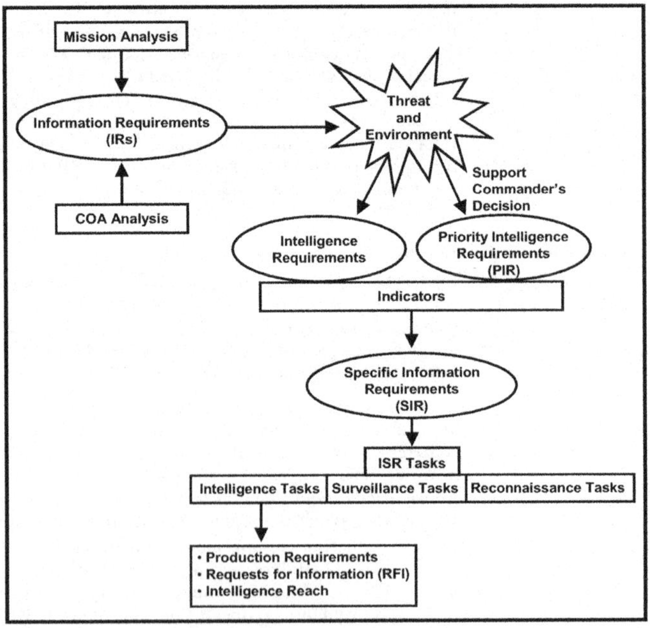

Figure 1-2. ISR Task Development Process.

1-29. **Intelligence Synchronization Considerations.** The G2/S2 generally follows six considerations in planning intelligence synchronization and ISR activities: anticipate, integrate, prioritize, balance, control, and reach. Refer to FM 2-01 for more information regarding intelligence synchronization.

- **Anticipate.** The intelligence staff must recognize when and where to shift collection or identify new intelligence requirements. The overall intent of this principle is to identify a new or adjust an existing requirement and present it to the commander for approval before waiting for the commander or staff to identify it.
- **Integrate.** The intelligence staff must be fully and continuously integrated into the unit's orders production and planning activities to ensure early identification of intelligence requirements. Early and continuous consideration of collection factors enhances the unit's ability to direct collection assets in a timely manner, ensures thorough planning, and increases flexibility in selecting assets.
- **Prioritize.** Prioritize each intelligence requirement based on its importance in supporting the commander's intent and decisions. Prioritization, based on the commander's guidance and the current situation, ensures that limited ISR assets and resources are directed against the most critical requirements.
- **Balance.** ISR capabilities complement each other. The intelligence staff should resist favoring or becoming too reliant on a particular unit, discipline, or system. Balance is simply planning redundancy, when required, eliminating redundancy when not desired, and ensuring an appropriate mix of ISR assets or types. The intelligence synchronization matrix (ISM) is useful in determining or evaluating balance.
- **Control.** To ensure timely and effective responses to intelligence requirements, a unit should first use ISR assets it controls. These assets usually are more responsive to their respective commander and also serve to lessen the burden on the ISR assets of other units, agencies, and organizations.
- **Reach.** Although usually not as responsive as a unit's own assets, intelligence reach may be the only way to satisfy an intelligence requirement. If at all possible, one should not depend solely on intelligence reach to answer a PIR.

1-30. **Develop IRs**. The intelligence staff develops a prioritized list of what information needs to be collected and produced into intelligence. Additionally, the intelligence staff dynamically updates and adjusts those requirements in response to mission adjustments and changes. This list is placed against a latest time intelligence is of value to ensure intelligence and information are reported to meet operational requirements. (See FM 7-15, ART 1.3.1) Figure 1-3 shows a comparison of IR, PIR, and intelligence requirements.

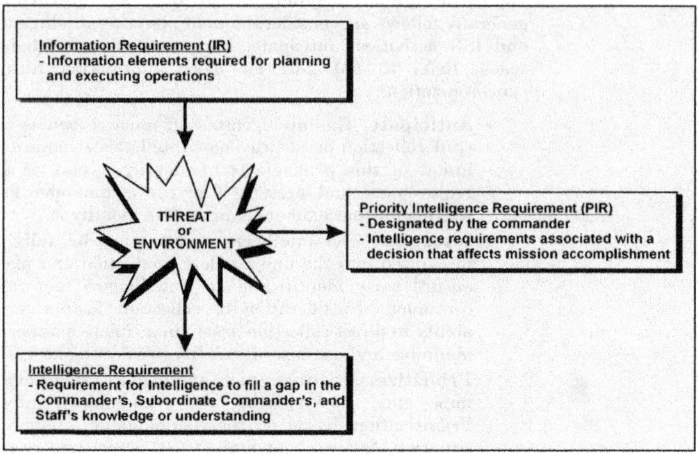

Figure 1-3. IR, PIR, and Intelligence Requirements Comparison.

1-31. An effective discussion of ISR has to include an understanding of the CCIRs. The CCIRs are elements of information required by commanders that directly affect decisionmaking and dictate the successful execution of military operations. The commander decides what information is critical based on the commander's experience, the mission, the higher commander's intent, and the staff's input (initial IPB, information, intelligence, and recommendations). Refer to FM 3-0 for more information regarding CCIRs.

1-32. Based on the CCIRs, two types of supporting IRs are generated: PIRs and FFIRs. However, commanders may determine that they need to know whether one or more essential elements of friendly information (EEFI) have been compromised or that the enemy is collecting against a designated EEFI. In those cases, commanders may designate that question as one of their CCIRs. Figure 1-4 shows the CCIR composition.

1-33. IRs are all of the information elements required by the commander and staff for the successful planning and execution of operations; that is, all elements necessary to address the factors of mission, enemy, terrain and weather, troops and support available, time available, civil considerations (METT-TC). Vetting by the commander or a designated representative turns an IR into either a PIR or an intelligence requirement. IRs are developed during COA analysis based on the factors of METT-TC.

1-34. PIRs are those intelligence requirements for which a commander has an anticipated and stated priority during the task of planning and decisionmaking. PIRs are associated with a decision based on an enemy action or inaction or the AO that will affect the overall success of the commander's mission. **The commander designates intelligence requirements tied directly to decisions as CCIR (PIR and FFIR).**

Answers to the PIRs help produce intelligence essential to the commander's situational understanding and decisionmaking. For information on PIR development, see FM 2-01.

1-35. The G2/S2 recommends to the commander those IRs produced during the MDMP that meet the criteria for PIR. They do not become CCIR (PIR and FFIR) until approved by the commander. Additionally, The commander may unilaterally designate PIRs. The IRs that are not designated by the commander as PIRs become intelligence requirements. The intelligence requirement is a gap in the command's knowledge or understanding of the AO or threat that the Intelligence warfighting function must fill.

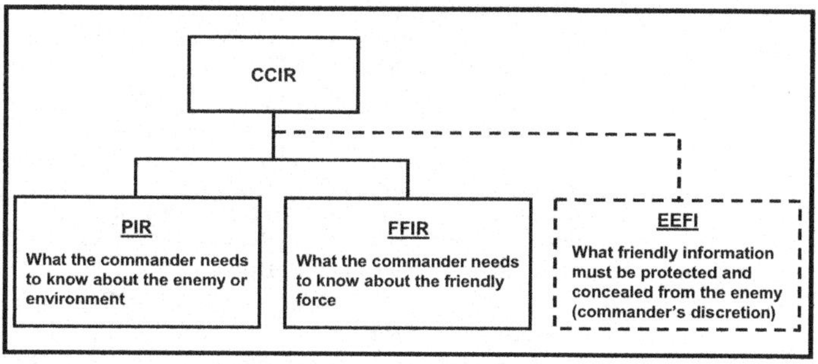

Figure 1-4. Commander's Critical Information Requirements Composition.

1-36. The G3/S3 then tasks the unit's assets to answer both the PIR and intelligence requirements through the ISR plan. PIR should—

- Ask only one question.
- Support a decision.
- Identify a specific fact, event, activity (or absence thereof) which can be collected.
- If linked to an ECOA, indicate an ECOA prior to, or as early as possible in, its implementation.
- Indicate the latest time the information is of value (LTIOV). The LTIOV is the absolute latest time the information can be used by the commander in making the decision the PIR supports. The LTIOV can be linked to time, an event, or a point in the battle or operation.

1-37. **Friendly Force IRs.** The staff also develops FFIRs which, when answered, provide friendly force information that the commander and staff need to achieve situational understanding and to make decisions.

1-38. **Essential Elements of Friendly Information**. EEFI establish information to protect, not information to obtain. However, commanders may determine that they need to know whether one or more EEFI have been compromised or that the enemy is collecting against a designated EEFI. In those cases, commanders may designate that question as one of their CCIRs, which generates PIRs and/or FFIRs. For example, a commander may determine that if the enemy discovers the location and movement of the friendly reserve, the operation is at risk. In this case, the location and movement of the friendly reserve are EEFI. The commander designates determining whether the enemy has discovered the location and movement of the friendly reserve as one of the CCIR. That CCIR, in turn, generates PIR and FFIR to support staff actions in determining whether the EEFI has been compromised.

1-39. **Develop the Intelligence Synchronization Plan.** The entire unit staff develops their IRs and determines how best to satisfy them. The staff uses reconnaissance and surveillance assets to collect information. The intelligence synchronization plan includes all assets that the operations officer can task or request and coordination mechanisms to ensure adequate coverage of the AOIs. (See FM 7-15, ART 1.3.1.2)

1-40. The intelligence synchronization plan, often presented in a matrix format as an ISM, aids in synchronizing the entire ISR effort with the overall operation and the commander's decisions and/or decision points (DPs). The intelligence synchronization plan is often produced in conjunction with the ISR plan. However, before performing intelligence synchronization and finalizing the intelligence synchronization plan, the G2/S2 must have—

- The CCIR (PIR and FFIR).
- A prioritized list of the remaining intelligence requirements.
- Evaluated ISR assets and resources.
- All of the assigned ISR tasks.

Perform ISR Integration

1-41. The operations officer, in coordination with the intelligence officer and other staff members, orchestrates the tasking and directing of available ISR assets to answer the CCIR. The operations officer, with input from the intelligence officer, develops tasks from the SIRs which coincide with the capabilities and limitations of the available ISR assets and the latest time information is of value (LTIOV). Intelligence requirements are identified, prioritized, and validated and an ISR plan is developed and synchronized with the scheme of maneuver. (See FM 7-15, ART 1.3.2, and FM 3-90)

1-42. The G3/S3, in coordination with the G2/S2 and other staff members, orchestrates the tasking and directing of available ISR assets to answer the CCIRs (PIRs and FFIRs) and IRs. The result of this process is the forming of the ISR plan. The ISR plan provides a list of all the ISR tasks to be accomplished. The G2/S2 and the G3/S3 develop tasks from the SIRs. These tasks are then assigned based on the capabilities and limitations of the available ISR assets and the LTIOV.

1-43. **Develop the ISR Plan.** The operations officer is responsible for developing the ISR plan. The entire unit staff analyzes each requirement to determine how best to satisfy it. The staff will receive orders and RFIs from both subordinate and adjacent units and higher headquarters. The ISR plan includes all assets that the operations officer can task or request and coordination mechanisms to ensure adequate coverage of the area of interest. (See FM 7-15, ART 1.3.2.1, and FM 3-90)

1-44. **The ISR Plan**. The ISR plan is usually produced as the ISR Annex to an OPORD (Annex L, Intelligence Surveillance, and Reconnaissance). Refer to FM 5-0 for specific and authoritative information on the ISR Annex. ISR is a continuous combined arms effort led by the operations and intelligence staffs in coordination with the entire staff that sets reconnaissance and surveillance in motion. The PIRs and other intelligence requirements drive the ISR effort. The commander takes every opportunity to improve situational understanding and the fidelity of the COP about the enemy and terrain through the deployment of ISR assets. Commanders integrate reconnaissance and surveillance to form an integrated ISR plan that capitalizes on their different capabilities. The ISR plan is often the most important part of providing information and intelligence that contributes to answering the CCIRs. For the G2/S2, an effective ISR plan is critical in answering the PIR. Also see FM 3-55 for more information on the ISR plan.

1-45. The ISR plan is not an MI-specific product—the G3/S3 is the staff proponent of the ISR plan— it is an integrated staff product executed by the unit at the direction of the commander. The G2/S2, however, must maintain situational understanding in order to recommend to the commander and G3/S3 changes or further development of the ISR plan. Based on the initial IPB and CCIRs, the staff—primarily the G2/S2—identifies gaps in the intelligence effort and develops an initial ISR plan based on available ISR assets. The G3/S3 turns this into an initial ISR Annex that tasks ISR assets as soon as possible to begin the collection effort.

1-46. The G3/S3, assisted by the G2/S2, uses the ISR plan to task and direct the available ISR assets to answer the CCIRs (PIRs and FFIRs) and intelligence requirements. Conversely, the staff revises the plan as other intelligence gaps are identified if the information is required to fulfill the CCIRs or in anticipation of future intelligence requirements. With staff participation, the G2/S2 intelligence officer synchronizes the ISR effort through a complementary product to the ISR plan—the intelligence synchronization plan.

1-47. **Execute and Update the ISR Plan**. The operations officer updates the ISR plan based on information received from the intelligence officer. The operations officer is the integrator and manager of the ISR effort through an integrated staff process and procedures. As PIRs are answered and new information requirements arise, the intelligence officer updates intelligence synchronization requirements and provides the new input to the operations officer who updates the ISR plan. He works closely with all staff elements to ensure the unit's organic collectors receive appropriate taskings. This ISR reflects an integrated collection strategy and employment, production and dissemination scheme that will effectively answer the commander's PIR. (See FM 7-15, ART 1.3.2.2, and FM 3-90)

Conduct Tactical Reconnaissance

1-48. To obtain, by visual observation or other detection methods, such as signals, imagery, measurement of signature or other technical characteristics, human interaction and other detection methods about the activities and resources of an enemy or potential enemy, or to secure data concerning the meteorological, hydrographic, or geographic characteristics and the indigenous population of a particular area. This task includes the conduct of NBC reconnaissance and the tactical aspects of SOF special reconnaissance. The five subtasks are—

- Conduct a Zone Reconnaissance.
- Conduct an Area Reconnaissance.
- Conduct a Reconnaissance in Force.
- Conduct a Route Reconnaissance.
- Conduct a Reconnaissance Patrol.

Note: This task branch includes techniques by which ART 1.1.4.1 (Collect Police Information) may be performed. (FM 7-15)

1-49. Reconnaissance is a mission undertaken to obtain by visual observation or other detection methods, information about the activities and resources of an enemy or potential enemy, and about the meteorological, hydrographic, or geographic characteristics of an AO. MI personnel and organizations can conduct reconnaissance through obtaining information derived from signals, imagery, measurement of signatures, technical characteristics, human interaction, and other detection methods. When performing reconnaissance, it is important to—

- Orient the reconnaissance asset on the named area of interest (NAI) and/or reconnaissance objective in a timely manner.
- Report all information rapidly and accurately.
- Complete the reconnaissance mission not later than (NLT) the time specified in the order.
- Answer the requirement that prompted the reconnaissance task.

Conduct Surveillance

1-50. To systematically observe the airspace, surface, or subsurface areas, places, persons, or things in the AO by visual, aural (audio), electronic, photographic, or other means. Other means may include but are not limited to space-based systems, and using special NBC, artillery, engineer, SOF, and air defense equipment. (See FM 7-15, ART 1.3.4, and FM 3-90)

Note: This task is a technique by which ART 1.1.4.1 (Collect Police Information) may be performed.

1-51. Conducting surveillance is systematically observing the airspace, surface, or subsurface areas, places, persons, or things in the AO by visual, aural, electronic, photographic, or other means. Surveillance activities include—

- Orienting the surveillance asset on the NAI and/or the surveillance objective in a timely manner.
- Reporting all information rapidly and accurately.
- Completing the surveillance mission NLT the time specified in the order.
- Answering the requirement that prompted the surveillance task.

PROVIDE INTELLIGENCE SUPPORT TO EFFECTS

1-52. The task of providing the commander information and intelligence support for targeting of the threat's forces, threat organizations, units and systems through lethal and non-lethal fires to include electronic attack and information operations. This task includes three subtasks: provide intelligence support to targeting, provide intelligence support to information operations, and provide intelligence support to combat assessment. (See FM 7-15, ART 1.4)

Provide Intelligence Support to Targeting

1-53. The intelligence officer, supported by the entire staff, provides the commander information and intelligence support for targeting the threat's forces and systems through employment of direct, indirect lethal and non-lethal fires. It includes identification of threat capabilities and limitations. The Intelligence warfighting function plays a crucial role by providing intelligence throughout the steps of the targeting process: decide, detect, deliver, assess. Support to targeting includes identifying threat capabilities and limitations. The G2/S2 supports the unit's lethal and non-lethal targeting effort through target development, target acquisition (TA), electronic warfare (EW), and combat assessment. This task has two subtasks:

- *Provide Intelligence Support to Target Development* is the systematic analysis of the enemy forces and operations to determine HVTs, systems, and system components for potential attack through maneuver, fires, or information.
- *Provide Intelligence Support to Target Detection* establishes procedures for dissemination of targeting information. The targeting team develops the sensor/attack system matrix to determine the sensor required to detect and locate targets. The intelligence officer places the following requirements into the integrated ISR plan.
  - Requires reconnaissance and surveillance operations to identify, locate, and track high-payoff targets (HPTs) for delivery of lethal or non-lethal effects.
  - Includes employing fires, offensive IO, and other attack capabilities against enemy C2 systems as part of the unit's FS plan and IO objectives.

Provide Intelligence Support to Information Operations

1-54. IO are actions taken to affect adversary information, influence other's decisionmaking processes and information systems while protecting one's own information and information systems. Overall operational continuity and mission success requires close, mutual coordination and synchronization of intelligence plans and operations with IO elements and related activities. Refer to Appendix A, this manual, for more information. This task has three subordinate tasks: (See FM 7-15, ART 1.4.2)

- Provide Intelligence Support to Offensive IO.
- Provide Intelligence Support to Defensive IO.
- Provide Intelligence Support to Activities Related to IO.

Provide Intelligence Support to Combat Assessment

1-55. Combat assessment is the determination of the overall effectiveness of force employment during military operations. The objective of combat assessment is to identify recommendations for the course of military operations. It answers the question, "Were the objectives met by force employment?" Although the assessment is primarily an intelligence responsibility, it requires input from and coordination with operations and fire support staffs. Combat assessment consists of conducting physical damage, functional damage, and target system assessments. (See FM 7-15, ART 1.4.3)

1-56. The staff determines how combat assessment relates to a specific target by conducting physical damage, functional damage, and target system assessments.

- **Conduct Physical Damage Assessment (PDA).** A PDA is an estimate of the extent of physical damage to a target based upon observed or interpreted damage. This post-attack target analysis is a coordinated effort among all units.
- **Conduct Functional Damage Assessment (FDA).** The FDA estimates the remaining functional or operational capability of a targeted facility or object. The staff bases FDA on the assessed physical damage and estimates of the threat's ability to recuperate (to include the time required to resume normal operations). Multiple echelons typically conduct this all-source analysis. The targeting or combat assessment cell integrates the initial target analyses with other sources, including intelligence, and then compares the original objective with the current status of the target to determine if the objective has been met.
- **Conduct Target System Assessment (TSA).** The TSA is an estimate of the overall impact of force employment against an adversary's target system. The unit, supported by higher echelon assets, normally conducts this assessment. The analyst combines all combat assessment reporting on functional damage to targets within a target system and assesses the overall impact on that system's capabilities. This process lays the groundwork for future recommendations for military operations in support of operational objectives.

1-57. **Munitions Effects Assessment (MEA).** MEA takes place concurrently and interactively with combat assessment since the same signatures used to determine the level of physical damage also give clues to munitions effectiveness. MEA is primarily the responsibility of operations and FS personnel, with input from the G2/S2. After the same weapon is used to attack several targets of a specific type, MEA should be accomplished to evaluate weapon performance. MEA analysts seek to identify through systematic trend analysis any deficiencies in weapon system and munitions performance or combat tactics by answering the question, "Did the weapons employed perform as expected?" Using a variety of input (targeting analysts, imagery analysts, structural engineers, and mission planners) analysts prepare a report assessing munitions performance. If combat troops capture attacked targets, it is then possible to collect detailed information on the target. Reports should detail weapon performance against specified target types. This information could have a crucial impact on future operations.

1-58. **Re-attack Recommendation**. Re-attack recommendations follow directly from both battle damage assessment (BDA) and MEA efforts. Basically re-attack recommendations answer the question, "Have we achieved the desired effects against our targeted objectives?" Evolving objectives, target selection, timing, tactics, weapons, vulnerabilities, and munitions are all factors in the new recommendations, combining both operations and intelligence functions.

1-59. **Targeting Meeting**. The role of the G2/S2 in targeting meetings is critical to ensuring intelligence supports the targeting process. The G2/S2 must come to the targeting meeting prepared. Additionally, the G2/S2 must understand the steps of the targeting process and tailor the intelligence products and G2/S2 participation according to the targeting process steps of: decide, detect, deliver, and assess.

1-60. There are two major areas in which the G2/S2 must prepare in order to support the targeting meeting.

- **Analysis:** Determine what the enemy is doing (current situation) and anticipated threat models; ECOA situation templates, ECOA sketches and statements, HVTs, and other associated products.
- **IPB Products:** Updated from the initial IPB effort, including analytical results (see above bullet), and tailored to the targeting requirements.

1-61. **Decide**. At the targeting meeting, the G2/S2 should be prepared to do the following:

- Brief the current and projected future enemy situation with an event template.
- Brief any combat assessments.
- Brief HVTs for potential selection of HPTs.
- Brief the current PIR portion of CCIRs, the PIRs of the CCIRs for the next phase or in accordance with the same timeframe of the targeting meeting.
- Brief the current ISM or intelligence synchronization plan.
- Participate in wargaming.
- Refine the initial ISR plan (in conjunction with the FS and G3/S3 representatives).

1-62. **Detect**. Based on information from the decide portion of the targeting meeting, the G2/S2 should be prepared to do the following:

- Identify specific ISR assets.
- Develop IRs and propose pertinent IRs to the commander for designation as PIRs.
- Complete the initial ISR plan; ensure it addresses the HPTs.

1-63. **Electronic Warfare**. EW provides targeting information to the commander based on transmitter location data. There are three major subdivisions within EW: electronic support (ES), electronic protection (EP), and electronic attack (EA).

- **ES** – Also referred to as electronic warfare support (EWS) in JP 1-02, involves actions to search for, intercept, identify, and locate or localize sources of intentional and unintentional radiated electromagnetic energy. In addition to supporting the overall ISR effort, ES is needed to produce intelligence required to support EA missions.
- **EP** – Involves passive and active means taken to protect personnel, facilities, and equipment from any effects of friendly or enemy employment of EW that degrades, neutralizes, or destroys friendly combat capabilities.
- **EA** – Involves the use of electromagnetic energy, directed energy, or anti-radiation weapons to attack personnel, facilities, or equipment with the intent of degrading, neutralizing, or destroying enemy combat capability and is considered a form of fires. EA includes actions taken to prevent or reduce an enemy's effective use of the electromagnetic spectrum (EMS), such as jamming and electromagnetic deception.

1-64. **Deliver**. The G2/S2 should be prepared to do the following:

- Brief how the Intelligence warfighting function will work to detect and track the target through the entire targeting process.
- Brief the implementation of the intelligence synchronization and ISR plans, including the LTIOV and communications structure.

1-65. **Assess**. Depending on the factors of METT-TC and the PIRs, the same ISR assets that detect and track the targets may be required to support combat assessments in order to determine if re-attack is required. Conversely, if re-attack is a decision that the commander will make based upon the effects on the enemy, combat assessment should be reflected in the PIR, and supported in the corresponding ISR plan.

THE OPERATIONAL ENVIRONMENT

1-66. The operational environment can be defined as a composite of the conditions, circumstances, and influences that affect the employment of capabilities and bear on the decision of the commander. The operational environment encompasses physical areas and factors of the air, land, maritime, and space domains. It also includes the information environment and enemy, adversary, friendly, and neutral systems.

OPERATIONAL VARIABLES (PMESII-PT)

1-67. Analysis of the operational environment—in terms of political, military, economic, social, infrastructure, and information with the addition of physical environment and time (PMESII-PT) variables—provides relevant information that senior commanders use to understand, visualize, and describe the operational environment. As a set, these operational variables are often abbreviated as PMESII-PT. A description of each of the PMESII-PT variables follows:

- **Political**. Describes the distribution of responsibility and power at all levels of governance or cooperation.
- **Military**. Explores the military capabilities of all relevant actors in a given operational environment.
- **Economic**. Encompasses individual behaviors and aggregate phenomena related to the production, distribution, and consumption of resources.
- **Social**. Describes the cultural, religious, and ethnic makeup within an operational environment.
- **Information**. Describes the nature, scope, characteristics, and effects of individuals, organizations, and systems that collect, process, disseminate, or act on information.
- **Infrastructure**. Is composed of the basic facilities, services, and installations needed for the functioning of a community or society.
- **Physical Environment**. Defines the physical circumstances and conditions that influence the execution of operations throughout the domains of air, land, sea, and space.
- **Time**. Influences military operations within an operational environment in terms of the decision cycles, operational tempo, and planning horizons. (FM 3-0)

1-68. For the intelligence portion, when identifying threats based on the systems approach to PMESII-PT, there are four primary components of the operational environment. These are the threat, terrain, weather, and civil considerations.

1-69. While the operational variables are directly relevant to campaign planning, they are too broad to be applied directly to tactical planning. That does not mean they are not of value at all levels; they are fundamental to developing the understanding of the operational environment necessary to conduct planning at any level, in any situation. The degree to which each operational variable provides useful information depends on the situation and echelon. For example, social and economic variables often receive close analysis as part of enemy and civil considerations at brigade and higher levels. They may affect the training and preparation of small units. However, they are not relevant to a small-unit leader's mission analysis. That leader may only be concerned with such questions as "Who is the tribal leader for this village?" "Is the electrical generator working?" "Does the enemy have antitank missiles?"

MISSION VARIABLES

1-70. Upon receipt of a warning order (WARNO) or mission, Army leaders narrow their focus to six mission variables. Mission variables are those aspects of the operational environment that directly affect a mission. They outline the situation as it applies to a specific Army unit. The mission variables are mission, enemy, terrain and weather, troops and support available, time available, and civil considerations (METT-TC). These are the categories of relevant information used for mission analysis. Army leaders use the mission variables to synthesize operational variables and tactical-level information with local knowledge about conditions relevant to their mission. The METT-TC categories are listed below:

- **Mission**. The mission is the task, together with the purpose, that clearly indicates the action to be taken and the reason.
- **Enemy**. Relevant information regarding the enemy may include the following:
  - Threat characteristics (previously order of battle [OB] factors).
    - Doctrine (or known execution patterns).
    - Personal habits and idiosyncrasies.
    - Equipment, capabilities, and vulnerabilities.
  - Threat COAs.
- **Terrain and Weather**. Terrain and weather are natural conditions that profoundly influence operations. Terrain and weather are neutral; they favor neither side unless one is more familiar with—or better prepared to operate in—the environment.
- **Troops and Support Available**. The number, type, capabilities, and condition of available friendly troops and support available. These include the resources available from joint, interagency, intergovernmental, multinational, host nation (HN), commercial (via contracting), and private organizations.
- **Time Available**. Time is critical to all operations. Controlling and exploiting time is central to initiative, tempo, and momentum. By exploiting time, commanders can exert constant pressure, control the relative speed of decisions and actions, and force exhaustion on enemy forces.
- **Civil Considerations**. Civil considerations comprise six characteristics expressed in the memory aid ASCOPE:
  - Areas.
  - Structures.
  - Capabilities.
  - Organizations.
  - People.
  - Events.

Note. For additional information on ASCOPE and the IPB process, see FM 2-01.3. Understanding the operational environment requires understanding the civil aspects of the AO. Civil considerations reflect how the manmade infrastructure, civilian institutions, and the attitudes and activities of the civilian leaders, populations, and organizations within an AO influence the conduct of military operations.

1-71. METT-TC enables leaders to synthesize operational level information with local knowledge relevant to their missions and tasks in a specified AO. Tactical and operational leaders can then anticipate the consequences of their operations before and during execution. See FM 3-0 for a detailed discussion of PMESII-PT and METT-TC.

THE INTELLIGENCE PROCESS

1-72. Intelligence operations generally include the five functions that constitute the intelligence process: **plan, prepare, collect, process,** and

produce. Additionally, there are three common tasks that occur across the five functions of the intelligence process: **analyze, disseminate,** and **assess**. The three common tasks are discussed after the last function, produce. The intelligence process functions are not necessarily sequential; this is what differentiates the Army's intelligence process from the Joint intelligence cycle. The intelligence process provides a common model with which to guide one's thinking, discussing, planning, and assessing about the threat or AOI environment. The intelligence process generates information about the threat, the AOI, and the situation, which allows the commander and staff to develop a plan, seize and retain the initiative, build and maintain momentum, and exploit success.

PLAN

1-73. The plan step consists of the activities that identify pertinent IRs and develop the means for satisfying those requirements. The CCIRs (PIRs and FFIRs) drive the planning of the ISR effort. The intelligence officer supports the G3/S3 in arranging the ISR effort, based on staff planning, to achieve the desired collection effects. Planning activities include, but are not limited to—

- Conducting IPB.
- Managing requirements.
- Submitting RFIs and using intelligence reach to fill information gaps.
- Evaluating reported information.
- Establishing the intelligence communications and dissemination architecture.
- Developing, managing, and revising the intelligence synchronization plan and the ISR plan as mission requirements change.

PREPARE

1-74. The failure of MI units or personnel to accomplish their tasks or missions can often be attributed to their failure to prepare. The prepare step includes those staff and leader activities which take place upon receiving the operation order (OPORD), operation plan (OPLAN), warning order (WARNO), or commander's intent in order to improve the unit's ability to execute tasks or missions.

1-75. The most habitual and egregious preparation failures committed by leaders (as evidenced by performance at the combat training centers) is not the conduct of specific tasks, but the failure to adequately coordinate for more generic combat requirements such as—

- Friendly forward unit liaison.
- Departing or reentering friendly lines or AOs.
- Fire support: enroute, at mission location, and return.
- Casualty evacuation procedures.
- Passwords (running, forward of friendly lines), recognition signals, call signs, and frequencies.
- Resupply.
- Movement through friendly AOs.

COLLECT

1-76. The collect step involves collecting and reporting information in response to ISR tasking. ISR assets collect information and data about threat forces, activities, facilities, and resources as well as information concerning the environmental and geographical characteristics of a particular AO. A successful ISR effort results in the timely collection and reporting of relevant and accurate information. This collected information forms the foundation of intelligence databases, intelligence production, and the G2's/S2's situational understanding. The requirements manager evaluates the reported information for its responsiveness to the CCIRs (PIRs and FFIRs).

PROCESS

1-77. Processing involves converting collected data and information into a form that is suitable for analyzing and producing intelligence. Examples of processing include enhancing imagery, translating a document from a foreign language, converting electronic data into a standardized report that can be analyzed by a system operator, and correlating information.

1-78. Processing data and information is performed unilaterally and cooperatively by both humans and automated systems.

PRODUCE

1-79. The produce step involves evaluating, analyzing, interpreting, synthesizing, and combining information and intelligence from single or multiple sources into intelligence or intelligence products in support of known or anticipated requirements. Production also involves combining new information and intelligence with existing intelligence in order to produce intelligence in a form that the commander and staff can apply to the MDMP and that supports and helps facilitate situational understanding. During the produce step, the intelligence staff manipulates information by—

- Analyzing the information to isolate significant elements.
- Evaluating the information to determine accuracy, timeliness, usability, completeness, precision, and reliability.
- Combining the information with other relevant information and previously developed intelligence.
- Applying the information to estimate possible outcomes.
- Presenting the information in a format that will be most useful to its eventual user.

1-80. The intelligence staff deals with numerous and varied production requirements based on PIRs and intelligence requirements; diverse missions, environments, and situations; and presentation requirements. Through analysis, collaboration, and intelligence reach, the G2/S2 and the staff use the collective intelligence production capability of higher, lateral, and subordinate echelons to meet the production requirements. Proficiency in these techniques and procedures facilitates the intelligence staff's ability to answer command and staff requirements regardless of the factors of METT-TC.

1-81. The three common tasks, discussed below, can occur throughout the intelligence process.

Analyze

1-82. The intelligence staff analyzes intelligence and information about the enemy's capabilities, friendly vulnerabilities, and the battlefield environment as well as issues and problems that arise within the intelligence process itself to determine their nature, origin, and interrelationships. This analysis enables commanders, staffs, and leaders to determine the appropriate action or reaction and to focus or redirect assets and resources to fill information gaps or alleviate pitfalls. It is also within the analyze function that intelligence analysts sort through large amounts of collected information and intelligence to obtain only that information which pertains to the CCIRs (PIRs and FFIRs), maintenance of the COP, and facilitates the commander's situational understanding.

Disseminate

1-83. Disseminating is communicating relevant information of any kind from one person or place to another in a usable form by any means to improve understanding or to initiate or govern action. Disseminating intelligence entails using information management techniques and procedures to deliver timely, relevant, accurate, predictive, and usable intelligence to the commander. Determining the product format and selecting the means to deliver it are key aspects of dissemination. Information presentation may be in a verbal, written, interactive, or graphic format. The type of information, the time allocated, and the individual preference of the commander all influence the information format.

Assess

1-84. Assessment plays an integral role in all aspects of the intelligence process. Assessing includes evaluating the effectiveness of intelligence in supporting the operation. Assessing the situation and available information begins upon receipt of the mission and continues throughout the intelligence process. The continual assessment of intelligence operations and ISR efforts, available information and intelligence, and various aspects of the factors of METT-TC are critical to ensure—

- The G2/S2 answers all CCIRs (PIRs and FFIRs).
- The G2/S2 and G3/S3 redirect ISR assets to support changing requirements.
- Using information and intelligence properly.

INTELLIGENCE DISCIPLINES

1-85. Intelligence disciplines are categories of intelligence functions. The Army's intelligence disciplines are All-Source Intelligence, Human Intelligence (HUMINT), Geospatial Intelligence (GEOINT), Imagery Intelligence (IMINT), Signals Intelligence (SIGINT), Measurement and Signatures Intelligence (MASINT), Open-Source Intelligence (OSINT), Technical Intelligence (TECHINT), and Counterintelligence (CI). For more information regarding the intelligence disciplines, see Part Three of this manual, as well as the respective manuals, which covers each individual intelligence discipline.

ALL-SOURCE INTELLIGENCE

1-86. All-source intelligence is defined as the intelligence products, organizations, and activities that incorporate all sources of information and intelligence, including open-source information, in the production of intelligence. All-source intelligence is a separate intelligence discipline, as well as the name of the task used to produce intelligence from multiple intelligence or information sources.

HUMAN INTELLIGENCE

1-87. HUMINT is the collection of foreign information—by a trained HUMINT Collector—from people and multimedia to identify elements, intentions, composition, strength, dispositions, tactics, equipment, personnel, and capabilities. It uses human sources as a tool, and a variety of collection methods, both passively and actively, to collect information.

GEOSPATIAL INTELLIGENCE

1-88. Title 10 U.S. Code § 467 establishes GEOINT. The definition of geospatial intelligence is—

> *The exploitation and analysis of imagery and geospatial information to describe, assess, and visually depict physical features and geographically referenced activities on the Earth. GEOINT consists of imagery, imagery intelligence, and geospatial information.*

1-89. The Army implementation of GEOINT is a result of the Army's organization, manning, and training. There are multiple types of data and information that various Army units and organizations collect, provide, and analyze in order to support the GEOINT enterprise. The two primary GEOINT service providers in the Army are MI units and organizations and Engineer (topographic) units and organizations. MI units and organizations provide imagery and IMINT to the enterprise. Engineer (topographic) units and organizations provide geospatial data and information to the enterprise. Therefore, while some of the collection, analysis, and exploitation of imagery and geospatial information occur within the intelligence warfighting function; some of the collection, analysis, and exploitation of imagery and geospatial information occur outside intelligence.

IMAGERY INTELLIGENCE

1-90. IMINT is intelligence derived from the exploitation of imagery collected by visual photography, infrared, lasers, multi-spectral sensors, and radar. These sensors produce images of objects optically, electronically, or digitally on film, electronic display devices, or other media.

SIGNALS INTELLIGENCE

1-91. SIGINT is a category of intelligence comprising either individually or in combination all communications intelligence (COMINT), electronic intelligence (ELINT), and foreign instrumentation signals intelligence (FISINT), however transmitted. SIGINT is derived from communications, electronics, and foreign instrumentation signals.

MEASUREMENT AND SIGNATURES INTELLIGENCE

1-92. MASINT is technically derived intelligence that detects, locates, tracks, identifies, and/or describes the specific characteristics of fixed and dynamic target objects and sources. It also includes the additional advanced processing and exploitation of data derived from IMINT and SIGINT collection. MASINT collection systems include but are not limited to radar, spectroradiometric, electro-optical (E-O), acoustic, radio frequency (RF), nuclear detection, and seismic sensors, as well as techniques for gathering chemical, biological, radiological, and nuclear (CBRN), and other material samples.

OPEN-SOURCE INTELLIGENCE

1-93. The *National Defense Authorization Act for Fiscal Year 2006* states:

> *Open source intelligence is produced from publicly available information that is collected, exploited, and disseminated in a timely manner to an appropriate audience for the purpose of addressing a specific intelligence requirement.*

1-94. Expressed in terms of the Army intelligence process, OSINT is relevant information derived from the systematic collection, processing, and analysis of publicly available information in response to intelligence requirements.

1-95. The Army does not have a specific military occupational specialty (MOS), additional skill identifier (ASI), or special qualification identifier (SQI) for OSINT. With the exception of the Asian Studies Detachment, the Army does not have base tables of organization and equipment (TOE) for OSINT units or staff elements. OSINT missions and tasks are imbedded within existing missions and force structure or accomplished through task organization.

TECHNICAL INTELLIGENCE

1-96. TECHINT is intelligence derived from the collection and analysis of threat and foreign military equipment and associated materiel for the purposes of preventing technological surprise, assessing foreign scientific and technical (S&T) capabilities, and developing countermeasures designed to neutralize an adversary's technological advantages.

COUNTERINTELLIGENCE

1-97. CI counters or neutralizes intelligence collection efforts through collection, counterintelligence investigations, operations, analysis, and production, and functional and technical services. CI includes all actions taken to detect, identify, track, exploit, and neutralize the multidiscipline intelligence activities of friends, competitors, opponents, adversaries, and enemies; and is the key intelligence community contributor to protect US interests and equities. CI assists in identifying EEFI, identifying vulnerabilities to threat collection, and actions taken to counter collection and operations against US forces.

UNIFIED ACTION

1-98. Unified action is the synchronization, coordination, and/or integration of the activities of governmental and nongovernmental entities with military operations to achieve unity of effort. (JP 1) Joint operations focus and maximize the complementary and reinforcing effects and capabilities of each service. Joint force commanders (JFCs) synchronize the complementary capabilities of the service components that comprise the joint force.

1-99. The employment of MI in campaigns and major operations must be viewed from a joint perspective, and the "mud-to-space" intelligence concept must establish a fully interoperable and integrated joint intelligence capability. ARFOR intelligence assets work with multinational and interagency partners to accomplish their missions. Ideally, multinational and interagency intelligence partners provide cultures, perspectives, and capabilities that reinforce and complement Army MI strengths and capabilities. Close intelligence coordination is the foundation of successful unified action. (See chapter 2.)

1-100. A critical part of current operations is the execution of the joint doctrinal concept of persistent surveillance. Joint doctrine defines persistent surveillance as a collection strategy that emphasizes the ability of some collection systems to linger on demand in an area to detect, locate, characterize, identify, track, target, and possibly provide battle damage assessment and re-targeting in real or near-real time. Persistent surveillance facilitates the formulation and execution of preemptive activities to deter or forestall anticipated adversary courses of action (JP 2-0).

THE NATURE OF LAND OPERATIONS

1-101. Landpower is the ability—by threat, force, or occupation—to gain, sustain, and exploit control over land, resources, and people (FM 3-0). Army operations reflect expeditionary and campaign capabilities that constantly adapt to each campaign's unique circumstances. Expeditionary capabilities require forces organized to be modular, versatile, and rapidly deployable. Rapidly deployed expeditionary force packages provide immediate options for seizing or retaining the operational initiative and also allow the conduct of sustained operations for as long as necessary. Army forces are organized, trained, and equipped for endurance. The Army's preeminent challenge is to balance expeditionary agility and responsiveness with the endurance and adaptability needed to complete a campaign successfully.

1-102. The capability to prevail in close combat is indispensable and unique to land operations. The outcome of battles and engagements depends on Army forces' ability to prevail in close combat. Many factors inherent in land combat combine to complicate the situation. These factors include chaos, complexity, insufficient intelligence, errors in understanding or planning, difficult terrain, the civilian population, and an adaptive and lethal enemy. The axiom "intelligence drives operations" continues to be true especially for land operations; operations and intelligence are inextricably linked.

1-103. Four considerations are preeminent in generating expeditionary capabilities and Army force packages.

- **Scope.** Considers and strives to understand an enemy throughout the depth of an operational area. Commanders rely on intelligence in order to use maneuver, fires, and other elements of combat power to defeat or destroy enemy forces.
- **Duration.** Forces routinely conduct missions prior to, during, and after the commitment of land combat forces. Intelligence is always engaged.
- **Terrain.** Missions occur among a complex variety of natural and manmade features. Employing forces in the complexity of the ground environment requires thorough planning.
- **Permanence.** Forces are integrated with, or assigned to, land combat forces as they seize or secure ground.

1-104. Several attributes of the land environment affect the application of landpower. These attributes include—

- The requirement to deploy and employ Army forces rapidly.
- The requirement for Army forces to operate for protracted periods.
- The nature of close combat.
- Uncertainty, chance, friction, and complexity.

1-105. Reconnaissance, Surveillance, and Target Acquisition (RSTA)/ISR is the means the Army uses to implement the joint doctrinal concept of persistent surveillance in support of tactical operations. Dependable technology and responsive intelligence lessen the effects of uncertainty, chance, friction, and complexity. Complex and dynamic Army tactical operations require extensive ISR capabilities to satisfy the commander's information requirements to detect, locate, characterize, identify, track, and target HPTs, and to provide combat assessment in real time within a very fluid operational environment.

1-106. *Reconnaissance, Surveillance, and Target Acquisition/ Intelligence, Surveillance, and Reconnaissance* is a full spectrum combined arms mission that integrates ground and air capabilities to provide effective, dynamic, timely, accurate, and assured combat information and multidiscipline actionable intelligence for lethal and non-lethal effects and decisions in direct support of the ground tactical commander.

FORCE PROJECTION OPERATIONS

1-107. Force projection is the military component of power projection. It is a central element of the national military strategy. Army organizations and installations, linked with joint forces and industry, form a strategic platform to maintain, project, and sustain ARFOR, wherever they deploy. Force projection operations are inherently joint and require detailed planning and synchronization. Force projection encompasses a range of processes—mobilization, deployment, employment, sustainment, and redeployment—discussed below.

1-108. The Army must change its mindset from depending on an "intelligence buildup" to performing intelligence readiness on a daily basis in order to meet the requirements for strategic responsiveness. MI personnel, even in garrison at the lowest tactical echelons, must use their analytic and other systems and prepare for possible operations on a daily basis.

1-109. Built on a foundation of intelligence readiness, the Intelligence warfighting function provides the commander with the intelligence needed to plan, prepare, and execute force projection operations. Successful intelligence during force projection operations relies on continuous collection and intelligence production before and during the operation. In a force projection operation, higher echelons will provide intelligence to lower echelons until the tactical ground force completes entry and secures the lodgment area. The joint force J2 must exercise judgment when providing information to subordinate G2s/S2s so not to overwhelm them. Figure 1-5 is an example of force projection intelligence.

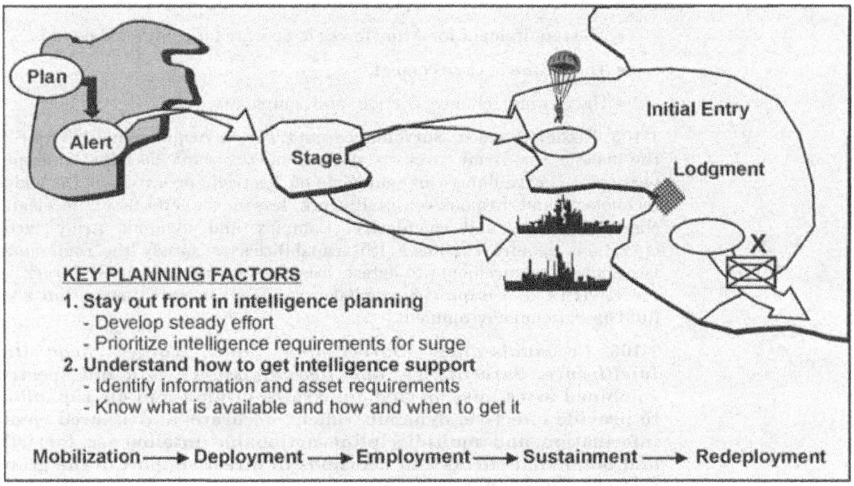

Figure 1-5. Force Projection Intelligence.

1-110. The G2/S2 must anticipate, identify, consider, and evaluate all threats to the entire unit throughout force projection operations. This is critical during the deployment and entry operations stages of force projection. During these stages, the unit is particularly vulnerable to enemy actions because of its limited combat power and knowledge of the AO. Intelligence personnel must, therefore, emphasize the delivery of combat information and intelligence products that indicate changes to the threat or AO developed during predeployment IPB.

1-111. The G2/S2 should—

- Review available databases on assigned contingency AOIs, conduct IPB on these AOIs, and develop appropriate IPB products.
- Comply with higher headquarters standing operating procedures (SOPs) and manuals for specific intelligence operations guidance.
- Coordinate for and rehearse electronic message transfers (for example, Internet Protocol addresses, routing indicators) using the same communications protocols with theater, higher headquarters, subordinate, and lateral units that the unit would use when deployed.
- Plan, train, and practice surging intelligence functions on likely or developing contingency crises.
- Prepare and practice coordination from predeployment through redeployment with other elements and organizations (for example, HUMINT, IMINT, SIGINT, MASINT, IO, staff weather officer [SWO], civil affairs [CA], psychological operations [PSYOP], and special operations forces [SOF] units, to include databases and connectivity).
- Include the following as a part of daily (sustainment) operations:
  - US Army Reserve and other augmentation.
  - A linguist plan with proficiency requirements. (Alert linguists through early entry phases of deployment.)
  - Training (individual and collective).
- Establish formal or informal intelligence links, relationships, and networks to meet developing contingencies.
- Forward all RFIs to higher headquarters in accordance with SOPs.
- Establish statements of intelligence interests (SIIs), other production, and I&W requirements.

1-112. To draw intelligence from higher echelons and focus intelligence downward, based on the commander's needs, the G2/S2 must—

- Understand the J2's multiple echelon and broadcast dissemination capability to ensure near-real time (NRT) reporting to all deployed, in transit, or preparing to deploy forces.
- Maintain or build intelligence databases on the environment and threats for each probable contingency.
- State and record the CCIR (as a minimum, list the PIRs and ISR tasks or requests).

1-113. Until the unit's collection assets become operational in the AO, the G2/S2 will depend upon intelligence from the ARFOR or JTF to answer the unit's intelligence needs. The following paragraphs describe the intelligence and ISR considerations during force projection.

MOBILIZATION

1-114. Mobilization is the process by which the armed forces or part of them are brought to a state of readiness for war or other national emergency. It assembles and organizes resources to support national objectives. Mobilization includes activating all or part of the US Army Reserve, and assembling and organizing personnel, supplies, and materiel. A unit may be brought to a state of readiness for a specific mission or other national emergency. This process, called mobilization, is where specific US Army

Active or Reserve units, capabilities, and personnel are identified and integrated into the unit. During mobilization, the G2/S2 must—

- Monitor intelligence reporting on threat activity and I&W data.
- Manage IRs and RFIs from their unit and subordinate units to include updating ISR planning.
- Establish habitual training relationships with their US Army Active and Reserve augmentation units and personnel as well as higher echelon intelligence organizations as identified in the existing OPLAN. Support the US Army Reserve units and augmentation personnel by preparing and conducting intelligence training and threat update briefings and by disseminating intelligence.
- Identify ISR force requirements for the different types of operations and contingency plans (CONPLANs).
- Identify individual military, civilian, and contractor augmentation requirements for intelligence operations. The Army, and the Intelligence warfighting function in particular, cannot perform its missions without the support of its Department of the Army Civilians (DACs) and contractors. The force increasingly relies on the experience, expertise, and performance of non-uniformed personnel and has fully integrated these non-uniformed personnel into the warfighting team.

1-115. During mobilization the G2/S2, in conjunction with the rest of the staff, must ensure the adequacy of training and equipping of US Army Active and Reserve MI organizations and individual augmentees to conduct intelligence operations.

1-116. The G2/S2 supports peacetime contingency planning with IPB products and databases on likely contingency areas. The G2/S2 establishes an intelligence synchronization plan that will activate upon alert notification. For smooth transition from predeployment to entry, the G2/S2 must coordinate intelligence synchronization and communications plans before the crisis occurs. The intelligence synchronization plan identifies the intelligence requirements supporting those plans, to include—

- ISR assets providing support throughout the AOI.
- Command and support relationships of ISR assets at each echelon.
- Report and request procedures not covered in unit SOPs.
- Sequence of deployment of ISR personnel and equipment. Early deployment of key ISR personnel and equipment is essential for force protection and combat readiness. Composition of initial and follow-on deploying assets is influenced by the factors of METT-TC, availability of communications, and availability of lift.
- Communications architecture supporting both intelligence staffs and ISR assets.
- Friendly vulnerabilities to hostile intelligence threats and plans for conducting FP measures. The staff must begin this type of planning as early as possible to ensure adequate support to FP of deploying and initial entry forces.
- Monitor time-phased force and deployment data (TPFDD) and recommend changes in priority of movement, unit, or capability to enable ISR operations.

1-117. The G2/S2 must continually monitor and update the OPLANs to reflect the evolving situation, especially during crisis situations. National intelligence activities monitor regional threats throughout the world and can answer some intelligence requirements supporting the development of OPLANs.

1-118. Upon alert notification, the G2/S2 updates estimates, databases, IPB products, and other intelligence products needed to support command decisions on force composition, deployment priorities and sequence, and the AOI. Units reassess their collection requirements immediately after alert notification. The G2/S2 begins verifying planning assumptions within the OPLANs. CI and ISR personnel provide FP support and antiterrorism measures.

1-119. Throughout mobilization, unit intelligence activities will provide the deploying forces with the most recent intelligence on the AO. The intelligence staff will also update databases and situation graphics. The G2/S2 must—

- Fully understand the unit, ARFOR, and joint force intelligence organizations.
- Revise intelligence and intelligence-related communications architecture and delete or integrate any new systems and software with the current architecture.
- Support 24-hour operations and provide continuous intelligence.
- Plan all required intelligence reach procedures.
- Determine transportation availability for deployment and availability when deployed.
- Determine all sustainability requirements.
- Determine intelligence release requirements and restrictions; releasability to multinational and HN sources.
- Review status of forces agreements (SOFAs), rules of engagement (ROE), international laws, and other agreements, emphasizing the effect that they have on intelligence collection operations. (Coordinate with the staff judge advocate [SJA] on these issues.)
- Ensure ISR force deployment priorities are reflected in the TPFDD to support ISR operations based upon the factors of METT-TC.
- Ensure intelligence links provide the early entry commander vital access to multi-source army and joint intelligence collection assets, processing systems, and databases.
- Review the supporting unit commanders' specified tasks, implied tasks, task organization, scheme of support, and coordination requirements with forward maneuver units. Address issues or shortfalls and direct or coordinate changes.
- Establish access to national HUMINT, IMINT, SIGINT, MASINT, and CI databases, as well as automated links to joint service, multinational, and HN sources.

DEPLOYMENT

1-120. Deployment is the movement of forces and materiel from their point of origin to the AO. This process has four supporting components:

predeployment activities, fort to port, port to port, and port to destination; these components are known collectively as reception, staging, onward movement, and integration (RSO&I) activities. Success in force projection operations hinges on timely deployment. The size and composition of forces requiring lift are based on the factors of METT-TC, availability of prepositioned assets, the capabilities of HN support, and the forward presence of US forces. Force or tactical tailoring is the process used to determine the correct mix and sequence of deploying units.

1-121. During deployment, intelligence organizations at home station or in the rear area take advantage of modern SATCOM, broadcast technology, and automated data processing (ADP) systems to provide graphic and textual intelligence updates to the forces enroute. Enroute updates help eliminate information voids and, if appropriate, allow the commander to adjust the plan prior to arrival in theater in response to changes in the OE or enemy actions.

1-122. Intelligence units extend established networks to connect intelligence staffs and collection assets at various stages of the deployment flow. Where necessary, units establish new communications paths to meet unique demands of the mission. The theater and corps analysis and control elements (ACEs) play a critical role in making communications paths, networks, and intelligence databases available to deploying forces.

1-123. Space-based systems are key to supporting intelligence during the deployment and the subsequent stages of force projection operations by—

- Monitoring terrestrial AOIs through ISR assets to help reveal enemy location and disposition, attempting to identify the enemy's intent.
- Providing communications links between forces enroute and in the continental United States (CONUS).
- Permitting MI collection assets to accurately determine their position through the Global Positioning System (GPS).
- Providing timely and accurate data on meteorological, oceanographic, and space environmental factors that might affect operations.
- Providing warning of theater ballistic missile launches.
- Providing timely and accurate weather information to all commanders through the Integrated Meteorological System (IMETS).

1-124. Situation development dominates intelligence operations activities during initial entry operations. The G2/S2 attempts to identify all threats to arriving forces and assists the commander in developing FP measures. During entry operations, echelons above corps (EAC) organizations provide intelligence. This support includes providing access to departmental and joint intelligence and deploying scalable EAC intelligence assets. The entire effort focuses downwardly to provide tailored support to deploying and deployed echelons in response to their CCIRs (PIRs and FFIR).

1-125. Collection and processing capabilities are enhanced, as collection assets build up in the deployment area, with emphasis on the build-up of the in-theater capability required to conduct sustained ISR operations. As the build-up continues, the G2/S2 strives to reduce total dependence on extended split-based intelligence from outside the AO. As assigned collection assets arrive into the theater, the G2/S2 begins to rely on them for tactical intelligence although higher organizations remain a source of intelligence.

1-126. As the ARFOR enter the theater of operations, the joint force J2 implements and, where necessary, modifies the theater intelligence architecture. Deploying intelligence assets establishes liaison with staffs and units already present in the AO. Liaison personnel and basic communications should be in place prior to the scheduled arrival of parent commands. ISR units establish intelligence communications networks.

1-127. CONUS and other relatively secure intelligence bases outside the AO continue to support deployed units. Systems capable of rapid receipt and processing of intelligence from national systems and high capacity, long-haul communications systems are critical to the success of split-based support of a force projection operation. These systems provide a continuous flow of intelligence to satisfy many operational needs.

1-128. The G2/S2, in coordination with the G3/S3, participates in planning to create conditions for decisive operations. The G2/S2 also adjusts collection activities as combat strength builds. During entry operations the G2/S2—

- Monitors FP indicators.
- Monitors the ISR capability required to conduct sustained intelligence operations.
- Monitors intelligence reporting on threat activity and I&W data.
- Develops measurable criteria to evaluate the results of the intelligence synchronization plan.
- Assesses—
  - Push versus pull requirements of intelligence reach.
  - Effectiveness of the intelligence communications architecture.
  - Reporting procedures and timelines.
  - Intelligence to OPLANs and OPORDs, branches, and sequels (to include planning follow-on forces).

EMPLOYMENT

1-129. Employment is the conduct of operations to support a JFC commander. Employment encompasses an array of operations, including but not limited to—

- Entry operations (opposed or unopposed).
- Shaping operations (lethal and non-lethal).
- Decisive operations (combat or support).
- Postconflict operations (prepare for follow-on missions or redeployment).

Entry Operations

1-130. Enemies often possess the motives and means to interrupt the deployment flow of ARFOR. Threats to deploying forces may include advanced conventional weaponry (air defense, mines, etc.) and WMD. Sea and air ports of debarkation (PODs) should be regarded as enemy HPTs because they are the entry points for forces and equipment. PODs are vulnerable because they are fixed targets with significant machinery and equipment that is vulnerable to attack; in addition to military forces and materiel, HN support personnel, contractors, and civilians may all be working there. An enemy attack, or even the threat of an enemy attack, on a POD can have a major impact on force projection momentum. Commanders at all levels require predictive intelligence so that they may focus attention on security actions that reduce vulnerabilities. To avoid, neutralize, or counter threats to entry operations, the commanders rely on the ability of the G2/S2 to support future operations by accurately identifying enemy reactions to US actions, anticipating their response to our counteractions and predicting additional ECOAs.

1-131. Predictive intelligence also supports the decisions the commander and staff must make about the size, composition, structure, and deployment sequence of the force in order to create the conditions for success. Commanders rely on predictive intelligence to identify potential friendly decisions before the actual event. While thorough planning develops friendly COAs to meet possible situations, the nature of an operation can change significantly before execution. The G2/S2 must provide timely, accurate, and predictive intelligence to ensure the commander can retain the initiative to implement the plan or make decisions before losing the opportunity to do so.

Shaping Operations

1-132. Shaping operations create and preserve conditions for the success of the decisive operation. Shaping operations include lethal and non-lethal activities conducted throughout the AO. They support the decisive operation by affecting enemy capabilities and forces, or by influencing enemy decisions. G2/S2 intelligence analysis and ISR activities support the development and execution of shaping operations by identifying threat centers of gravity and decisive points on the battlefield. The G2/S2 also ensures the intelligence process focuses on the CCIRs (PIRs and FFIRs). It is critical that the G2/S2 provide timely, accurate, and predictive intelligence to the commander to facilitate the execution of shaping operations. Predictive intelligence should provide sufficient time for the commander to understand how the enemy will react to US COAs so that appropriate shaping operations can be implemented.

Decisive Operations

1-133. Decisive operations are those that directly accomplish the task assigned by the higher headquarters. Decisive operations conclusively determine the outcome of major operations, battles, and engagements. Continuously synchronized ISR activities, coupled with predictive intelligence results and products during all stages of force projection, combine to ensure the commander is prepared to employ the right forces with the right support to conduct decisive operations at the most appropriate place and time. In addition to coordinating with the G3/S3 to synchronize all ISR activities, the G2/S2 must answer the CCIRs (PIRs and FFIRs) and provide continuous assessments of the enemy's current situation and predict the enemy's subsequent COA, branches, and/or sequels.

Postconflict Operations

1-134. Upon cessation of hostilities or truce, deployed forces enter a new stage of force projection operations. Postconflict operations focus on restoring order, reestablishing HN infrastructure, preparing for redeployment of forces, and planning residual presence of US forces. While postconflict operations strive to transition from conflict to peace, there remains a possibility of resurgent hostilities by individuals and forces. The ISR effort, particularly predictive intelligence analysis, remains just as critical in postconflict operations as in the other employment operations. ISR operations support the postconflict emphasis on restoration operations exemplified by commanders redirecting their CCIRs (PIRs and FFIRs) and IRs to support units conducting these missions. These operations might include—

- Engineer units conducting mine clearing or infrastructure reconstruction operations.
- Medical and logistics units providing humanitarian relief.
- MP units providing law and order assistance.
- CA units reestablishing local control and preparing for the orderly transition of local governments and to support civil-military operations (CMO).

SUSTAINMENT

1-135. Sustainment involves providing and maintaining levels of personnel and materiel required to sustain the operation throughout its duration. It is essential to generating combat power. CSS may be split-based between locations within and outside CONUS. These operations include ensuring units have the MI assets required to accomplish the mission, such as personnel (including linguists), communications systems, ISR systems, and appropriate maintenance support.

REDEPLOYMENT

1-136. Redeployment is the process by which units and materiel reposture themselves in the same theater; transfer forces and materiel to support another JFC's operational requirements; or return personnel and materiel to the home or demobilization station upon completion of the mission. Redeployment operations encompass four phases:

- Recovery, reconstitution, and pre-deployment activities.
- Movement to and activities at the port of embarkation (POE).
- Movement to the POD.
- Movement to home station.

1-137. As combat power and resources decrease in the AO, FP and I&W become the focus of the commander's intelligence requirements. This in turn drives the selection of those assets that must remain deployed until the end of the operation and those that may redeploy earlier. The S2—

- Monitors intelligence reporting on threat activity and I&W data.
- Continues to conduct intelligence to FP.
- Requests ISR support (theater and national systems) and intelligence in support of redeployment.

1-138. After redeployment, MI personnel and units recover and return to predeployment activities. ISR units resume contingency-oriented peacetime intelligence operations. US Army Reserve ISR units demobilize and return to peacetime activities. G2/S2s must—

- Monitor intelligence reporting on threat activity and I&W data.
- Update or consolidate databases.
- Maintain intelligence readiness.
- Provide their input into the Force Design Update (FDU) process to refine modified table of organizations and equipment (MTOE) and evaluate the need for Individual Mobilization Augmentee (IMA) personnel.

Chapter 2

Intelligence and Unified Action

UNIFIED ACTION

2-1. As discussed in JP 1-02 and Chapter 1, unified action describes the wide scope of actions (including the synchronization of activities with governmental and NGO agencies) taking place within unified commands, subordinate unified commands, or JTFs under the overall direction of the commanders of those commands. Under unified action, commanders integrate joint, single-service, special, and supporting intelligence operations with interagency, nongovernmental, and multinational operations. The ARFOR often brings unique ISR capabilities to unified action operations.

2-2. In a unified action, the intelligence staff takes on additional responsibilities, relationships, and infrastructure to execute multi-service and multinational intelligence operations. This chapter discusses the synchronization of Army intelligence efforts with joint and other national and international partners to achieve unity of effort and to accomplish the commander's intent.

THE LEVELS OF WAR

2-3. The levels of war are doctrinal perspectives that clarify the links between strategic objectives and tactical actions. Although there are no finite limits or boundaries between them, the three levels of war are strategic, operational, and tactical (see Figure 2-1).

- The strategic level is that level at which a nation, often as one of a group of nations, determines national and multinational security objectives and guidance and develops and uses national resources to accomplish them.
- The operational level is the level at which campaigns and major operations are conducted and sustained to accomplish strategic objectives within theaters or AOs. It links the tactical employment of forces to strategic objectives.
- The tactical level is the employment of units in combat. It includes the ordered arrangement and maneuver of units in relation to each other, the terrain, and the enemy to translate potential combat power into victorious battles and engagements.

2-4. Understanding the interdependent relationship of all three levels of war helps commanders visualize a logical flow of operations, allocate resources, and assign tasks. Actions within the three levels are not associated with a particular command level, unit size, equipment type, or force or component type. The concept of strategic, operational, and tactical intelligence operations aids JFCs and their J2s in visualizing the flow of intelligence from one level to the next. The concept facilitates allocating required collection, analysis, production, and dissemination resources; and facilitates the assignment of appropriate intelligence tasks to national, theater, component, and supporting intelligence elements.

STRATEGIC

2-5. The President and the Secretary of Defense use strategic intelligence to develop national strategy and policy, monitor the international situation, prepare military plans, determine major weapon systems and force structure requirements, and conduct strategic operations.

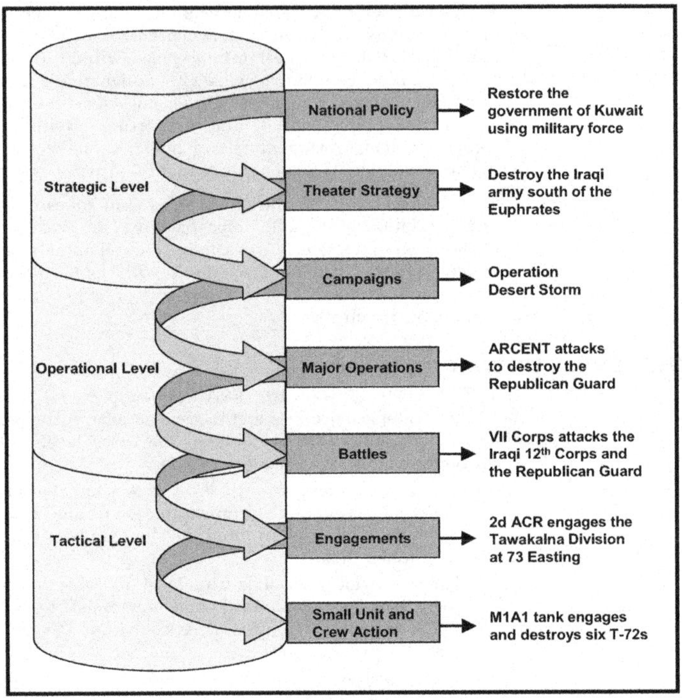

Figure 2-1. The Levels of War.

2-6. Intelligence supports joint operations across the full spectrum of military operations. It determines the current capabilities, and forecasts future developments, of adversaries or potential adversaries that could affect the national security and interests. A strategic operation also produces the intelligence required by combatant commanders to prepare strategic estimates, strategies, and plans to accomplish missions assigned by higher authorities. Theater intelligence includes determining when, where, and in what strength the adversary will stage and conduct theater level campaigns and strategic unified operations. The intelligence staff should also focus predictive analysis efforts on identifying strategic threat events and how these events will impact US actions at the strategic, operational, and tactical levels. Intelligence operations support IO as well. Intelligence operations support strategic planning by—

- Developing strategic intelligence policy.
- Preparing the strategic collection plan.
- Allocating national intelligence resources.

2-7. Intelligence requirements are those requirements generated from the staff's IRs regarding the enemy and environment that are not a part of the CCIRs (PIRs and FFIRs). Intelligence requirements require collection and can provide answers in order to identify indicators of enemy actions or intent, which reduce the uncertainties associated with an operation. Significant changes (that is, branches and sequels) with an operation usually lead to changes in intelligence requirements. Of particular importance is information relating to enemy or threat strategic vulnerabilities, strategic forces, strategic centers of gravity, and any capabilities relating to the development and employment of CBRNE.

2-8. Global and regional issues and threats are identified and reported to the President and the Secretary of Defense, as well as to the senior military leadership and the combatant commanders. Intelligence requirements include any foreign developments that could threaten the US, its citizens abroad, or multinational military, political, or economic interests. Intelligence also includes identifying hostile reactions to US reconnaissance activities and indications of impending terrorist attacks (I&W). For a complete listing of strategic tasks, refer to CJCSM 3500.04C.

OPERATIONAL

2-9. Combatant commanders and subordinate JFCs and their component commanders are the primary users of intelligence. At the operational echelons, intelligence—

- Focuses on the military capabilities and intentions of enemies and threats.
- Provides analysis of events within the AOI and helps commanders determine when, where, and in what strength the adversary might stage and conduct campaigns and major operations.
- Supports all phases of military operations, from mobilization all the way through redeployment of US forces, and continues during sustainment.
- Supports all aspects of the joint campaign.
- Identifies adversary centers of gravity and HVTs.
- Provides critical support to friendly IO.

2-10. The JFC and staff allocate intelligence resources and request support from national agencies, other theaters, and multinational partners. During stability operations and support operations, operational intelligence includes training and assisting multinational partners in conducting intelligence operations.

2-11. Combatant commanders use intelligence concerning the nature and characteristics of the battlespace to determine the type and scale of operations. Intelligence also aids in determining the impact of significant regional features and hazards on the conduct of both friendly and adversary operations. Significant regional factors include the natural environment, political, informational, economic, industrial, infrastructure, geographic, demographic, topographic, hydrographic, climatic, populace, cultural, medical, lingual, historical, and psychological aspects of the AOI. Intelligence analysis also assists in determining the ROE and other restrictions which will affect operations in the JFC's AO.

2-12. Intelligence relating to the adversary's military and nonmilitary capabilities assists in determining the adversary's ability to conduct military operations. Factors that operational intelligence addresses include mobilization potential, force structure (including alliance forces), force dispositions, equipment, military doctrine, C2 structure, and their MDMP. Intelligence includes the continuous refinement of the OBs for the entire array of the enemy's forces in the AOI. For a complete listing of operational tasks, refer to CJCSM 3500.04C.

US Army Intelligence and Security Command (INSCOM)

2-13. The U.S. Army has vested its intelligence at the operational level with INSCOM, the major Army command (MACOM) responsible for the Army's intelligence forces above Corps. INSCOM's mission is to conduct and support dominant intelligence, security, and information operations for military commanders and national decisionmakers. The INSCOM strategy is to provide superior information and information capabilities to Army commanders, while denying the same to adversaries. Headquarters (HQ), INSCOM, in coordination with its major subordinate commands (MSCs), provides a myriad of general intelligence support operations. INSCOM is providing a globally focused, rapidly deployable, knowledge based, adaptively force packaged capability, supporting commanders and leaders with actionable intelligence at the point of decision. INSCOM serves as the national to tactical Intelligence bridge.

Army Space Program Office (ASPO)

2-14. ASPO executes the Army's Tactical Exploitation of National Capabilities Program (TENCAP). The program focuses on exploiting current and future tactical potential of national systems and integrating the capabilities into the Army's tactical decisionmaking process. Army TENCAP systems enable the tactical commander maximum flexibility to satisfy intelligence needs under a wide range of operational scenarios. ASPO is the point of contact (POC) for all tactical activities between MACOMs or users and the National Reconnaissance Office (NRO).

TACTICAL

2-15. Tactical commanders use intelligence for planning and conducting battles and engagements. Relevant, accurate, predictive, and timely intelligence allows tactical units to achieve an advantage over their adversaries. Precise and predictive intelligence, on the threat and targets, is essential for mission success. Predictive intelligence also enables the staff to better identify or develop ECOAs. Tactical intelligence—

- Identifies and assesses the enemy's capabilities, COAs, and vulnerabilities, as well as describes the battlespace.
- Seeks to identify when, where, and in what strength the enemy will conduct tactical level operations.
- Provides the commander with information on imminent threats to the force including those from terrorists, saboteurs, insurgents, and foreign intelligence collection.
- Provides critical support to friendly IO.
- Develops and disseminates targeting information and intelligence.

2-16. Intelligence provides the tactical commander with the information and intelligence he requires to successfully employ combat forces against enemy forces. Thus, intelligence tasks support the execution of battles and engagements. These intelligence tasks are different from those at other levels due to their ability to immediately influence the outcome of the tactical commander's mission. They include information gathered from tactical sources, such as combat information, interrogations, debriefings, and eliciting information from captured or misplaced personnel. For a complete listing of tactical tasks, refer to FM 7-15.

CATEGORIES OF INTELLIGENCE

2-17. Directly or indirectly, the unit receives intelligence from throughout the US intelligence community. The intelligence operations and products of national, joint, and service organizations that make up the intelligence community fall into one of six categories:

- I&W.
- Current intelligence.
- GMI.
- Target intelligence.
- S&T intelligence.
- CI.

2-18. The categories of intelligence are distinguishable from each other primarily by the purpose of the intelligence product. The categories can overlap and some of the same intelligence is useful in more than one category. Depending upon the echelon, intelligence organizations use specialized procedures to develop each category of intelligence. The following information describes each category.

INDICATIONS AND WARNINGS

2-19. Indications and warnings (I&W) are those intelligence activities intended to detect and report time-sensitive intelligence information on foreign developments that could involve a threat to the US or allied and/or coalition military, political, or economic interests or to US citizens abroad. I&W includes forewarning of enemy actions or intentions; the imminence of hostilities; insurgency; nuclear or non-nuclear attack on the US, its overseas forces, or alliled and/or coalition nations; hostile reactions to US reconnaissance activities; terrorist attacks; and other similar events. (JP 3-13). I&W comes from time-sensitive information and analysis of developments that could involve a threat to the US and multinational military forces, US political or economic interests, or to US citizens. While the G2/S2 is primarily responsible for producing I&W intelligence, each element, such as the MPs conducting PIO, within every unit contributes to I&W through awareness of the CCIRs and reporting related information.

CURRENT INTELLIGENCE

2-20. Current intelligence involves the integration of time-sensitive, all-source intelligence and information into concise, accurate, and objective reporting on the battlespace and current enemy situation. One of the most important forms of current intelligence is the enemy situation portion of the COP. The G2/S2 is responsible for producing current intelligence for the unit. Current intelligence supports ongoing operations across full spectrum operations. In addition to the current situation, current intelligence should provide projections of the enemy's anticipated situations (estimates) and their implications on the friendly operation.

GENERAL MILITARY INTELLIGENCE

2-21. GMI is intelligence concerning military capabilities of foreign countries or organizations or topics affecting potential US or multinational military operations relating to armed forces capabilities, including OB, organization, training, tactics, doctrine, strategy, and other factors bearing on military strength and effectiveness and area and terrain intelligence. This broad category of intelligence is normally associated with long-term planning at the national level. However, GMI is also an essential tool for the intelligence staff and should be in place long before the start of preparations for a particular military operation. An up-to-date, comprehensive intelligence database is critical to the unit's ability to plan and prepare rapidly for the range of operations and global environments in which it may operate. GMI supports the requirement to quickly respond to differing crisis situations with corresponding intelligence spanning the globe. One of the many places to get information for GMI is the medical intelligence database. The G2/S2 planner develops his initial IPB from GMI products.

2-22. The G2/S2 develops and maintains the unit's GMI database on potential threat forces and environments based on the commander's guidance. As an essential component of intelligence readiness, this database supports the unit's planning, preparation, execution, and assessment of operations. The G2/S2 applies and updates the database as it executes its intelligence production tasks.

TARGET INTELLIGENCE

2-23. Target intelligence is the analysis of enemy units, dispositions, facilities, and systems to identify and nominate specific assets or vulnerabilities for attack, re-attack, or exploitation (for intelligence). It consists of two mutually supporting production tasks: target development and combat assessment.

- **Target development** is the systematic evaluation and analysis of target systems, system components, and component elements to determine HVTs for potential attack through maneuver, fires, or non-lethal means.
- Once attacked, **combat assessment** provides a timely and accurate estimate of the affects of the application of military force (lethal or non-lethal) an IO on targets and target systems based on predetermined objectives.

SCIENTIFIC AND TECHNICAL INTELLIGENCE

2-24. Scientific and technical intelligence (S&TI) is the product resulting from the collection, evaluation, analysis, and interpretation of foreign S&T information which covers foreign developments in basic and applied research and in applied engineering techniques and S&T characteristics, capabilities, and limitations of all foreign military systems, weapons, weapon systems, and materiel, the research and development (R&D) related thereto, and the production methods employed for their manufacture.

2-25. S&T intelligence concerns foreign developments in basic and applied sciences and technologies with warfare potential. It includes characteristics, capabilities, vulnerabilities, and limitations of all weapon systems, subsystems, and associated materiel, as well as related R&D. S&T intelligence also addresses overall weapon systems and equipment effectiveness. Specialized organizations—such as the Defense Intelligence Agency (DIA) Missile and Space Intelligence Center (MSIC), INSCOM, Air Missile Defense (AMD), Area Air and Missile Defense Command (AAMDC), and National Ground Intelligence Center (NGIC)—produce this category of intelligence. The G2/S2 establishes instructions within SOPs, orders, and plans for handling and evacuating captured enemy materiel (CEM) for S&T intelligence (S&TI) exploitation.

COUNTERINTELLIGENCE

2-26. As previously defined at the end of Chapter 1, CI attempts to identify and recommend countermeasures to the threat posed by foreign intelligence security services (FISS) and the ISR activities of non-state entities such as organized crime, terrorist groups, and drug traffickers. CI analysis incorporates information from all sources as well as the results of CI collection, investigations, and operations to analyze the multidiscipline threat posed by foreign intelligence services and activities.

INTELLIGENCE COMMUNITY

2-27. There are many organizations in the intelligence community that support military operations by providing specific intelligence products and services. The J2/G2/S2 and his staff must be familiar with these organizations and the methods of obtaining information from them as necessary. Figure 2-2 shows organizations that compose the intelligence community.

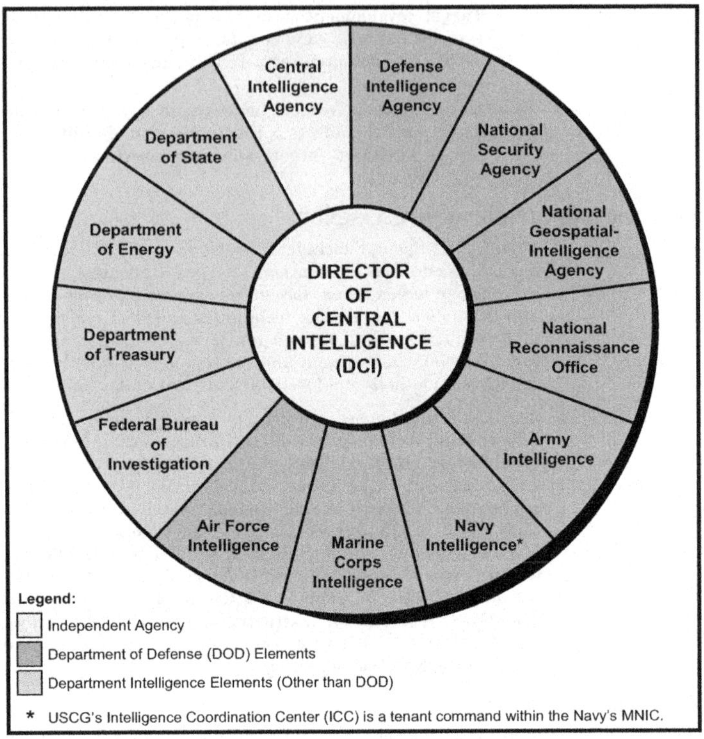

Figure 2-2. Intelligence Community Membership.

DOD AGENCIES

Defense Intelligence Agency (DIA)

2-28. DIA is a combat support agency and a major collector and producer in the defense intelligence community. The DIA supports the full spectrum of operations to include basic MI, counterterrorism, counterdrugs, medical intelligence, WMD and proliferation, United Nations (UN) peacekeeping and multinational support, missile and space intelligence, noncombatant evacuation operations (NEOs), targeting, combat assessment, and battle damage assessment (BDA).

National Security Agency (NSA)

2-29. NSA ensures cryptologic planning and support for joint operations. Working with the tactical cryptologic units of a command, the NSA provides SIGINT and information security (INFOSEC), encompassing communications security (COMSEC) and computer security, as well as telecommunications support and operations security (OPSEC). The people and equipment providing SIGINT, INFOSEC, and OPSEC constitute the United States Cryptologic System (USCS). The NSA, through the USCS, fulfills cryptologic command and/or management, readiness, and operational responsibilities in support of military operations according to the Secretary of Defense tasking, priorities, and standards of timeliness.

National Geospatial-Intelligence Agency (NGA)

2-30. The NGA (formerly National Imagery and Mapping Agency [(NIMA)] mission is to provide timely, relevant, and accurate intelligence and geospatial information in support of national security objectives of the United States. The Director of NGA advises the Secretary of Defense, Director Central Intelligence (DCI), Chairman of the Joint Chiefs of Staff (JCS), and the combatant commanders on imagery, IMINT, and geospatial information. The Operations Directorate, Customer Support Office, is the focal point for interface with external customers, including the JCS, combatant commands, services, and national and defense agencies.

National Reconnaissance Office (NRO)

2-31. The NRO develops and operates unique and innovative space reconnaissance systems and conducts intelligence-related activities essential for US security. The role of the NRO is to enhance US government and military information superiority across full spectrum operations. NRO responsibilities include supporting I&W, monitoring arms control agreements, and performing crisis support to the planning and conduct of military operations. The NRO accomplishes its mission by building and operating IMINT and SIGINT reconnaissance satellites and associated communications systems

US Navy (USN)

2-32. Naval intelligence products and services support the operating naval forces, the Department of the Navy, and the maritime intelligence requirements of national level agencies. Naval intelligence operates the National Maritime Intelligence Center (NMIC). Naval intelligence responsibilities include maritime intelligence on global merchant affairs, counter-narcotics, fishing issues, ocean dumping of radioactive waste, technology transfer, counter-proliferation, cryptologic related functions, CI, I&W support, management of Coast Guard collection, and development of new weapons systems and countermeasures.

US Marine Corps (USMC)

2-33. USMC intelligence provides pre-deployment training and force contingency planning for requirements that are not satisfied by theater, other service, or national capabilities. The Marine Corps Intelligence Agency (MCIA) handles the integration, development, and application of GMI, technical information, all-source production, and open-source information.

US Air Force (USAF)

2-34. USAF ISR assets fill a variety of roles to meet US national security requirements. The USAF operates worldwide ground sites and an array of airborne ISR platforms to meet national level intelligence requirements. To support day-to-day USAF operations and to meet specific USAF requirements, intelligence professionals at the wing and squadron levels use suites of interoperable analysis tools and dissemination systems to tailor information received from all levels and agencies in the intelligence community. USAF responsibilities include all-source information on aerospace systems and potential adversaries' capabilities and intentions, cryptologic operations, I&W, IO, and criminal investigative and CI services.

NON-DOD AGENCIES

2-35. Although the primary focus of non-DOD members of the intelligence community is strategic intelligence and support to the President and the Secretary of Defense, these agencies also produce intelligence and intelligence products that support operational and tactical ARFOR. This responsibility includes assessing potential issues and situations that could impact US national security interests and objectives. These agencies identify global and regional issues and threats. Some of the intelligence products and services these agencies provide are essential to accurate assessment of the threat and battlespace, particularly during stability operations and support operations.

Central Intelligence Agency (CIA)

2-36. The CIA's primary areas of expertise are in HUMINT collection, imagery, all-source analysis, and the production of political and economic intelligence. CIA and military personnel staff the CIA's Office of Military Affairs (OMA). As the CIA's single POC for military support, OMA negotiates, coordinates, manages, and monitors all aspects of agency support for military operations. This support is a continuous process, which the agency enhances or modifies to respond to a crisis or developing operation. Interaction between OMA and the DCI representatives to the Office of the Secretary of Defense (OSD), the Joint Staff, and the combatant commands facilitates providing national level intelligence in support of joint operations, contingency and operations planning, and exercises.

Department of State, Bureau of Intelligence and Research

2-37. The Bureau of Intelligence and Research coordinates programs for intelligence, analysis, and research; it produces intelligence studies and current intelligence analyses essential to foreign policy determination and execution. Its subordinate Bureau of International Narcotics Matters develops, coordinates, and implements international narcotics control assistance activities. It is the principal POC and provides policy advice on international narcotics control matters for the Office of Management and Budget (OMB), the NSC, and the White House Office of National Drug Control Policy (ONDCP). The Bureau also oversees and coordinates the international narcotics control policies, programs, and activities of US agencies.

Department of Energy, Office of Nonproliferation and National Security

2-38. The Department of Energy's Office of Nonproliferation and National Security assists in the development of the State Department's policy, plans, and procedures relating to arms control, nonproliferation, export controls, and safeguard activities. Additionally, this office is responsible for—

- Managing the department's R&D program.
- Verifying and monitoring arms implementation and compliance activities.
- Providing threat assessments and support to headquarters and field offices.

Department of the Treasury

2-39. The US Treasury Department's intelligence-related missions include producing and disseminating foreign intelligence relating to US economic policy and participating with the Department of State in the overt collection of general foreign economic information.

Federal Bureau of Investigation (FBI)

2-40. The FBI is the principal investigative arm of the Department of Justice (DOJ) and has primary responsibility for CI and counterterrorism operations conducted in the US. CI operations contemplated by any other organizations in the US must be coordinated with the FBI. Any overseas CI operation conducted by the FBI must be coordinated with the CIA.

Department of Homeland Security

2-41. The mission of the Department of Homeland Security is to develop and coordinate the implementation of a comprehensive national strategy to secure the US from terrorist threats or attacks. The organizational construct of homeland security mission is a framework of prepare, deter, preempt, defend, and respond. Component agencies will analyze threats and intelligence, guard our borders and airports, protect our critical infrastructure, and coordinate the response of our nation for future emergencies. The component agencies of the department include—

- US Coast Guard (USCG).
- US Customs Service.
- US Border Patrol.
- Federal Emergency Management Agency (FEMA).
- Immigration and Naturalization Service (INS).
- Transportation Security Administration.
- Federal Protective Service.
- Office of Domestic Preparedness.

United States Coast Guard (USCG)

2-42. The USCG, subordinate to the Department of Homeland Security, has unique missions and responsibilities as both an armed force and a law enforcement agency (LEA), which makes it a significant player in several national security issues. The USCG intelligence program supports counter-drug operations, mass seaborne migration operations, alien migration interdiction operations, living marine resource enforcement, maritime intercept operations, port status and/or safety, counterterrorism, coastal and harbor defense operations, and marine safety and/or environmental protection.

Other Agencies

2-43. There are a number of US Government agencies and organizations, not members of the intelligence community, that are responsible for collecting and maintaining information and statistics related to foreign governments and international affairs. Organizations such as the Library of Congress, the Departments of Agriculture and Commerce, the National Technical Information Center, US Information Agency, US Information Service, and the US Patent Office are potential sources of specialized information on political, economic, and military-related topics. The intelligence community may draw on these organizations to support and enhance research and analysis and to provide relevant information and intelligence for planners and decisionmakers. Many other US Government agencies may become directly involved in supporting DOD especially during stability operations and support operations. (See JP 2-02 for a description of agency support to joint operations and intelligence.) These organizations include—

- Department of Transportation.
- Disaster Assistance Response Team within the Office of Foreign Disaster.
- US Agency for International Development.
- NGA.

UNIFIED ACTION INTELLIGENCE OPERATIONS

2-44. In unified action, ARFOR synchronize their actions with those of other participants to achieve unity of effort and to accomplish the combatant commander's objectives. Unified action links subordinates to the combatant commander under combatant command (command authority) (COCOM). Multinational, interagency, and nonmilitary forces work with the combatant commander through cooperation and coordination. Combatant commanders form theater strategies and campaigns, organize joint forces, designate operational areas, and provide strategic guidance and operational focus to subordinates. The aim is to achieve unity of effort among many diverse agencies in a complex environment. Subordinate JFCs synchronize joint operations in time and space, direct the action of foreign military forces (multinational operations), and coordinate with governmental and NGOs (interagency coordination) to achieve the same goal.

2-45. The J2 staff provides intelligence promptly, in an appropriate form, and by any suitable means to those who need it. Intelligence personnel ensure that the consumers understand the intelligence and assist them as they apply the intelligence to their operations.

2-46. Dissemination requires establishing appropriate communications systems and procedures. The J2 and other intelligence personnel must fully participate in all operation planning and execution, and develop close working relationships with the JFC and other staff elements.

2-47. The commander and staff assess intelligence operations to determine their effectiveness and to make any necessary improvements. The intelligence process functions presented in Chapter 4 provide the criteria for evaluating intelligence operations.

JOINT INTELLIGENCE OPERATIONS

2-48. The JTF commanders and his intelligence staff must—

- Understand the intelligence requirements of superior, subordinate, and component commands.
- Identify organic intelligence capabilities and shortfalls.
- Access theater and national systems to ensure appropriate intelligence and CI products are available to the JTF.

2-49. The JTF's intelligence effort focuses on integrating multi-source information and multi-echelon intelligence into all-source intelligence products that provide clear, relevant, and timely knowledge of the enemy and battlespace. These products must be in formats that are readily understood and directly usable by the recipient in a timely manner. They must neither overload the user nor the communications architecture.

2-50. The J2 is the JTF commander's focal point for intelligence. The J2 directly supports the JFC's responsibilities for determining objectives, directing operations, and evaluating the effects of those operations. The J2 supports the execution of the plan with the intelligence needed to sustain the operation, attain joint force objectives, provide support to subordinate commands, and continually support FP efforts. The J2 analyzes the potential threat situation and provides assessments to support friendly opportunities. The J2 then supports the execution of the plan with the operational intelligence needed to sustain the operations, attain joint force objectives, and support FP. To maintain the initiative, the JFC will seek to get inside the adversary's decisionmaking cycle; that is, the JFC will seek to develop procedures and an organization in order to receive new and accurate intelligence and respond to the new situation faster than the adversary. The J2 must help in identifying the adversary's decisionmaking cycle and identifying vulnerabilities that may be exploited. The J2 also—

- Ensures the provision of the required ISR support to the JTF and its subordinate functional and service components.
- Assists the JTF commander in defining intelligence responsibilities and PIRs.
- Actively participates in joint staff planning and in planning, preparing, executing, and assessing ISR efforts. This includes leading the JTF's joint collection board and providing representation at the joint targeting board.

2-51. Figure 2-3 shows a typical JTF J2 organization. The overall organization of the JTF and operations will dictate actual composition of the J2. At a minimum, a core element of analytical, ISR management, and administrative capabilities is required.

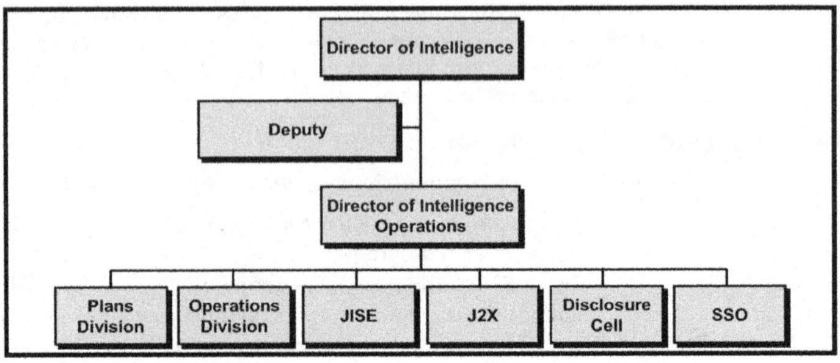

Figure 2-3. Typical Joint Task Force J2 Organization.

Considerations In Joint Intelligence Operations

2-52. The Army, Air Force, Navy, Marine Corps, and SOF must provide globally responsive assets to support the combatant commanders' theater strategies and the national security strategy. The capabilities of the other armed forces complement those of ARFOR. During joint operations, they provide support consistent with JFC-directed missions. When conducting joint intelligence operations, there are a number of unique problems that can arise due to the complexity of integrating the efforts of the different services and commands. Elements affecting joint intelligence operations among the different services include the following:

- Intelligence liaison is critical to the success of intelligence operations and requires early establishment, particularly between units that have not routinely trained together and possess differing capabilities. As a minimum, organizations exchange liaison teams with the higher echelon organization. Additional liaison may be necessary to facilitate joint force collection, production, or dissemination requirements. Liaison teams—
  - Support planning and C2 of intelligence operations.
  - Ensure timely two-way flow of intelligence between commands.
  - Manage intelligence and resource requirements of the subordinate command.
  - Advise the commander on service ISR capabilities, limitations, and employment.
- Commanders and staffs use IPB to understand the battlespace and develop or refine plans and orders. IPB products exchanged between echelons ensure a common picture of the battlespace and estimate of the situation.
- Communications considerations for joint operations include—
  - Planning for intelligence communications transition to facilitate execution of branches or sequels to the plan or to accommodate shifting of the main effort from one force to another.

  - Identifying the initial communications architecture to include establishing procedures and protocols for information exchanges (databases, text, imagery, voice, and video).
  - Balancing the availability of service-unique intelligence systems between echelons or services. This may require each service providing additional resources. The senior commander is responsible for allocating resources.
  - Disseminating intelligence between commands and services. Additional communications equipment, intelligence terminals, and personnel may be required to balance capabilities between services.
- Identifying the databases each service possesses or has access to; determining which databases will support the operation and, if necessary, merge them into a single database; and ensure access by the entire force.
- Providing a focal point for subordinate command access to national or joint intelligence is essential. The senior commander will request and allocate resources required to support this access.

2-53. The JFC's intelligence requirements, concept of operation, and intent drive the ISR effort. The different organizations and services participating in joint intelligence operations must continuously share information, intelligence, and products to satisfy requirements. See FM 2-01 for details on intelligence requirements and RM. Activities required to facilitate an effective joint collection effort include—

- Coordinating intelligence and ISR operations to optimize capabilities of collection assets and reduce duplication of effort.
- Integrating supporting national and theater intelligence collection assets into the intelligence synchronization plan.
- Establishing procedures for tracking and handing off HPTs between services and echelons.
- Establishing procedures for cueing Army and other service collection assets.
- Coordinating airspace for ISR assets.
- Maximizing available linguist capabilities. Shortages of military linguists trained in target languages may require cross-leveling Army and other service linguists.

2-54. Reporting and intelligence production considerations in joint operations include—

- Establishing production criteria and thresholds that produce timely and relevant intelligence keyed to the commander's intelligence and targeting requirements. The ISM, attack guidance matrix (AGM), and HPT list are examples of tools used to support joint intelligence and targeting efforts. The sharing of these products to all echelons and services is crucial.
- Establishing common methodology and criteria for producing the BDA and supporting the combat assessment function.
- Establishing set standards for the number and frequency of periodic reports such as intelligence summaries (INTSUMs).

Joint Intelligence Architecture

2-55. In addition to the J2 staffs at every joint level of command, the key organizations in the joint intelligence architecture are the National Military Joint Intelligence Center (NMJIC) (see Figure 2-4), the Joint Intelligence Centers (JICs) or Joint Analysis Centers (JACs) of the unified commands, and, when formed, the JTF's joint intelligence support element (JISE). Working together, these organizations play the primary role in managing and controlling joint intelligence operations. The formal relationships which link these organizations facilitate information management and optimize complementary intelligence functions by echelon without obstructing the timely flow of intelligence up, down, or laterally.

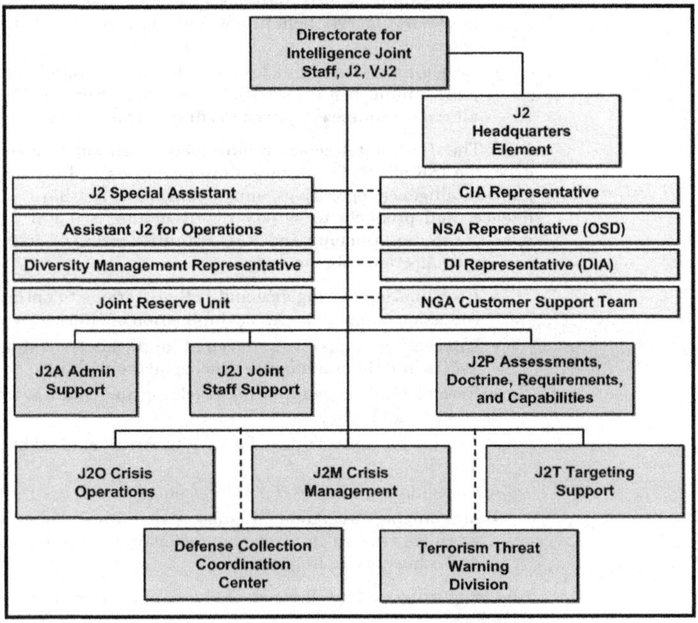

Figure 2-4. National Military Joint Intelligence Center.

2-56. The NMJIC is the focal point for intelligence activities in support of joint operations. The NMJIC is the CJCS J2 intelligence watch within the National Military Command Center (NMCC) in the Pentagon. When the assigned or attached assets cannot satisfy the combatant commander and the JTF commander's crisis-related and time-sensitive intelligence requirements, the NMJIC, as part of the J2's staff, sends tasks to appropriate agencies to fill the requirements. During crises, the NMJIC expands as necessary to establish a working group, intelligence task force, or an expanded intelligence task force. The NMJIC participates in targeting by developing national level target lists. When requested, the NMJIC supports the theater in performing BDA. Besides supporting the combatant commands and JTFs, the NMJIC supports any multinational partners and prescribed international organizations.

2-57. The JIC is organized in accordance with the combatant commander's prerogatives, but normally performs the general functions described in JP 2-0 and specific unified command intelligence TTP. The JIC is responsible for providing intelligence to JTF and its subordinates during military operations. If the JIC cannot meet the combatant commander's requirements, the JIC forwards requests to the NMJIC or to subordinate command levels through established channels, using standard command procedures. In some cases, the JIC may also ensure timely support by approving a direct communication path, in advance, between requesters such as the JTF JISE and outside producers. In this case, the parties must inform the JIC of all requests as they occur. This method is most appropriate when the parties require products and services, which the JIC does not routinely produce.

2-58. At the discretion of the JTF commander, the J2 can establish a JISE during the crisis or preparation stage for operations to augment the J2 staff. The JISE is a tailored subset of the JIC or the intelligence organization of the service component designated as the JTF headquarters.

2-59. Under the direction of the J2, the JISE normally manages the intelligence collection, production, and dissemination for the JTF. The JISE provides intelligence to JTF operational forces and performs common intelligence functions. By design, the JISE is scaleable and can expand to meet the needs of the JTF and the operating environment. It is composed of analytical experts, analysis teams, and ISR managers that provide services and products, which the JTF, JTF staff, and subordinate components require. These experts, mentioned above, focus on solving the operational intelligence problems of concern to the JTF commander. The JISE's capability to perform all-source analysis and ISR synchronization is key to producing operational intelligence that is timely, relevant, and complete. Figure 2-5 illustrates the features of a typical JISE.

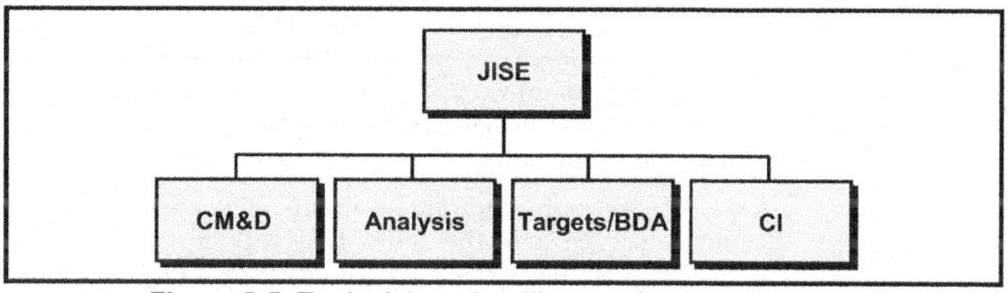

Figure 2-5. Typical Joint Intelligence Support Element.

Joint Task Force Intelligence Organizations

2-60. In addition to the JISE, the JTF commander and J2 may require other supporting JICs or teams based on projected operations. The JTF commander may make a request to the NMJIC for specific national intelligence agency capabilities. The NMJIC evaluates and coordinates the JTF commander's requirements with the J3, J5, and national intelligence agencies and tailors the composition of the deployment packages to meet those needs. The deployment packages, such as the National Intelligence Support Team (NIST), provide access to the entire range of capabilities resident in the national intelligence agencies and can focus those capabilities on the JTF commander's intelligence requirements. The J2X manages and coordinates the HUMINT and CI activities of national, theater, and service components operating within the JTF's JOA. The Joint Captured Materiel Exploitation Center (JCMEC) assists in management of recovery, exploitation, and disposal of captured enemy equipment (CEE). The JTF commander's requirements dictate the composition and tailoring of such deployment packages.

Augmentation Considerations

2-61. Depending on the scale of the operations, the intelligence organizations described above and those of the JTF's subordinate command may require personnel augmentation. Optimum use of available intelligence assets is essential to ensure quality support in meeting the JTF commander's requirements.

2-62. The demand for additional intelligence increases significantly during crisis and wartime operations. As the need arises for more intelligence personnel, the intelligence presence also increases at all command levels. Locating additional intelligence personnel and knowing how to integrate those personnel into the operation is vital. The JTF J2 should identify intelligence personnel augmentation requirements in accordance with the Chairman, Joint Chiefs of Staff (CJCS) Instruction 1301.01. The combatant commander's joint table of mobilization and distribution (JTMD) also contains the standing augmentation requirements. The JTMD should reflect the need for either IMA or Individual Ready Reserve (IRR) personnel. The combatant commander and the JTF refine personnel requirements and initiate requests when they anticipate or start an operation.

2-63. A consideration for the JTF when requesting support or augmentation is that these national level teams and individual augmentees are not totally self-contained elements; rather they require logistic, information, and other support from the supported command. Each deployment is unique based on mission, duration, team composition, and capabilities required. A full NIST, for example, requires a private, access-controlled area within a sensitive compartmented information facility (SCIF) work environment and dedicated secure communications.

2-64. For more information on intelligence operations as they apply to other armed services, see the individual service intelligence doctrine. See also JP 2-0 series, JP 3-0, JP 3-55, and JP 5-0 for more details on joint intelligence operations and considerations.

Multinational Intelligence

2-65. Multinational intelligence operations take place within the structure of an alliance or coalition. Some multinational military organizations, such as the North Atlantic Treaty Organization (NATO) and the UN Command in the Republic of Korea (ROK), are highly structured and enduring. Others, such as the coalition formed during the Gulf War, are less formal and temporary.

2-66. In multinational operations, the multinational force commander (MNFC) exercises command authority over a military force composed of elements from two or more nations. Therefore, in most multinational operations, the JTF must share intelligence, as necessary, for mission accomplishment with foreign military forces and coordinate exchange of intelligence with those forces.

2-67. In some circumstances, the JTF may need to seek authority to go outside the usual political-military channels to provide information to NGOs. The JTF must tailor intelligence policy and dissemination criteria to each multinational operation.

INTELLIGENCE REACH

2-68. Military forces use intelligence reach to rapidly access information, receive support, and conduct collaboration and information sharing with other units and organizations unconstrained by geographic proximity, echelon, or command.

2-69. Intelligence reach supports minimizing the deployed footprint of ISR assets. By providing enhanced information and tailored intelligence products, intelligence reach can greatly enhance the intelligence capabilities of the unit and play a significant role in improving the commander's ability to make decisions on the battlefield. It allows the commander to harness national, joint, foreign, and other military organization resources. Table 2-1 shows examples of partners and sources for intelligence reach.

2-70. Detailed planning and training are critical to the success of intelligence reach operations. The following are steps that the staff can take to ensure optimal use, operability, and effectiveness of intelligence reach:

- Establish data exchange methods and procedures.
- Establish electronic message transfer procedures.
- Establish homepages for identified forces.
- Establish POCs for I&W centers, production centers, Theater JICs, DIA, INSCOM, and their major subordinate commands such as NGIC and the theater intelligence brigades and groups.
- Ensure the intelligence staff has the necessary personnel, training, automated systems, bandwidth, and resources to conduct intelligence reach.
- Determine IRs through staff planning. Develop production requirements for identified intelligence gaps.
- Order geospatial products for the projected joint AOI.
- Establish and maintain a comprehensive directory of intelligence reach resources before deployment and throughout operations. The value of intelligence reach will greatly increase as the staff develops and maintains ready access to rich information resources. These resources are numerous and may include, for example, Army, Joint, DOD, non-DOD, national, commercial, foreign, and university research programs. Know what types of information the resources can provide. Continuously expand the resource directory through identification of new resources.

- Use intelligence reach first to fill intelligence gaps and requirements and answer RFIs. This technique can preclude unnecessary tasking or risk to limited ISR assets.
- Maintain continuous situational understanding and anticipate intelligence requirements. Use intelligence reach to fulfill these requirements and provide the results to the commander and staff for planning and decisionmaking.
- Exchange intelligence reach strategies with other units.
- Present the information retrieved through intelligence reach in a usable form. Share the information derived from intelligence reach with subordinate, lateral, and higher echelons. Ensure follow-on forces have all information as well.

Table 2-1. Examples of Intelligence and Reach Partners and Sources.

ARMY	SERVICES	JOINT	DOD
ACE ISE TSE (TIB/TIG) 902d NGIC ATCAE ARISCs	ONI NMIC AIA NAIC MCIA	USEUCOM JAC USSOUTHCOM JIC USSOCOM JIC USSPACECOM CIC USJFCOM AIC USCENTCOM JIC USTRANSCOM JIC USSTRATCOM USPACOM JIC USNORTHCOM JIC	DIA CMO MSIC AFMIC DHS NGA NSA RSOCs
NON-DOD	**NATIONAL**	**COMMERICAL**	**FOREIGN**
DOE FBI DOS DEA FEMA INS	DCI NIC CIA NRO INR DOT, Office of Intel Support	RAND Jane's Defence Weekly Economic Intelligence Unit CNN Reuters Associated Press United Press International	DIS NFHQ DIO

INTELLIGENCE REACH COMPONENTS

2-71. Intelligence reach requires the G2/S2 to develop a strategy on how best to support the unit's mission with intelligence reach capabilities. There are six basic components of the strategy:

- Push
- Database Access.
- Pull.
- Broadcast Services.
- Collaborative Tools.
- Requirements Management.

Push

2-72. Push occurs when the producers of intelligence or information are knowledgeable of the customer's requirements and are able to send the desired intelligence to the customer without further requests. Push is accomplished through the most appropriate or efficient dissemination means. Unless coordinated as an acceptable dissemination method, push is not accomplished solely by posting intelligence products on a web page. The entity that posts the document must ensure the intended recipient is aware of the product's location or has received the product.

2-73. Push begins with the statement of intelligence interest (SII). The SII establishes the unit's intelligence and IRs. The SII is prepared by intelligence staff organizations to register their interest in receiving recurring hardcopy and softcopy reports, studies, and publications covering a wide variety of intelligence subjects. The J2/G2/S2 works with the Department of the Army Production and Dissemination Management (DAPDM) to establish a profile for information transfer. Intelligence reports or information meeting the unit's requirements are then automatically sent directly to the unit's classified or unclassified networks or communication systems.

2-74. Organizations that push information and data down to subordinate units must be careful of the amount, detail, and focus of the information they are sending. Too much information will overwhelm the subordinate unit. Crucial information may be lost in the midst of an overabundant flow of information, much of which is of little use. Intelligence reach often runs more efficiently if the user pulls the information or at least focuses the intelligence producers to send the appropriate information with the correct level of detail at the appropriate time.

Database Access

2-75. Access to local, theater, DOD, non-DOD, and commercial databases allows analysts to leverage stored knowledge on topics ranging from basic demographics to OB information. A validated DIA Customer Number (acquired by the J2/G2/S2) in combination with Secret Internet Protocol Network (SIPRNET) and Joint Worldwide Intelligence Communications System (JWICS) connectivity establishes access to most of the databases online. The challenge for an analyst is to gain an understanding of the structure, contents, strengths, and weaknesses of the database regardless of database type. Additionally, the procedures are often difficult for extracting portions or downloading and transferring the database to unit level automated information systems.

2-76. Each intelligence discipline has unique databases established and maintained by a variety of agencies. Database access is accomplished through unit or agency homepages via SIPRNET (Intelink-S) and JWICS (Intelink). The DAPDM office, upon approving the unit's SII, validates the requirement for access to the majority of these databases. Units coordinate with the DA dissemination and program manager (DPM) and/or the agency for access to those databases requiring passwords and permissions beyond normal Intelink-S and Intelink access.

Pull

2-77. Pull occurs when the requestor is familiar enough with existing databases and products to be able to anticipate the location of the desired information. Knowledge of both the types and locations of intelligence databases can greatly increase the efficiency of the Intelligence BOS by saving time and effort on the part of analysts at every echelon.

2-78. The G2/S2 must also ensure that intelligence reach capability extends to multinational forces, augmentees, and other services or organizations working as part of the JTF. The G2/S2 must forward relevant intelligence to units which do not possess the necessary automation to conduct intelligence reach.

Broadcast Services

2-79. Broadcast services are an integrated, interactive dissemination system, focusing on tactical user's IRs using joint message formats. The theater broadcast's data prioritization and processing occur through information management. Selected tactical, theater, DOD, and national collectors broadcast their messages using prescribed message formats. The analyst decides which types of messages are required. Broadcast service data is responsive to the operational commander's information needs and allows tactical users to construct successively detailed pictures of the battlespace.

2-80. Tactical users require intelligence and information to be available at the lowest classification level possible (for example, collateral SECRET) and releasable, as necessary, to multinational partners. Broadcast service information management will adhere to national policy as it relates to the provision of data to NATO or multinational partners.

Collaborative Tools

2-81. Collaboration is the sharing of knowledge, expertise, and information, normally online, and may take numerous forms. Collaborative tools are computer-based tools (groupware) that help individuals work together and share information. They allow for virtual online meetings and data sharing. As an example of the use of collaborative tools, the President or the Secretary of Defense and DOD during a crisis situation will establish a number of crisis action teams (CATs) or joint interagency working groups (JIAWGs). These groups or teams are formed to address specified subjects or topics in support of the warfighter or decisionmaker. The groups will focus on the crisis and normally publish their products on a homepage. Analysts with online access can participate in a number of ways. This includes passively accessing the homepages to study the products, sending queries to the identified POCs, or by having their organization join the CAT or JIAWG. Table 2-2 shows some examples of collaborative tools.

Requirements Management

2-82. The RM system provides a mechanism for submitting RFIs, tasking, and managing ISR assets. Analysts who are trained and familiar with the RM process and the various tasking procedures can leverage its systems for refined information. Each intelligence discipline has established procedures for requesting specific information. It is therefore advantageous to have someone familiar with each discipline's procedures to participate in synchronizing the ISR effort. For example, an analyst receives a HUMINT-based intelligence information report, which provides information on an event or subject. If he requires more refined data or clarification, he should submit a time-sensitive requirement or a source-directed requirement. This creates a two-way communication from the field to the collectors without creating additional standing requirements. The more familiar the analyst is with the RM process, the better he can leverage it.

2-83. The intelligence reach component of RM includes the ability of an intelligence officer at any level to request information, which is beyond what is available at his location. The normal procedure for obtaining intelligence or information not obtainable through the use of available ISR assets is to submit an RFI to the next higher echelon. Users enter RFIs into the RM system. See ST 2-33.5 for more information on intelligence reach operations and FM 2-01 for more information on RM.

Table 2-2. Examples of Collaborative Tools.

Tool	Description
Chat (Audio and Text)	Used to conduct conversations online.
Whiteboard	Permits real-time display of drawings, pictures, or documents for group discussion and comment. Participants can annotate in real time as well.
Bulletin Board	Used to post notices and facilitate discussions on any topic.
Video	Used at a desktop computer or a video teleconferencing (VTC) center to allow the person or group to see with whom they are communicating.
Discussion Groups (Newsgroups)	Topics are posted to a website for discussion and comment where participants can follow a line of discussion on a topic.
File-Sharing Tools	Virtual file cabinets allow information to be stored on web servers, and are available to anyone having access to the site and electronic permission to use the files.
Presentation Tools	Used in a virtual auditorium to allow lectures and briefings to be given to an audience.
Application Sharing	An entire team can use an application running on one computer to revise documents.
Text Tools	Allows live text input and editing by group members. Once complete, the text document can be copied into word processing software.
E-mail	Electronic mail.
Persistent Capability	The ability to preserve files, briefings, or other team or project material for future reference. Properly organized, it becomes an information management device and is invaluable to a long-term effort.
Instant Messaging	This allows real-time exchange of notes and messages.

PART TWO

Intelligence in Full Spectrum Operations

Part Two provides a primer of what and who constitutes both MI and the Intelligence BOS and the process that these entities use in order to provide warfighters and decisionmakers with the intelligence they require in order to accomplish their respective missions.

Chapter 3 discusses the role of MI and the Intelligence BOS within full spectrum operations. It provides an overview of intelligence readiness, particularly the intelligence requirements associated with force projection. The concept of MI asset technical control is discussed as a complement to, not a replacement of, the Army's command and support relationships.

Chapter 4 presents the Intelligence Process—the Intelligence BOS methodology that accomplishes the primary focus of intelligence in full spectrum operations, which provides the warfighter with effective intelligence.

Chapter 3

Fundamentals in Full Spectrum Operations

FULL SPECTRUM OPERATIONS

3-1. Army commanders at all echelons may combine different types of operations simultaneously and sequentially to accomplish missions. Full spectrum operations include offensive, defensive, stability, and support operations. Missions in any environment require ARFOR that are prepared to conduct any combination of these operations:

- **Offensive operations** aim to destroy or defeat an enemy. Their purpose is to impose US will on the enemy and achieve decisive victory.
- **Defensive operations** defeat an enemy attack, buy time, economize forces, or develop conditions favorable for offensive operations. Defensive operations alone normally cannot achieve a decision. Their purpose is to create conditions for a counteroffensive that allow ARFOR to regain the initiative.
- **Stability operations** promote and protect US national interests by influencing the threat, political, and information dimensions of the OE through a combination of peacetime developmental, cooperative activities, and coercive actions in response to crises. Regional security is supported by a balanced approach that enhances regional stability

and economic prosperity simultaneously. ARFOR presence promotes a stable environment.

- **Support operations** employ ARFOR to assist civil authorities, foreign or domestic, as they prepare for or respond to crises and relieve suffering. Domestically, ARFOR respond only when the President or Secretary of Defense direct. ARFOR operate under the lead federal agency and comply with provisions of US law, to include the Posse Comitatus and Stafford Acts. (See FM 3-0.)

3-2. Intelligence supports the commander across full spectrum operations. It helps the commander decide when and where to concentrate sufficient combat power to overwhelm the enemy. ISR is essential for the commander to preclude surprise from the enemy, maintain the initiative on the battlefield, and win battles. Commanders at all levels synchronize intelligence with the other BOSs to maximize their ability to see and strike the enemy simultaneously throughout the AO.

3-3. Every soldier in the command is responsible for detecting and reporting enemy activities, dispositions, and capabilities. This task is critical because the environment we operate in is characterized by violence, uncertainty, complexity, and asymmetric methods by the threat. The increased situational awareness that soldiers develop through personal contact and observation is a critical element of that unit's ability to more fully understand the OE. However, soldiers collect information, they are not intelligence collectors. While medical personnel cannot be assigned ISR tasks due to their Geneva Convention category status, medical personnel who gain information through casual observation of activities in plain view while discharging their humanitarian duties will report the information to their supporting intelligence element.

OFFENSIVE OPERATIONS

3-4. Offensive operations at all levels require effective intelligence to help the commander avoid the enemy's main strength and to deceive and surprise the enemy. During offensive operations, intelligence must provide the commander with the composition, disposition, limitations, employment characteristics, and anticipated enemy actions in a timely enough manner for the commander to significantly affect the enemy commander's decision cycle. It ensures commanders have the intelligence they need to conduct offensive operations with minimum risk of surprise.

DEFENSIVE OPERATIONS

3-5. The immediate purpose of defensive operations is to defeat an enemy attack. Commanders defend to buy time, hold key terrain, hold the enemy in one place while attacking in another, or destroy enemy combat power while reinforcing friendly forces. Intelligence should determine the enemy's strength, COAs, and location of enemy follow-on forces. The defending commander can then decide where to arrange his forces in an economy-of-force role to defend and shape the battlefield. Intelligence support affords him the time necessary to commit the striking force precisely.

3-6. Intelligence in area defensive operations identifies, locates, and tracks the enemy's main attack and provides the commander time to allocate sufficient combat power to strengthen the defense at the point of the enemy's main effort. Intelligence should also identify where and when the commander can most decisively counterattack the enemy's main effort or exploit enemy vulnerabilities.

3-7. Although the battlefield is normally organized as decisive, shaping, and sustaining operations, commanders conducting defensive operations may delineate intelligence in battlefield organizational terms of deep area, close area, and rear area.

STABILITY OPERATIONS

3-8. The environment is often much more complex during stability operations and as a result intelligence is often more complex. In fact, intelligence is even more important a factor (or operational multiplier) during stability operations. As a result, the commander must be even more involved in and knowledgeable of ISR (to include ISR operations the commander controls and other higher level ISR operations that may be occurring within his AO) during stability operations. For example, the commanders must understand the complex details of HUMINT and special access program (SAP) operations.

3-9. The commander requires the appropriate intelligence and IPB products in order to determine how best to influence the threat, political and information dimensions of the operational environment, and enhance regional stability. The identification and analysis of characteristics of the terrain and weather, politics, infrastructure, health status, civilian press, attitudes, and culture of the local populace and all possible threats are important in conducting stability operations. A lack of knowledge concerning local politics, customs, and culture could lead to US actions which attack inappropriate targets or which may offend or cause mistrust among the local population.

SUPPORT OPERATIONS

3-10. Support operations are usually nonlinear and noncontiguous. Commanders designate the decisive, shaping, and sustaining operations necessary for mission success. However, determining the intelligence requirements that drive the ISR effort intending to identify any potential threat's centers of gravity and decisive points may require a more complex and unorthodox thought process than that used in offensive and defensive operations. The G2/S2 may have to define the enemy differently. In support operations, the adversary is often the effects of disease, hunger, or disaster on a civilian population. US forces conducting support operations must also fully understand the organization and identity of key figures or groups within the region where they are operating, as these figures may influence greatly the actions of the population—both civilian and military.

ELEMENTS OF COMBAT POWER

3-11. Combat power is the ability to fight. It is the total means of destructive or disruptive force, or both, that a military unit or formation can apply against the adversary at a given time. Commanders combine the elements of combat power to meet constantly changing requirements and defeat an enemy. The elements of combat power are—

- **Manuever** is the employment of forces, through movement combined with fire or fire potential, to achieve a position of advantage with respect to the enemy to accomplish the mission. Maneuver is the means by which commanders concentrate combat power to achieve surprise, shock, momentum, and dominance.
- **Firepower** provides the destructive force essential to overcoming the enemy's ability and will to fight. Firepower magnifies the effects of maneuver by destroying enemy forces and restricting his ability to counter friendly actions; maneuver creates the conditions for the effective use of firepower.
- **Leadership** is the most dynamic element of combat power. Confident, audacious, and competent leadership focuses on the other elements of combat power and serves as the catalyst that creates conditions for success. Leadership provides purpose, direction, and motivation in all operations.
- **Protection** is the preservation of fighting potential of a force so the commander can apply maximum force at the decisive time and place. Protection is neither timidity nor risk avoidance.
- **Information** enhances leadership and magnifies the effects of maneuver, firepower, and protection. Today, ARFOR use information collected to increase their situational understanding before engaging the enemy. Information from the COP and running estimates, transformed into situational understanding, allows commanders to combine the elements of combat power in new ways.

THE FOUNDATIONS OF ARMY OPERATIONS

3-12. Understanding the principles of war and tenets of Army operations is fundamental to operating successfully across the range of military operations. The principles of war and tenets of Army operations form the foundation of Army operational doctrine. (Refer to FM 3-0 for a full description of the Principles of War and Tenets of Army Operations.)

- **The Principles of War** provide general guidance for conducting war and military operations other than war at the strategic, operational, and tactical levels. The principles are the enduring bedrock of Army doctrine.
- **The Tenets of Army Operations** build upon the principles of war. They further describe the characteristics of successful operations. These tenets are essential to victory.

THE OPERATIONAL FRAMEWORK

3-13. The operational framework consists of the arrangement of friendly forces and resources in time, space, and purpose with respect to each other and the enemy or situation. It consists of the **AO, battlespace**, and the **battlefield organization**. The framework establishes an area of geographic and operational responsibility and provides a way for commanders to visualize how to employ forces against the enemy. Commanders design an operational framework to accomplish their mission by defining and arranging its three components. They use the framework to focus combat power. The operational framework provides a mechanism through which the commander can focus the Intelligence BOS effort. Understanding the operational framework, AOIR, and intelligence coordination line (ICL) and their relationship to each other is key to planning and executing ISR operations.

- **AO** is an operational area defined by the JFC for land and naval forces. AOs do not typically encompass the entire operational area of the JFC but should be large enough for component commanders to accomplish their missions and protect their forces. AOs should also allow component commanders to employ their available systems to the limits of their capabilities. The AO is the basic control measure for all operations that defines the geographical area for which a particular unit is responsible. The commander—
  - Assumes responsibility for intelligence, maneuver, fires, terrain management, security, and movement within his AO.
  - Establishes control measures within his AO to assign responsibilities, coordinate intelligence, fires, and maneuver, and to control other activities.
- **Battlespace** is the environment, factors, and conditions that must be understood to successfully apply combat power, protect the force, or complete the mission. This includes air, land, sea, space, and the included enemy and friendly forces; facilities; weather; terrain; the EMS; and the information environment within the operational areas and AOIs. The G2/S2 performs IPB and synchronizes ISR activities throughout the battlespace as determined by the commander's METT-TC considerations.
  - Area of influence is the geographical area in which a commander can directly influence operations by maneuver of FS systems normally under the commander's command or control. Areas of influence surround and include the associated AO.
  - AOI is that area of concern to the commander, including the area of influence, areas adjacent thereto, and extending into enemy territory to the objectives of current or planned operations. This area also includes areas occupied by enemy forces who could jeopardize the accomplishment of this mission. (See JP 1-02.)
  - Information environment refers to a commander's battlespace that encompasses information activity affecting the operation. To envision that part of the information environment that is within their battlespace, commanders determine the information activities that affect their own operational capabilities and opposing C2 and information systems.
  - Force Projection Bases are the intermediate staging bases and power projection platforms.
  - Home Stations are the permanent locations of active and RC units. To a significant degree, events occurring at home station affect the morale and performance of deployed forces. Thus, the commander's battlespace encompasses all home station functions.

- **Battlefield Organization** is the allocation of forces in the AO by purpose. It consists of three all-encompassing categories of operations:
  - Decisive Operations are those that directly accomplish the task assigned by higher headquarters. Decisive operations conclusively determine the outcome of major operations, battles, and engagements.
  - Shaping Operations are operations at any echelon that create and preserve conditions for the success of decisive operations.
  - Sustaining Operations are operations at any echelon that enable shaping and decisive operations by providing CSS, rear area and base security, movement control, terrain management, and infrastructure development.

3-14. AOIs extend into enemy territory, to the objectives of current or planned operations. They include areas occupied by enemy forces that could jeopardize the accomplishment of the mission. AOIs also serve to focus intelligence development and IO directed at factors outside the AO that may affect the operation. (FM 3-0, para 4-79) The scheme of support includes the coordination of reconnaissance and surveillance missions and AOIs with a joint force or higher headquarters and lateral units to answer the intelligence requirements within the AOI. The G2/S2 monitors enemy activities within the AOI and provides intelligence on enemy activities that may affect the unit.

3-15. The AOIR is an area allocated to a commander in which the commander is responsible for the provision of intelligence within the means at the commander's disposal. It is an area allocated to the commander in which the commander is responsible for the collection of information concerning the threat and the environment and the analysis of that information in order to produce intelligence. Higher headquarters also ensure through intelligence handovers, collection management, and deconfliction that problems with duplication, confliction, and command and control do not occur in the AOIR. (FM 2-19.402/FM 34-80-2) They include the available ISR assets, capability of the G2/S2 section, available intelligence architecture, and METT-TC considerations. The AOIR cannot extend beyond a unit's AOI; however, it can be smaller than its AO as well as vary (expand or contract) during an operation. An example of when a unit's AOIR is smaller than its AO is when a higher headquarters ISR effort covers an area within the unit's AO.

3-16. ICLs designate the boundary between AOIRs. The G2/S2 establishes ICLs to facilitate coordination between higher, lateral, and subordinate units; coordinates with the G3/S3 to direct subordinates to track enemy units and HPTs in their areas; and hands over intelligence responsibility for areas of the battlefield. The establishment of ICLs ensures that there are no gaps in the collection effort; that all echelons are aware of the location, mission, and capabilities of other assets. The G2/S2 keeps abreast of collection activities in progress (all echelons) and battlefield developments through the ICLs. ICLs help—

- Facilitate coordination between higher, lateral, and subordinate units.

- Ensure higher, lateral, and subordinate units share information and intelligence as enemy entities and HPTs move into, within, or transition between AOIRs.
- Specify intelligence responsibility for areas of the battlefield.

ARMY CAPABILITIES

3-17. Commanders combine AC and RC ARFOR—consisting of different types of units with varying degrees of modernization—with multinational forces and civilian agencies to achieve effective and efficient unified action. A broad range of organizations makes up the institutional Army that supports the field Army. Institutional Army organizations design, staff, train, and equip the force. The institutional Army assists in effectively integrating Army capabilities. It does this through leadership and guidance regarding force structure, doctrine, modernization, and budget. (See FM 100-11.)

TASK ORGANIZATION

3-18. The Army supports JFCs by providing tailored force packages to accomplish joint missions and dominate enemies and situations on land. Trained and equipped AC and RC units comprise these force packages. Within these force packages, Army commanders organize groups of units for specific missions. They reorganize for subsequent missions when necessary. This process of allocating available assets to subordinate commanders and establishing their command and support relationships is called task organizing. A temporary grouping of forces designed to accomplish a particular mission is a task organization. The ability of ARFOR to tailor (select forces based upon a mission) and task organize (temporarily organize units to accomplish a tactical mission) gives them extraordinary agility. It allows operational and tactical level commanders to organize their units to make the best use of available resources. The ability to task organize means ARFOR can shift rapidly among offensive, defensive, stability, and support operations.

COMBINED ARMS

3-19. The fundamental basis for the organization and operations of ARFOR is combined arms. Combined arms is the synchronized or simultaneous application of several arms—such as infantry, armor, field artillery, engineers, air defense, and aviation—to achieve an effect on the enemy that is greater than if each arm was used against the enemy separately or sequentially. The ultimate goal of Army organization for operations remains success in joint and combined arms warfare. Its combined arms capability allows commanders to form Army combat, combat support (CS), and CSS forces into cohesive teams focused on common goals.

COMMAND AND SUPPORT RELATIONSHIPS

3-20. Establishing clean command and support relationships is fundamental in organizing for all operations. These relationships can achieve clear responsibilities and authorities among subordinate and supporting units. The commander designates command and support relationships within his authority to weigh the decisive operation and support his scheme of maneuver. Some forces available to a commander are given command or support relationships that limit his authority to prescribe additional relationships. Command and support relationships carry with them varying responsibilities to the subordinate unit by parent and gaining units. By knowing the inherent responsibilities of each command and support relationship, a commander may organize his force to establish clear relationships.

3-21. Command relationships establish the degree of control and responsibility commanders have for forces operating under their tactical control (TACON). When commanders establish command relationships, they determine if the command relationship includes administrative control (ADCON). Figure 3-1 shows command and support relationships.

3-22. Support relationships define the purpose, scope, and effect desired when one capability supports another. Support relationships establish specific responsibilities between supporting and supported unit. Army support relationships are direct support (DS), general support (GS), general support-reinforcing (GSR), and reinforcing.

3-23. While not an actual C2 function, technical control often affects certain intelligence operations. Technical control ensures adherence to existing policies or regulations and provides technical guidance for MI activities, particularly HUMINT, SIGINT, and CI operations. Commanders direct operations but often rely on technical expertise to plan, prepare, execute, and assess portions of the unit's collection effort. Technical control also involves translating ISR tasks into the specific parameters used to focus highly technical or legally sensitive aspects of the ISR effort. Technical control includes, but is not limited to—

- Defining, managing, or guiding the employment of specific ISR assets.
- Identifying critical technical collection criteria such as technical indicators.
- Recommending collection techniques, procedures, or assets.
- Conducting operational reviews.
- Conducting operational coordination.
- Conducting specialized training for specific MI personnel or units.

3-24. An example of technical control is the Prophet control team converting the PIR and SOR sets from the MDMP process and assigning times and anticipated enemy frequencies for subordinate Prophets to collect.

	IF RELATIONSHIP IS:	INHERENT RESPONSIBILITIES ARE: Has Command Relationship with:	May Be Task Organized by:	Receives CSS from:	Assigned Position or AO By:	Provides Liaison To:	Establishes/ Maintains Communications with:	Has Priorities Established by:	Gaining Unit Can Impose Further Command or Support Relationship of:
COMMAND	Attached	Gaining unit	Gaining unit	Gaining unit	Gaining unit	As required by gaining unit	Unit to which attached	Gaining unit	Attached; OPCON; TACON; GS; GSR; R; DS
COMMAND	OPCON	Gaining unit	Parent unit and gaining unit; gaining unit may pass OPCON to lower HQ. Note 1	Parent unit	Gaining unit	As required by gaining unit	As required by gaining unit and parent unit	Gaining unit	OPCON; TACON; GS; GSR; R; DS
COMMAND	TACON	Gaining unit	Parent unit	Parent unit	Gaining unit	As required by gaining unit	As required by gaining unit and parent unit	Gaining unit	GS; GSR; R; DS
COMMAND	Assigned	Parent unit	Parent unit	Parent unit	Gaining unit	As required by parent unit	As required by parent unit	Parent unit	Not Applicable
SUPPORT	Direct Support (DS)	Parent unit	Parent unit	Parent unit	Supported unit	Supported unit	Parent unit; Supported unit	Supported unit	Note 2
SUPPORT	Reinforcing (R)	Parent unit	Parent unit	Parent unit	Reinforced unit	Reinforced unit	Parent unit; reinforced unit	Reinforced unit: then parent unit	Not Applicable
SUPPORT	General Support Reinforcing (GSR)	Parent unit	Parent unit	Parent unit	Parent unit	Reinforced unit and as required by parent unit	Reinforced unit and as required by parent unit	Parent unit; then reinforced unit	Not Applicable
SUPPORT	General Support (GS)	Parent unit	Parent unit	Parent unit	Parent unit	As required by parent unit	As required by parent unit	Parent unit	Not Applicable

NOTE 1. In NATO, the gaining unit may not task organize a multinational unit (see TACON).

NOTE 2. Commanders of units in DS may further assign support relationships between their subordinate units and elements of the supported unit after coordination with the supported commander.

Figure 3-1. Army Command and Support Relationships and Inherent Responsibilities.

Chapter 4

Intelligence Process in Full Spectrum Operations

THE INTELLIGENCE PROCESS

4-1. Commanders use the operations process of plan, prepare, execute, and assess to continuously design and conduct operations (see Figure 4-1). The commander cannot successfully accomplish the activities involved in the operations process without information and intelligence. The design and structure of intelligence operations support the commander's operations process by providing him with intelligence regarding the enemy, the battlefield environment, and the situation.

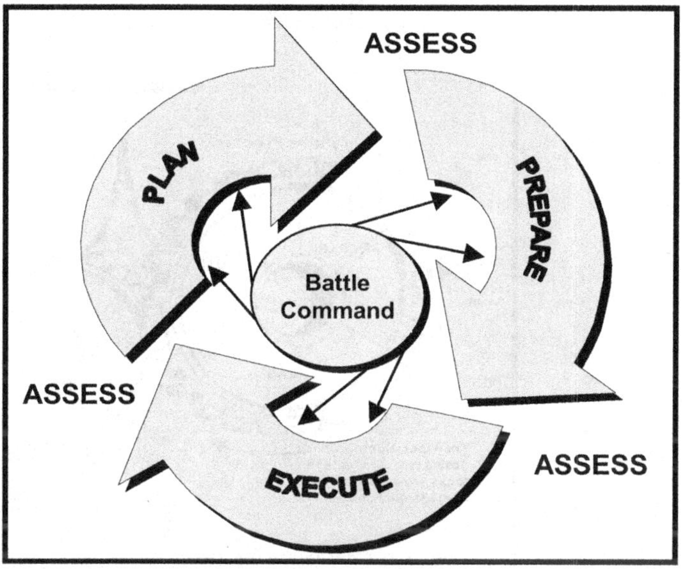

Figure 4-1. The Operations Process.

4-2. Intelligence operations consist of the functions that constitute the intelligence process: **plan, prepare, collect, process, produce**, and the three common tasks of **analyze, disseminate,** and **assess**. Just as the activities of the operations process overlap and recur as circumstances demand, so do the functions of the intelligence process. Additionally, the analyze, disseminate, and assess functions of the intelligence process occur continuously throughout the intelligence process.

4-3. The operations process and the intelligence process are mutually dependent. The commander, through the operations process, provides the guidance and focus through CCIRs and PIRs that drives the intelligence process; the intelligence process provides the continuous intelligence essential to the operations process. Intelligence about the enemy, the battlefield environment, and the situation allows the commander and staff to develop a plan, seize and retain the initiative, build and maintain momentum, and exploit success (see Figure 4-2). The intelligence process is just one of the mechanisms that provides input to build the COP and facilitate the commander's situational understanding.

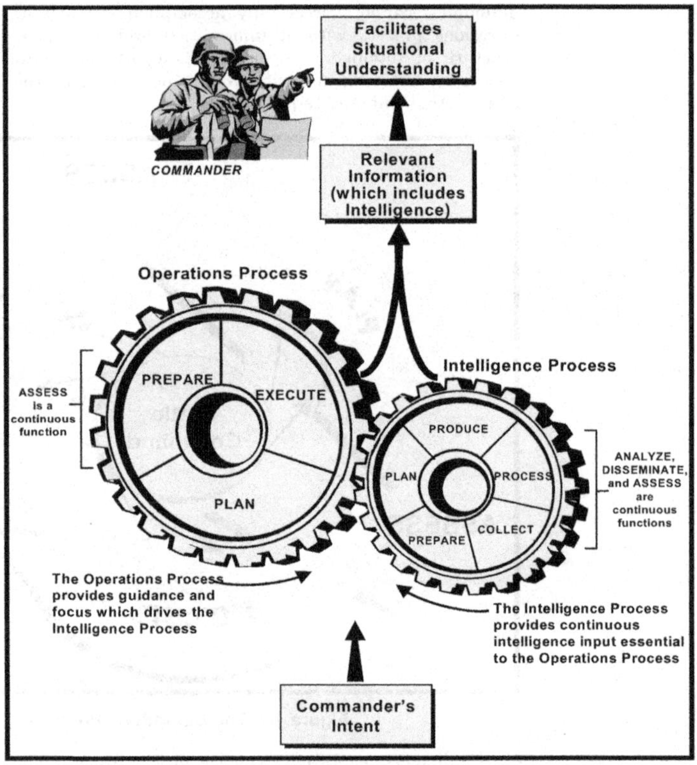

Figure 4-2. The Relationship Between the Operations and Intelligence Processes.

PLAN

4-4. The planning step of the intelligence process consists of activities that include assessing the situation, envisioning a desired outcome (also known as setting the vision), identifying pertinent information and intelligence requirements, developing a strategy for ISR operations to satisfy those requirements, directing intelligence operations, and synchronizing the ISR effort. The commander's intent, planning guidance, and CCIRs (PIRs and FIRs) drive the planning of intelligence operations. Planning, managing, and coordinating these operations are continuous activities necessary to obtain information and produce intelligence essential to decisionmaking.

COORDINATE

4-5. Staff and leaders coordinate with various elements, units, and organizations to ensure the necessary resources, linguist support (see Appendix B), information, intelligence, training, and procedures are in place to facilitate effective intelligence operations.

- Coordination for Movement of ISR Assets. All ISR assets at one time or another will move through or near another unit's AO. To avoid fratricide, ISR elements must coordinate with units, G3/S3, G2/S2, and each other, as well as coordinate with the fire support officer (FSO) to establish no-fire areas and/or other control measures around ISR assets and the air defense officer (ADO) in reference to aerial ISR assets in order to establish the appropriate weapons control status.
- Coordination for Information and Intelligence. The intelligence staff must prepare and practice coordination with personnel from all MI units, non-MI units, other service components, and multinational organizations that may contribute to or facilitate the collection effort. This coordination enables the G2/S2 to share and update databases, information, and intelligence and ensures connectivity with those organizations. All units are sources of relevant information regarding the enemy and the operational environment.
- Liaison. In order to accomplish the mission, exchange information and intelligence, move through certain areas and ensure FP, it may be necessary to coordinate with many different elements, organizations, and local nationals of the country in which friendly forces are conducting operations. Local nationals include police, town officials, foreign military forces, and political and other key figures within the AO. Operations may also necessitate coordination with other US and multinational forces; for example, the International Police Task Force (IPTF), Joint Commission Observers (JCO), Organization for Security and Cooperation in Europe (OSCE), Allied Military Intelligence Battalion (AMIB), and Defense HUMINT Service (DHS).
- Movement. Coordination with the G3/S3 ensures ease of movement and safe passage of friendly forces through an area. Coordinating movement also helps avoid fratricide.

INTELLIGENCE BOS CONSIDERATIONS

4-6. The Intelligence BOS is a unified system that anticipates and satisfies intelligence needs. Commanders ensure its proper employment by clearly articulating intent, designating CCIRs (PIRs and FFIRs), and prioritizing targets. Commanders must, however, understand the limitations of the Intelligence BOS to preclude unrealistic expectations of the system. The following are Intelligence BOS limitations:

- Intelligence only reduces uncertainty on the battlefield; it does not eliminate it entirely. The commander will always have to determine the presence and degree of risk involved in conducting a particular mission.
- The Intelligence BOS is composed of finite resources and capabilities. Intelligence systems and soldiers trained in specific ISR skills are limited in any unit. Once lost to action or accident, these soldiers and systems are not easily replaceable; for some, it may not be possible to replace them during the course of the current operation. The loss of soldiers and equipment can result in the inability to detect or analyze enemy actions. The loss of qualified language-trained soldiers, especially soldiers trained in low-density languages or skills, could adversely affect intelligence operations as well.
- The Intelligence BOS cannot effectively and efficiently provide intelligence without adequate communications equipment, capacity, and connectivity. Commanders and G2/S2s must ensure communications support to intelligence has the appropriate priority.
- Commanders and G2/S2s cannot expect that higher echelons will automatically send them everything they need. While intelligence reach is a valuable tool, the push of intelligence products from higher echelons does not relieve subordinate staffs from conducting detailed analysis and focusing the efforts of higher headquarters. Nor can they expect products pushed to them to be always at the level of detail they require. Commanders and G2/S2s must focus higher echelons by clearly articulating and actively pursuing intelligence requirements. By providing higher echelons with a clear picture of the required intelligence products, commanders can also narrow the flow of intelligence and information and preclude being overwhelmed by too much information.

4-7. Commanders should be aware that intelligence collection is enabled by, and subject to, laws, regulations, and policies to ensure proper conduct of intelligence operations. While there are too many to list here specifically, categories of these legal considerations include United States Codes (USCs), Executive Orders, National Security Council Intelligence Directives (NCSIDs), Army Regulations, United States Signal Intelligence Directives (USSIDs), SOFAs, ROE, and other international laws and directives.

PREPARE

4-8. The prepare step includes those staff and leader activities which take place upon receiving the OPORD, OPLAN, WARNO, or commander's intent to improve the unit's ability to execute tasks or missions and survive on the battlefield. These activities include—

- Effecting necessary coordination in accordance with the OPORD, METT-TC, unit SOP.
- Establishing and testing the intelligence architecture. This activity includes complex and technical issues like hardware, software, communications, COMSEC materials, network classification, technicians, database access, liaison officers (LNOs), training, funding, and TTP.
- Establishing an intelligence team attitude. This activity includes knowing different unit's and organization's capabilities, training the necessary collective skills, establishing effective relationships with different units and organizations, developing mutual battle rhythms and TTP, and leveraging the right architectures and collaboration tools.
- Coordinating effective analytic collaboration. Effective analytic collaboration is necessary to maximize the complementary analytic capabilities of different units and organizations that produce intelligence within the same theater of operations. Coordinating this collaboration is an effort-intensive activity that requires careful mutual planning, division of labor, defined responsibilities, and procedures for adapting to changing circumstances as they develop.
- Establishing reporting procedures.
- Conducting IPB.
- Producing Intelligence Estimates.
- Presenting briefings.
- Ensuring staff and personnel are trained. If personnel are not adequately trained at this point, they must be trained or the leader must evaluate the risk they bring to the operation.
- Planning refinement, brief-backs, SOP reviews, rehearsals, and coordinating with various elements and organizations.
- Establishing other troop-leading procedures (TLPs) or coordination, as necessary, in accordance with METT-TC factors.

G2/S2 PREPARATION ACTIVITIES

4-9. The G2/S2 takes numerous steps before mission execution to ensure intelligence operations run smoothly and effectively. These steps include, but are not limited to, the following:

- Conduct rehearsals.
- Conduct communication rehearsals.
- Review and update available databases and IPB products.
- Review applicable SOPs, Army Regulations, DA Pamphlets, Field Manuals, and ROE for guidance in conducting intelligence operations.
- Plan and practice actions supporting likely contingencies, or the branches or sequels to an operation.
- Ensure coordination measures are still in effect.
- Ensure training (individual and collective).

- Verify communications protocols with theater and higher headquarters and subordinate and lateral units.
- Update intelligence databases.
- Update the forces with the most recent intelligence on the AO immediately before mission execution.

VERIFICATION

4-10. Coordination for or requesting provisions or services is only the first step in acquiring them. It is crucial that staff and leaders check to verify that procedures, personnel, equipment, and services are in place and ready for mission execution.

REHEARSALS

4-11. Rehearsals help units prepare for operations by either verifying that provisions and procedures are in place and functioning or identifying inadequacies, which staff and leaders must remedy. They allow participants in an operation to become familiar with and to translate the plan into specific actions that orient them to their environment and other units when executing the mission. They also imprint a mental picture of the sequence of key actions within the operation and provide a forum for subordinate and supporting leaders and units to coordinate. (FM 6-0)

REPORTING PROCEDURES

4-12. The timely and accurate reporting of CCIRs (PIRs and FFIRs) and IRs is key to successful operations. All assets should know when, how often, and what format to use when reporting. The G2/S2 must verify the frequencies, alternate frequencies, and reactions during jamming, as well as the LTIOV for specific information to be reported. Unit SOPs provide the proper reporting procedures.

4-13. The G2/S2 coordinates with the unit staff, subordinate and lateral commands, and higher echelon units to ensure that specific reporting assets, personnel, equipment (especially communications), and procedures are in place. The G2/S2 requests or establishes the appropriate message addresses, routing indicators, mailing addresses, and special security office (SSO) security accreditation for units.

COMMUNICATIONS

4-14. Staff and leaders must work closely with the G6/S6 or signal officer (SIGO) to coordinate for the required communication links. The unit may require classified and unclassified network connections for their equipment. If elements of the unit will be working outside the range of the unit's communications systems, then it is necessary to coordinate for global or extended range communications. Leaders must obtain the required type and amount of communications equipment and related components as well as the latest fills and frequencies. They must possess and be familiar with all the instructions, passwords, policies, regulations, and directives conducive to OPSEC. They must also ensure soldiers are trained in the use and procedures involved in operating communications equipment.

SITUATION UPDATES

4-15. Each staff section and element conducts activities to maximize the operational effectiveness of the force. Coordination and preparation are just as important, if not more important, as developing the plan. Staff preparation includes assembling and continuously updating estimates. For example, continuous IPB provides accurate situational updates for commanders.

INTELLIGENCE HANDOFF

4-16. A well-executed intelligence hand-off will ensure a smooth and seamless transition between units. It is important that the incoming unit becomes familiar with the operation as soon as possible to avoid compromising the intelligence production and flow of the mission. The following are points to consider during a mission hand-off:

- Briefings and reports (learn what briefings are required and when as well as report formats and requirements).
- Past, present, and planned activities within the AOI.
- Established SOPs (know procedures for reporting; intelligence contingency funds [ICFs] and incentive use if applicable; emplacement and use of ISR equipment).
- Key personalities (introductions are required; establish rapport and a good working relationship with all key personalities).
  - Key personnel on the base or camp (their responsibilities; how to contact them).
  - Key personnel in other US and multinational service components (coordinate for exchange of information and intelligence).
  - Key personalities from surrounding towns (local figures).
  - Key national level political and military figures.
- Supporting units (know where to go for provisions, information, or assistance and POCs within those organizations).
- Current attitudes (understand current attitudes and perspectives of the local populace).
- Equipment operation and idiosyncrasies (equipment may run on different applications; personnel may need to train on specific equipment and procedures).
- Area familiarization (identify NAIs, key terrain, minefields, and boundaries; know camp locations, routes and route names, checkpoints, towns, and troubled resettlement areas).

RULES OF ENGAGEMENT

4-17. Although ROE training was presented during the plan function of the intelligence process, leaders at all levels can take the opportunity during the prepare function to ensure their subordinates completely understand the ROE. It is also during this function that commanders may need to consider exceptions to, or modifications of, the ROE to facilitate HUMINT and CI collection or to enable the placement of ISR assets.

COLLECT

4-18. Recent ISR doctrine necessitates that the entire staff, especially the G3/S3 and G2/S2, must change their reconnaissance and surveillance mind-set to conducting ISR. The staff must carefully focus ISR on the CCIR (PIR and FFIR) but also enable the quick retasking of units and assets as the situation changes. This doctrinal requirement ensures that the enemy situation not just our OPLAN "drives" ISR operations. Well-developed procedures and carefully planned flexibility to support emerging targets, changing requirements, and the need to support combat assessment is critical. The G2/S2 and G3/S3 play a critical role in this challenging task that is sometimes referred to as "fighting ISR" because it is so staff intensive during planning and execution (it is an operation within the operation). Elements of all units on the battlefield obtain information and data about enemy forces, activities, facilities, and resources as well as information concerning the environmental and geographical characteristics of a particular area.

ISR TASKS

4-19. ISR tasks are the actions of the intelligence collection effort. ISR tasks consists of three categories:

- Intelligence.
- Surveillance.
- Reconnaissance.

4-20. Intelligence tasks are included in Annex B of the OPORD under Scheme of Intelligence. They include the following:

- **Intelligence Production**. Intelligence production includes analyzing information and intelligence and presenting intelligence products, conclusions, or projections regarding the OE and enemy forces in a format that enables the commander to achieve situational understanding.
- **Request for Information**. Submitting an RFI to the next higher or lateral echelon is the normal procedure for obtaining intelligence information not available through the use of available ISR assets. Users enter RFIs into an RFI management system where every other user of that system can see it. Hence, an echelon several echelons above the actual requester becomes aware of the request and may be able to answer it. A G2/S2 who receives an RFI from a subordinate element may use intelligence reach to answer RFIs.
- **Intelligence Reach**. Intelligence reach allows the commander to access the resources of national, joint, foreign, and other military organizations and units. Requestors can acquire information through push and pull of information, databases, homepages, collaborative tools, and broadcast services. (See Chapter 2 for more information on intelligence reach.)

4-21. For information on reconnaissance or surveillance tasks, refer to FM 7-15 and FM 3-55.

SPECIAL RECONNAISSANCE

4-22. Special reconnaissance (SR) is the complementing of national and theater intelligence collection assets and systems by obtaining specific, well-defined, and time-sensitive information of strategic or operational significance. It may complement other collection methods where there are constraints of weather, terrain, hostile countermeasures, and/or other systems availability. SR places US or US-controlled personnel conducting direct observation in hostile, denied, or politically sensitive territory when authorized. SOF may conduct these missions unilaterally or in support of conventional operations. (See JP 3-05 and FM 101-5-1.)

4-23. Army Special Operations Forces (ARSOF) elements conduct SR missions to obtain information not available through other means. SR operations encompass a broad range of collection activities to include reconnaissance, surveillance, and TA. SR complements national and theater collection systems that are more vulnerable to weather, terrain masking, and hostile countermeasures. SR missions provide intelligence or information that is often not available through other means. Typical SR missions include—

- TA and surveillance of hostile C2 systems, troop concentrations, deep-strike weapons, lines of communication (LOCs), WMD systems, and other targets.
- Location and surveillance of hostage, prisoner of war, or political prisoner detention facilities.
- Post-strike reconnaissance for BDA.
- Meteorologic, geographic, or hydrographic reconnaissance to support specific air, land, or sea operations.

4-24. For more information on special reconnaissance, see FM 3-05.102.

REPORTING

4-25. The most critical information collected is worthless if not reported in a timely manner. Collectors may report information via verbal, written, graphic, or electronic means. Unit SOPs must clearly state the transmission means of different types of reports (for example, sent by voice frequency modulated [FM] radios or by automated means). In general, the transmission of reports for enemy contact and actions, CCIRs, exceptional information, and NBC reports is by voice FM, and then followed up with automated reports. Commanders and staffs must remember that timely reporting, especially of enemy activity, is critical in fast-moving operations. Collectors must report accurate information as quickly as possible. Commanders and staff must not delay reports for the sole purpose of editing and ensuring the correct format. This is particularly true for reporting information or intelligence that answers the PIR.

TIME-SENSITIVE REPORTING

4-26. Intelligence and time-sensitive combat information that affects the current operation is disseminated immediately upon recognition. Combat information is unevaluated data, gathered by or provided directly to the tactical commander which, due to its highly perishable nature or the criticality of the situation, cannot be processed into tactical intelligence in time to satisfy the user's tactical intelligence requirements. Thus combat information is provided directly to the tactical commander (see JP 1-02). The routing of combat information proceeds immediately in two directions: directly to the commander and through routine reporting channels, which include intelligence analysis and production elements.

4-27. Time-sensitive information usually includes reports concerning enemy contact and actions and CCIRs.

PROCESS

4-28. The process function converts relevant information into a form suitable for analysis, production, or immediate use by the commander. Processing also includes sorting through large amounts of collected information and intelligence (multidiscipline reports from the unit's ISR assets, lateral and higher echelon units and organizations, and non-MI elements in the battlespace). Processing identifies and exploits that information which is pertinent to the commander's intelligence requirements and facilitates situational understanding. Examples of processing include developing film, enhancing imagery, translating a document from a foreign language, converting electronic data into a standardized report that can be analyzed by a system operator, and correlating dissimilar or jumbled information by assembling like elements before the information is forwarded for analysis.

4-29. Often collection assets must collect and process their data prior to disseminating it. MI systems have their own reporting and processing systems, the details of which are in the appropriate MI system manuals and technical manuals. Some collection assets, particularly air reconnaissance and ground scouts, can report relevant information that is immediately usable by the tactical commander (for example, for targeting purposes). However, the personnel in the reporting chain still process these reports by evaluating their relevancy and accuracy. In many cases, the output of a collection asset is data, or information of limited immediate use to a commander. Also, in certain situations ROE dictate a requirement for target confirmation by other sources.

4-30. The intelligence staff processes information collected by the unit's assets as well as that received from higher echelons. The intelligence staff processes many types of information and data from intelligence reach, unmanned aerial vehicle (UAV) imagery, radar imagery, mobile target indicators (MTIs), and HUMINT and SIGINT reports.

PRODUCE

4-31. In the production step, the G2/S2 integrates evaluated, analyzed, and interpreted information from single or multiple sources and disciplines into finished intelligence products. Like collection operations, the G2/S2 must ensure the unit's information processing and intelligence production are prioritized and synchronized to support answering the CCIRs (PIRs and FFIRs).

4-32. Intelligence products must be timely, relevant, accurate, predictive, and usable. The accuracy and detail of every intelligence product has a direct effect on how well the unit plans and prepares for operations. However, the G2/S2 and unit must use intelligence (no matter what form the intelligence is in) that meets the requirements but might not be as detailed or refined as possible or in a better form. A good answer on time is better than a more refined answer that is late.

4-33. The G2/S2 produces intelligence for the commander as part of a collaborative process. The commander drives the G2/S2's intelligence production effort by establishing intelligence and IRs with clearly defined goals and criteria. Differing unit missions, environments, and situations impose numerous and varied production requirements on the G2/S2 and his staff.

4-34. The G2/S2 must employ collaborative analysis techniques and procedures that leverage intelligence production capability of higher and subordinate echelons to meet these requirements. Proficiency in these techniques and procedures enables the G2/S2 to answer the commander's and staff's requirements regardless of the mission, environment, and situation. The G2/S2 and staff intelligence products enable the commander to—

- Plan operations and employ maneuver forces effectively.
- Recognize potential COAs.
- Employ effective tactics and techniques.
- Take appropriate security measures.
- Focus ISR.

COMMON INTELLIGENCE PROCESS TASKS

ANALYZE

4-35. Analysis occurs at various stages throughout the intelligence process. Personnel conducting intelligence operations at all levels analyze intelligence, information, and problems to produce intelligence, solve problems and, most importantly, answer the PIRs. Leaders at all levels conduct analysis to assist in making many types of decisions. An example is a HUMINT collector analyzing an intelligence requirement in order to determine the best possible collection strategy to use against a specific source.

4-36. Analysis in RM is critical to ensuring the IRs receive the appropriate priority for collection. The intelligence staff analyzes each requirement to determine its feasibility, whether or not it supports the commander's intent, and to determine the best method of satisfying the IRs. The staff also analyzes collected information to determine if it satisfies requirements.

4-37. During the produce function, the intelligence staff analyzes information from multiple sources to develop all-source intelligence products. The intelligence staff analyzes information and intelligence to ensure the focus, prioritization, and synchronization of the unit's intelligence production is in accordance with the PIRs.

4-38. In situation development, the intelligence staff analyzes information to determine its significance relative to predicted ECOAs and the CCIRs (PIRs and FFIRs). Through predictive analysis, the staff attempts to identify enemy activity or trends that represent opportunities or risks to the friendly force.

They use the indicators developed for each ECOA and CCIRs (PIRs and FFIRs) during the MDMP as the basis for their analysis and conclusions.

DISSEMINATE

4-39. Successful operations at the tactical and operational levels require an increased ability to synchronize fires, have faster access to intelligence, and enhance situational understanding and effective FP. Timely and accurate dissemination of intelligence is key to the success of these and other operations. Commanders must receive combat information and intelligence products in time and in an appropriate format to support decisionmaking. Additionally, sharing the most current all-source information and intelligence at all echelons is essential for commanders to maintain situational understanding.

4-40. To achieve this, it is imperative that the commander and staff establish and support a seamless intelligence architecture—including an effective dissemination plan—across all echelons to ensure information and intelligence flow in a timely manner to all those who need them. Intelligence and communications systems continue to evolve in their sophistication, application of technology, and accessibility to the commander. Their increasing capabilities also create an unprecedented volume of information available to commanders at all echelons. Finally, the commander and staff must have a basic understanding of these systems and how they contribute to the Intelligence BOS. A dissemination plan can be a separate product, or integrated into existing products such as the ISR synchronization plan or ISM, the decision support template (DST), or decision support matrix (DSM).

Dissemination Procedures

4-41. The G2/S2 and intelligence personnel at all levels assess the dissemination of intelligence and intelligence products.

4-42. Disseminating intelligence simultaneously to multiple recipients is one of the most effective, efficient, and timely methods. This can be accomplished through various means; for example, push, broadcast. However, within the current tactical intelligence architecture, reports and other intelligence products move along specific channels. The staff helps streamline information distribution within these channels by ensuring dissemination of the right information in a timely manner to the right person or element. There are three channels through which commanders and their staffs communicate: command, staff, and technical.

4-43. **Command Channel**. The command channel is the direct chain-of-command link that commanders, or authorized staff officers, use for command-related activities. Command channels include command radio nets (CRNs), video teleconferences (VTC), and the Maneuver Control System (MCS).

4-44. **Staff Channel**. The staff channel is the staff-to-staff link within and between headquarters. The staff uses the staff channel for control-related activities. Through the staff channel, the staff coordinates and transmits intelligence, controlling instructions, planning information, provides early warning information, and other information to support C2. Examples of staff channels include the operations and intelligence radio net, telephone, the staff huddle, VTC, and the BOS-specific components of the Army Battle Command System (ABCS).

4-45. **Technical Channel.** Staffs typically use technical channels to control specific combat, CS, and CSS activities. These activities include fire direction and the technical support and sensitive compartmented information (SCI) reporting channels of intelligence and ISR operations. The SIGINT tasking and reporting radio net, intelligence broadcast communications, and the wide area networks (WANs) supporting single intelligence discipline collection, processing, and production are examples of technical channels.

Presentation Techniques and Procedures

4-46. The staff's objective in presenting information is to provide the commander with relevant information. Table 4-1 lists the three general methods that the staff uses to present information and meet its information objective. Systems within the ABCS contain standard report formats, maps, and mapping tools that assist the staff in presenting information in written, verbal, and graphic form. Audio and video systems such as large format displays and teleconferencing systems enable the staff to use a combination of the methods in multimedia presentations.

Table 4-1. Presentation Methods and Products.

Method	Products
Written Narrative	Reports, Estimates, and Studies
Verbal Narrative	Briefing (information, decision, mission, and staff)
Graphic	Charts, Overlays, and Electronic Displays

Intelligence Communications Architecture

4-47. The intelligence communications architecture transmits intelligence and information to and from various ISR elements, units, and agencies by means of automation and communication systems. With the continued development of sensors, processors, and communications systems, it is increasingly important to understand the requirements of establishing an effective communications architecture. The G2/S2 must identify the Intelligence BOS specific requirements of the unit's overall communications architecture. Refer to FM 2-33.5 for more information on intelligence communications reach. The following are some (but not all) of the questions which the staff must answer in order to establish the intelligence communications architecture:

- Where are the unit's collectors?
- What and where are the unit's processors?
- Where are the unit's intelligence production elements?
- Where are the unit's decisionmakers?
- How does the unit disseminate information from its producers to its decisionmakers and/or consumers?

- Are the systems which the unit's collectors, producers, processors, and consumers use compatible with each other? If not, what is the plan to overcome this challenge?
- How can the unit access databases and information from higher and other agencies?

ASSESS

4-48. Assessment is the continuous monitoring—throughout planning, preparation, and execution—of the current situation and progress of an operation, and the evaluation of it against criteria of success to make decisions and adjustments. Assessment plays an integral role in all aspects of the intelligence process. Assessing the situation and available information begins upon receipt of the mission and continues throughout the intelligence process. The continual assessment of intelligence operations and ISR assets, available information and intelligence, the various aspects of the battlefield environment, and the situation are critical to—

- Ensure the CCIRs (PIRs and FFIRs) are answered.
- Ensure intelligence requirements are met.
- Redirect collection assets to support changing requirements.
- Ensure operations run effectively and efficiently.
- Ensure proper use of information and intelligence.
- Identify enemy efforts at deception and denial.

4-49. During planning, the intelligence staff conducts a quick initial assessment of the unit's intelligence posture and holdings, status of intelligence estimates, and any other available intelligence products. From this assessment the commander issues his initial guidance and a WARNO.

4-50. While the majority of the unit is engaged in preparation, the ISR effort should already have begun. It is during this period when the prepare and execute activities of the operations process overlap, that the G2/S2 assesses the current situation as well as the progress of ISR operations.

4-51. During execution the intelligence staff continues assessing the effectiveness of the ISR effort while at the same time assessing the results and products derived from the collection effort. The critical aspects of assessment at this point include determining whether the PIRs have been answered, will be answered with the current ISR operations, or which ISR operations to adjust in order to answer the CCIRs (PIRs and FFIRs). This type of assessment requires sound judgment and a thorough knowledge of friendly military operations, characteristics of the AO and AOI, and the threat situation, doctrine, patterns, and projected future COAs.

PART THREE

Military Intelligence Disciplines

Part Three provides a more detailed explanation of the intelligence disciplines introduced in Part One of this manual.

Chapter 5 defines and discusses the roles and fundamentals of the all-source intelligence discipline.

Chapter 6 defines and discusses the roles, fundamentals, and generic organization of the HUMINT discipline.

Chapter 7 defines and discusses the roles, fundamentals, and generic organization of the IMINT discipline.

Chapter 8 defines and discusses the roles, fundamentals, and generic organization of the SIGINT discipline.

Chapter 9 defines and discusses the roles, fundamentals, and generic organization of the MASINT intelligence discipline.

Chapter 10 defines and discusses the roles, fundamentals, and generic organization of the TECHINT discipline.

Chapter 11 defines and discusses the roles, fundamentals, and generic organization of the CI discipline.

Chapter 5

All-Source Intelligence

DEFINITION

5-1. All-source intelligence is defined as the intelligence products, organizations, and activities that incorporate all sources of information and intelligence, including open-source information, in the production of intelligence. All-source intelligence is a separate intelligence discipline, as well as the name of the function used to produce intelligence from multiple intelligence or information sources.

ROLE

5-2. The operational environment provides an ever-growing volume of data and information available from numerous sources, from which the commander can use to achieve situational understanding. His situational understanding enables him to make decisions in order to influence the outcome of the operation; prioritize and allocate resources; assess and take risks; and understand the needs of the higher and subordinate commanders. The commander depends upon a skilled G2/S2 working within his intent to support his ISR effort and provide all-source intelligence analysis conclusions and projections of future conditions or events.

FUNDAMENTALS

5-3. All-source intelligence production satisfies intelligence requirements. It provides an overall picture of the adversary and the battlespace. It reduces the possibility of error, bias, and misinformation through the use of multiple sources of information and intelligence.

PLAN

5-4. The utilization of all-source intelligence facilitates the development of accurate and concise contingency-specific plans, orders, and intelligence products. Additionally, the G2/S2 must retrieve, update, or develop any required intelligence databases. The most important all-source products are the modified combined obstacle overlay (MCOO), event templates, ECOA sketches, and the HPT list provided during initial IPB in support of the plan function. Through all-source intelligence, the commander can make informed decisions about COAs and adversary capabilities.

5-5. The intelligence staff coordinates its efforts with other elements such as the engineer terrain team, the unit surgeon, the Air Force weather team, and other assets or elements that can support the analytical effort.

5-6. IPB plays a significant role in the planning phase of the intelligence process. It is a systematic, continuous process of analyzing the threat and environment and is designed to support staff estimates and the MDMP. The IPB is led by the G2/S2, with participation by the entire staff. IPB allows the commander and staff to make informed decisions, develop COAs, and focus ISR efforts where they are most needed. IPB continuously assists the commander and staff in focusing ISR assets on the appropriate targets at critical points in time and space.

5-7. The time available for the IPB process may not permit the luxury of conducting each step in detail. Overcoming time limitations requires the commander to identify those parts of IPB that are most important to the commander in planning and executing his mission. Applying the specific steps or degree of detail performed varies according to METT-TC. There are four steps in the IPB process: define the battlefield environment, describe the battlefield's effects, evaluate the threat, and determine ECOAs.

Define the Battlefield Environment

5-8. In this step, the G2/S2—

- Identifies characteristics of the battlefield that will influence friendly and threat operations including terrain (mobility), weather, hydrological data, infrastructure, and civilian demographics. Produces accurate, timely, and predictive IPB products that depict the aspects of the battlefield.

- Identifies the limits of the command's AO and AOI.
- Identifies gaps in current intelligence holdings, identifies IRs, and recommends CCIRs (PIRs and FFIRs).

5-9. Defining the battlefield environment includes identifying enemy forces (their location, mobility, general capabilities, and weapon ranges) and all other aspects of the environment that could have an effect on the unit's ability to accomplish the mission. Depending on the situation, these considerations may also include—

- Geography, terrain, and weather of the area.
- Information environment to include but not limited to computer and communications systems and capabilities, data acquisition systems and capabilities, media access and distribution, areas prone to electromagnetic interferences, and systems to generate electromagnetic interference.
- Population demographics (ethnic groups, religious sects, age distribution, health status, income groups).
- Medical threat to include but not limited to endemic and epidemic diseases, occupational and environmental health hazards, and poisonous and toxic plants and animals.
- Political or socio-economic factors, including factions, clans, and gangs.
- Infrastructures, such as transportation or telecommunications, and critical decisionmaking infrastructures and supporting information systems.
- ROE or legal restrictions, such as international treaties or agreements.

Describe the Battlefield's Effects

5-10. This IPB step deals with the effects of the battlefield environment on the current operations and potential enemy and friendly COAs. It begins with the assessment of existing and projected conditions of the battlefield environment, which the staff accomplishes through terrain analysis, weather analysis, and analysis of other characteristics of the battlefield. To conduct terrain analysis, the staff uses maps, reconnaissance, and other specialized terrain products (maps, overlays, databases, software). These products address such factors as wet or dry cross-country mobility, transportation systems (road and bridge information), vegetation type and distribution, surface drainage and configuration, surface materials (soils), ground water, manmade structures, and obstacles. The results of evaluating the terrain's effects should be expressed by identifying areas of the battlefield that influence each COA.

Evaluate the Threat

5-11. Evaluating the threat involves determining the threat force capabilities and the doctrinal principles and TTP that threat forces prefer to employ. The result of this evaluation produces a threat model, which portrays how threat forces normally execute operations and how they have reacted to similar situations in the past. The threat model includes an evaluation of the threat's strengths, weaknesses, and vulnerabilities, including an evaluation of typical HVTs.

Determine ECOAs

5-12. Determining ECOAs involves the identification and development of likely ECOAs that will influence accomplishing the friendly mission. Developing ECOAs is a form of predictive intelligence analysis and production. The procedures for this step include—

- Identifying the enemy's likely objectives and predicting the desired end-state.
- Identifying the full set of COAs available to the enemy.
- Predicting the enemy's most likely and most dangerous COAs.
- Evaluating and prioritizing each COA.
- Developing each COA in the amount of detail that time permits.
- Identifying initial ISR and collection requirements.

5-13. After conducting the initial IPB, the staff, primarily the G2/S2, identifies gaps in the available intelligence, develops the initial PIRs and IRs, and develops the initial input to the ISR plan based on the commander's guidance.

5-14. Indicators are the basis for situation development. Indicators are activities that will confirm or deny the event specified in an intelligence requirement. They are any positive or negative evidence of enemy activity or characteristic of the AO that points toward enemy capabilities, vulnerabilities, or intentions. Individual indicators cannot stand alone. Each indicator is integrated with other factors and indicators before patterns are detected and enemy intentions established. Indicators are developed by the analysts in the G2/S2 section. The event matrix shows the threat activities, or indicators, to look for in each NAI, and the timelines during which each NAI should be active. All indicators are developed to answer the commander's PIRs and IRs. The analyst uses indicators to correlate particular events or activities with probable ECOAs and to determine what events or activities must occur for an enemy to follow a particular COA. The ability to read indicators (including recognition of enemy deception indicators) contributes to the success of friendly operations. The analyst integrates information from all sources to confirm indicators of enemy activities. As indicators are detected and confirmed, PIRs are answered.

PREPARE

Activities

5-15. All-source activities during the prepare function include—

- Conducting rehearsals.
- Conducting communications rehearsals and verifiying communications protocols with higher, lateral, and subordinate units.
- Planning and practicing actions that support likely contingencies, branches, or sequels.
- Reviewing and updating available databases and IPB products.
- Ensuring control and coordination measures are still in effect.

- Updating the force with the most recent intelligence before mission execution.

Intelligence Estimates

5-16. The intelligence estimate is a continuous process that is the product of all actions the intelligence staff performs throughout the MDMP. The intelligence estimate provides a timely and accurate evaluation of the enemy and the AO (and often the AOIR) at a given time. It provides the background the G2/S2 uses to portray enemy actions during COA analysis.

5-17. The G2/S2 must clearly understand the weather and terrain effects and the ability to visualize the battlespace before producing the intelligence estimate. This understanding facilitates accurate assessments and projections regarding the enemy: enemy situation (including strengths and weaknesses), enemy capabilities and an analysis of those capabilities (COAs available to the enemy), and conclusions drawn from that analysis. The intelligence estimate's conclusion identifies the enemy's most likely COA and most dangerous COA. ECOAs must include a sketch of the COA with a task and purpose for the enemy's actions. Any gaps in the intelligence estimate are identified as information requirements.

5-18. The intelligence estimate may be written or oral. At the tactical level, especially during operations and exercises, intelligence estimates are usually delivered orally, supported by graphic displays and other decision support tools. During contingency planning, especially at corps level and above, estimates are usually written. During deliberate planning at joint headquarters, estimates are always written (see JP 5-00.2). However, the intelligence estimate should always be prepared as thoroughly as time and circumstances permit. A comprehensive intelligence estimate considers both the tangible (quantifiable) and the intangible aspects of the enemy's operations. It translates enemy strengths, weapon systems, training, morale, and leadership into combat capabilities and projections of future enemy actions.

5-19. Different sections of the intelligence estimate receive more emphasis during different activities within the operations process.

- During planning, the most important decision the commander makes is selecting a COA on which to base the plan. Thus, the sections of the intelligence estimate that focus on the commander's selecting the most appropriate COA are the most important at that time.
- During preparation, the intelligence estimate must focus on the commander's decisions that affect the ability of the unit to execute the upcoming operation. The intelligence estimate—and functions—that supports these decisions must focus on answering the CCIRs (PIRs and FFIRs) and guiding the ISR effort.
- During execution, the intelligence estimate focus is to support the anticipated command decisions. The most important action of the intelligence estimate—and the ISR effort— is to answer the PIR. However, it is also during the execution phase that the intelligence estimate must look ahead: anticipating branches or sequels to the

current operation, transition to other operations, changes in the ECOAs, and required adjustments to the ISR effort.

5-20. Refer to FM 2-01 for examples of intelligence estimates.

COLLECT

5-21. The collect intelligence process function within the all-source intelligence discipline is limited to gathering the necessary information, intelligence, and intelligence products required to perform analysis. Thus, all-source intelligence depends upon the other intelligence disciplines to perform collection. This is also the primary reason why OSINT is defined as a source of information and intelligence and not a separate intelligence discipline.

PROCESS

5-22. The process function converts relevant information into a form suitable for analysis, production, or immediate use by the commander. Processing also includes sorting through large amounts of collected information and intelligence to identify and exploit the information which is pertinent to the commander's intelligence requirements and facilitates situational understanding.

PRODUCE

5-23. All-source intelligence analysis and production—

- Drives collection to answer the PIR.
- Provides the enemy situation.
- Provides INTSUMs and other intelligence reports.
- Supports situational understanding.
- Provides predictive estimates of enemy actions; specifically, ECOAs.
- Provides continuously updated IPB.
- Provides all-source target packages (or folders).

5-24. As previously stated, the intelligence estimate is a continuous process. The commander and staff constantly collect, process, store, display, and disseminate information. Staff members update their estimates as they receive new information, such as—

- When they recognize new facts.
- When they replace assumptions with facts or find their assumptions invalid.
- When they receive changes to the mission or when changes are indicated.

5-25. Technological advances and NRT information allow estimates to be continuously updated; the running estimate and its intelligence component, the intelligence running estimate, exemplify this.

Running Estimate

5-26. Running estimates are continuously updated estimates based on new information as the operation proceeds. They serve as a staff technique to support the commander's visualization and decisionmaking, as well as the staff's tool for assessing during preparation and execution. In the running estimate, staff officers continuously update their conclusions and recommendations as they evaluate the impact of new facts. Current doctrine emphasizes the COP as the primary tool that provides the current situation and, when merged with the running estimates, facilitates commanders achieving situational understanding. All staff sections (BOSs) provide their respective input to the COP. The COP as used today, combined with the running estimate, is predictive and enhances our current ability to collect, process, store, display, and disseminate information.

5-27. The portion of the COP that depicts the enemy situation is currently limited to displaying the locations and dispositions of enemy and threat forces in a relatively static manner, sometimes referred to as snapshots in time. The enemy situation portion of the COP requires analysis to provide the required level of detail. The COP and running estimates sufficiently provide effective support for battle command of knowledge-based organizations.

5-28. The main differentiation between the running estimate and the old staff estimates is the emphasis on not only continuously updating the facts of the estimate but also continuously updating the conclusions and recommendations while including projections of future conditions of the entire battlespace. While the COP is primarily a display of current intelligence and information, the running estimate requires the merging of the staff's cognitive processes with automation applications. The primary focus of the staff's cognitive process is to present predictive or anticipatory intelligence in support of the commander's decisionmaking or situational understanding. See Figure 5-1.

5-29. The running estimate is a product of the entire battle staff. Just as all BOSs contribute to the COP, they will also contribute their portion of the running estimate. The running estimate integrates the running estimates from each BOS. The Intelligence BOS input to the running estimate is the intelligence running estimate.

Intelligence Running Estimate

5-30. The intelligence running estimate is a continuous flow and presentation of relevant information and predictive intelligence that, when combined with the other staff running estimates, enables the decisionmaker's visualization and situational understanding of the AOI in order to achieve information superiority. The intelligence running estimate requires constant verification to support situational understanding of the current situation as well as predictive assessments for future operations.

5-31. While it is possible to employ the concept of the intelligence running estimate today, the true seamless and continuous update feature will be achieved with technological enablers that are not yet present in the force. In the future, technology should allow the information from the running estimate to become implanted as a part of the COP. However, there will always be a place for some separate context, orally or written, separate from the COP display to add fidelity and assist in the commander's visualization and decisionmaking. Additional training must be implemented not just within MI but also throughout the force in order to institute the running estimate concept.

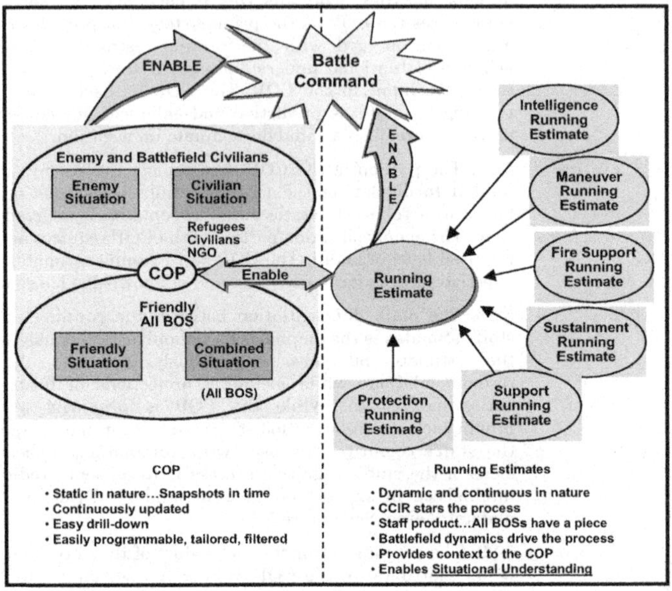

Figure 5-1. Example of the COP and a Running Estimate.

ANALYZE

5-32. Analysis is key in converting combat information and all intelligence from each discipline into all-source intelligence products and targeting information. From the multiple sources of information received, the staff analyzes and identifies critical information; determines the relevancy and accuracy of the information; and reaches conclusions about the information. These conclusions are either immediately disseminated or are used to form the basis of intelligence analysis products.

DISSEMINATE

5-33. The intelligence staff extracts and disseminates pertinent intelligence from products developed as a result of all-source analysis. The intelligence staff—

- Rapidly disseminates time-sensitive, all-source analysis intelligence and intelligence products to higher, lateral, and subordinate commands in order to keep all commanders abreast of current developments in the situation and battlespace in accordance with unit SOPs.

- Disseminates all-source analysis intelligence to other G2/S2 intelligence activities for additional processing or detailed analysis and exploitation in accordance with unit SOPs.

5-34. For intelligence reach operations all-source intelligence products are available and disseminated in a variety of forms. It is incumbent on the requestor to ensure that the all-source product can be transmitted over the available communications systems. This includes verifying the appropriate security level of the communication system.

ASSESS

5-35. The commander, the intelligence staff, and the intelligence consumers assess the analysis, production, and dissemination of all-source intelligence. The staff identifies intelligence gaps, emerging operational requirements, duplication of effort, or new targets or threats and redirects all-source activities as appropriate to meet intelligence requirements. Throughout the assessment process, the G2/S2, the intelligence staff, and intelligence users should be alert for evidence of possible enemy denial and deception efforts. The staff assesses intelligence and information for—

- Consistency with the current situation and threat trends.
- Accuracy and confirmation by other intelligence sources or disciplines.
- Source reliability and credibility.
- Pertinence to PIRs, intelligence requirements, RFIs, and other intelligence tasks or requests.

Chapter 6

Human Intelligence

DEFINITION

6-1. HUMINT is the collection by a trained HUMINT Collector of foreign information from people and multimedia to identify elements, intentions, composition, strength, dispositions, tactics, equipment, personnel, and capabilities. It uses human sources as a tool and a variety of collection methods, both passively and actively, to gather information to satisfy the commander's intelligence requirements and cross-cue other intelligence disciplines.

ROLE

6-2. The role of HUMINT Collectors is to gather foreign information from people and multiple media sources to identify adversary elements, intentions, composition, strength, dispositions, tactics, equipment, personnel, and capabilities. It uses human sources and a variety of collection methods to gather information to satisfy the commander's intelligence requirements, and cross-cue other intelligence disciplines.

HUMINT FUNCTIONS

6-3. HUMINT functions are interrelated, mutually supporting, and can be derived from one another. No single function or technical capability can provide a full understanding of our adversaries. HUMINT functions are defined below and shown in Figure 6-1.

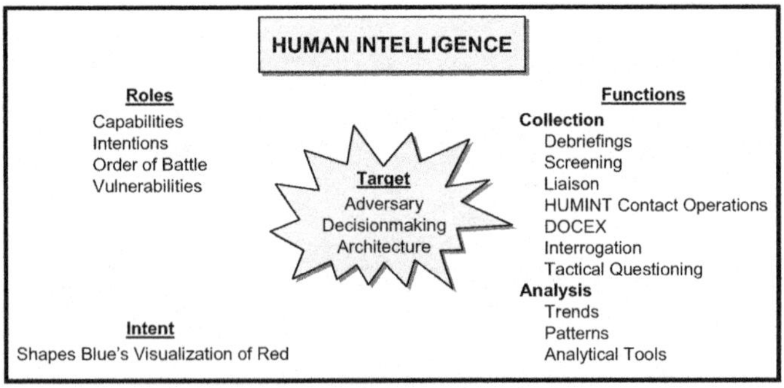

Figure 6-1. HUMINT Roles and Functions.

COLLECTION

6-4. HUMINT collection methods and operations include—

- **Debriefings.** Debriefings are the systematic questioning of individuals to procure information to answer specific collection requirements by direct and indirect questioning techniques. Sources for debriefings are categorized as friendly forces, US and non-US civilians to include refugees, displaced persons, and local inhabitants.
- **Screening.** Screening is the process of evaluating and selecting human and media sources for the prioritized collection of information in support of commander's PIRs and IRs. Screening categorizes and prioritizes sources based on the probability of a particular source having priority information and the degree of difficulty of extracting that information from the source. While screening is not in itself an information collection technique, it is vital to the rapid collection of information.
- **Liaison.** HUMINT elements conduct liaison with US, multinational, and HN military and civilian agencies, to include NGOs. Liaison is conducted to obtain information of interest and to coordinate or deconflict HUMINT activities.
- **HUMINT Contact Operations.** HUMINT contact operations are tactically oriented, overt collection activities that use human sources to identify attitude, intentions, composition, strength, dispositions, tactics, equipment, target development, personnel, and capabilities of those elements that pose a potential or actual threat to US and multinational forces. These forces provide early warning of imminent danger to US and multinational forces and contribute to the MDMP.
- **Document Exploitation (DOCEX).** DOCEX is the systematic extraction of information from all media formats in response to collection requirements.
- **Interrogation.** Interrogation is the systematic effort to procure information to answer specific collection requirements by direct and indirect questioning techniques of persons who are in the custody of the forces conducting the questioning.

UNIT SUPPORT TO HUMINT COLLECTION

6-5. Small units contribute to HUMINT collection through a number of different ways.

Tactical Questioning

6-6. Tactical questioning can provide critical information for situational understanding. Tactical questioning is the expedient initial questioning for information of immediate tactical value. Soldiers conduct tactical questioning based on the unit's SOP, ROE, and the order for that mission. Unit leaders must include specific guidance for tactical questioning in the order for appropriate missions. The unit S3 and S2 must also provide specific guidance down to the unit level to help guide tactical questioning. For a more detailed discussion of tactical questioning, see ST 2-91.6.

Combat Patrolling

6-7. Through observing and interacting with the local environment during the conduct of missions, handling EPWs/detainees, and handling captured documents, soldiers serve as the commander's "eyes and ears" whether—

- Performing traditional offensive or defensive operations.
- Performing a patrol in a stability operation.
- Manning a TCP or a roadblock in a support operation.
- Occupying an observation post.
- Passing through an area in a convoy.
- Performing any operation that involves observing and reporting on elements of the environment and activities in the AO.

ISR Operations

6-8. The information that the soldier reports as a result of completed missions up the chain of command forms a vital part of planning and operations. Careful and quick handling of EPWs/detainees and documents also helps the ISR effort. Unit headquarters must ensure that the reports they receive and forward contain all the subtle detail provided by the small unit. For tactical operations, there are four levels of reporting:

- Immediate reporting of information of critical tactical value, based either on predetermined criteria or common sense.
- Normal reporting, submitted before the unit S2 section performs the debriefing.
- Debriefing performed by the S2 section.
- Follow-up reporting, submitted after the unit S2 section performs the debriefing.

6-9. The four levels of reporting facilitate the unit S2 section capturing all the fine details of the small unit's activities for all-source analysis and future planning. The unit S2 must ensure that information of HUMINT and CI value is reported to the G2X.

ANALYSIS

6-10. IPB, all-source, and single-source analysis are used to template adversarial OB. HUMINT elements analyze operational taskings to determine the best collection methods to employ in satisfying CCIRs (PIRs and FFIRs) and IRs. As information is collected, it is analyzed for completeness and accuracy to identify significant facts for subsequent interpretation and inclusion in intelligence products, and passing to other collectors. This analysis also helps to identify collection gaps and focus or refocus collection efforts. Analysis provides the commander with situational understanding of the battlespace to execute operations. Raw information, open-source material, and finished intelligence products are analyzed in response to local and national requirements. Analysis occurs at all levels from tactical to strategic.

- At the tactical level, HUMINT teams focus their efforts on supporting mission requirements and contributing to the all-source COP.

- Operational analysis is used to assess adversary intentions, capabilities, dispositions, and their regional impacts.
- Strategic analysis supports national and Army programs through compilation of local and regional analysis into global assessments; identifies technology development trends and patterns in military activities and capabilities.

6-11. HUMINT products consist of, but are not limited to, target nomination, input to threat and vulnerability assessments, intelligence estimates, and intelligence information reports. Finalized intelligence derived from HUMINT activities is incorporated into joint and national intelligence databases, assessments, and analysis products. HUMINT products are also incorporated into the COP to support situational awareness. HUMINT production takes place at all levels.

- Operational and tactical production includes tactical alerts, spot reports, and current intelligence; input to threat and/or vulnerability assessments tailored to specific activities, units, installations, programs or geographic areas, and target studies to support contingency planning and major exercises; studies of military activities and capabilities.
- Strategic products include assessments supporting national and Army information requirements on foreign technology development; worldwide assessments of the organization, location, funding, training, operations capabilities and intentions of terrorist organizations; analyses of the capabilities of international narcotics trafficking organizations.

OPERATIONAL EMPLOYMENT

6-12. Army HUMINT must support the full spectrum of military operations. Commanders require a well-trained HUMINT force consisting of AC, RC, civilian government, and contractor personnel. The HUMINT force is focused on and dedicated to the collection of data and information relevant to the commander's PIRs and IRs. Effective employment of Army HUMINT elements in all phases of operations and at all levels from tactical to strategic will remain paramount to ensuring that commanders have the best pictures possible of their adversaries. HUMINT Collectors use the following collection methods.

LEVELS OF EMPLOYMENT

6-13. **Strategic and Departmental.** Strategic and departmental operations will be conducted by HUMINT elements supporting national, DOD, and DA required missions (for example, support to NATO and special operations and missions). The Army strategic HUMINT force resides in DHS. Trained Army HUMINT professionals must be available to augment DOD HUMINT efforts and coordinate between Army and Joint or DOD elements during crisis and war. Strategic HUMINT activities are generally carried out by departmental level HUMINT elements or Army HUMINT assets directly assigned to DOD and Joint positions. The Army overt HUMINT capability at the strategic level will be deployed and under operational control (OPCON) of the Army Service Component Commander (ASCC).

6-14. **Operational.** HUMINT operational missions must support combatant commanders, generally in geographic AORs. Operational HUMINT assets provide capabilities to support theater requirements for HUMINT collection. Operational HUMINT elements will focus on threat identification and capabilities. As with all levels of employment, HUMINT activities and functions will include DOCEX, collection, operations, analysis, and production. HUMINT elements must be capable of quickly transitioning from a peacetime mission to crisis operations to support combatant commander requirements. Operational elements may also be deployed to support or reinforce tactical forces in CONOPS.

6-15. **Tactical.** Army HUMINT play a crucial role in supporting tactical forces. HUMINT teams conduct operations throughout the battlespace during CONOPS. CONOPS support activities include debriefings, screenings, contact operations, and interrogations. HUMINT activities in CONOPS focus on the threat and assisting the senior intelligence officer (SIO) and commander in understanding the threat's decisionmaking process. During peacetime, organic tactical HUMINT teams conduct activities pursuant with approved regulations and command guidance.

SUPPORT TO CONTINGENCY OPERATIONS

6-16. **Initial Phase.** The initial phase of operations from peacetime military engagement (PME) to major theater war (MTW) lays the foundation of future team operations. In general, the priority of effort focuses inward on security of operating bases, areas of troop concentration, and C2 nodes to identify the collection threat to US forces that could be used by adversarial elements to plan hostile acts against US activities and locations.

6-17. **Continuation Phase.** Once security of the operating bases has been established, the operational focus of HUMINT teams shifts outside the operating base to continue to detect, identify, and neutralize the collection threat to US forces as well as to provide I&W of hostile acts targeting US activities. The HUMINT team uses several collection methods, to include HUMINT contact operations, elicitation, and liaison, to answer the supported commander's requirements.

6-18. **HUMINT Team.** A key element to the HUMINT team's success is the opportunity to spot, assess, and develop relationships with potential sources of information. Operating as independent teams, without being tied to combat assets, enables the HUMINT team's maximum interaction with the local population, thereby maximizing the pool of potential sources of information. A second key element of a HUMINT team's success is its approachability to the local population. A soft posture enables a HUMINT team to appear as non-threatening as possible. Experience has shown that the local population in general is apprehensive of fully and openly armed patrols and soldiers moving around population centers. During some operations civilian attire or non-tactical vehicles may be used to lower the

HUMINT team profile. NOTE: In some special situations, these measures are taken to make the operation less visible to the casual observer. Also, in some cultures, sharing food and beverages among friends is expected; exceptions to restrictions or general orders should be considered to facilitate successful HUMINT team operations, many of which are geared towards developing relationships with potential sources of information.

SUPPORT TO INSTALLATIONS AND OPERATING BASES

6-19. Commanders may restrict personnel to base camps and installations during the initial stages of operations (PME to MTW), when the operational environment is being assessed, or as a temporary expedient when the threat level exceeds the ability to provide reasonable FP. Operational restrictions minimize the risk to the HUMINT team, but minimizing its collection potential may increase the risk to the force as a whole. While confined to an installation or a base camp, the HUMINT team can maintain a limited level of information collection by—

- Screening locally employed personnel.
- Debriefing combat and ISR patrols.
- Debriefing friendly force personnel who are in contact with the local population.
- Conducting limited local open-source information collection.
- Contributing to the threat and vulnerability assessments of the base camp.

TACTICS, TECHNIQUES, AND PROCEDURES

6-20. At the HUMINT team level, team members conduct mission analysis and planning specific to their AO. Backwards planning and source profiling are used extensively to choose HUMINT targets. To verify adequate area coverage, the HUMINT team may periodically develop and use HUMINT target overlays and other HUMINT analytical tools that illustrate the HUMINT situation, identify gaps, and help refocus the collection effort.

6-21. The HUMINT team is also in constant contact with the supported S2 and the other ISR assets (Scouts, PSYOP, CA, and MPs) in order to coordinate and deconflict operations and to cross-check collected information. The supported unit S2, with the help of the HUMINT team, regularly and systematically debriefs all ISR assets.

6-22. The HUMINT team must be integrated into the supported unit's ISR plan. The HUMINT operational management team (OMT) chief will advise the supported unit on the specific capabilities and requirements of the team to maximize mission success.

OPERATIONAL RISK MITIGATION

6-23. The employment of HUMINT teams includes varying degrees of contact with the local population. As the degree of contact with the population increases, both the quantity and quality of collection increases. In many instances, however, there is a risk to the team inherent with increased exposure to the local population. The decision at what level to employ a team is METT-TC dependent. The risk to HUMINT assets must be balanced with the need to collect priority information and to protect the force as a whole. ROE, SOFA, direction from higher headquarters, and the overall threat level may also restrict the deployment and use of HUMINT teams. The commander should consider exceptions to the ROE to facilitate HUMINT collection.

6-24. Risks are minimized through the situational awareness of HUMINT team members. They plan and rehearse to readily react to any situation and carry the necessary firepower to disengage from difficult situations. If it becomes necessary to call for assistance, adequate and redundant communications equipment is critical. These scenarios and actions should be trained prior to deployment into a contingency area and rehearsed continuously throughout the deployment.

6-25. A supported unit commander is often tempted to keep the HUMINT team "inside the wire" when the threat condition (THREATCON) level increases. The supported commander must weigh the risk versus potential information gain when establishing operational parameters of supporting HUMINT teams. This is necessary especially during high THREATCON levels when the supported unit commander needs as complete a picture as possible of the threat arrayed against US and multinational forces.

6-26. When it is not expedient to deploy the HUMINT team independently due to threat levels or other restrictions, the team can be integrated into other ongoing operations. The HUMINT team may be employed as part of a combat, ISR, or MP patrol or used to support CA, PSYOP, engineer, or other operations. This method reduces the risk to the team while allowing a limited ability to collect information. It has the advantage of placing the team in contact with the local population and allowing it to spot, assess, and interact with potential sources of information. However, this deployment method restricts collection by subordinating the team's efforts to the requirements, locations, and timetables of the unit or operation into which it is integrated and does not allow for the conduct of sensitive source operations. **This method of employment should be considered a last resort.**

HUMINT EQUIPMENT

6-27. Basic C2, transportation, and weapons requirements do not differ significantly from most soldier requirements and are available as unit issue items. However, HUMINT teams have unique communications, collection, processing, and mission-specific requirements.

COMMUNICATIONS

6-28. Dedicated and secure long-range communications are key to the success of the HUMINT team mission. HUMINT team operations require a secure, three-tiered communications architecture consisting of inter/intra-team radios, vehicle-based communications, and a CI and HUMINT base station.

6-29. The HUMINT team must have access to existing communications networks such as the tactical local area network (LAN). The HUMINT team must also be equipped with its own COMSEC devices. It is imperative that the HUMINT team acquire access to the public communication system of the HN. This can be in the form of either landlines or cellular telephones. Such access enables the HUMINT team to develop leads which can provide early indicators to US forces.

Interoperability

6-30. Communications systems must be equipped with an open-ended architecture to allow for expansion and compatibility with other service elements, government organizations, NGOs, and multinational elements to effectively communicate during CONOPS. All ISR systems must be vertically and horizontally integrated to be compatible across all BOSs and with Legacy and Interim Force elements.

Satellite Communications On-The-Move (SOTM)

6-31. To provide real time or NRT information reporting, HUMINT elements must have the capability to transmit voice, data, imagery, and video while on the move. HUMINT teams must be able to transmit while geographically separated from their parent unit while operating remotely. This broadband requirement can only be achieved through a SATCOM capability and must be achievable while mobile.

HUMINT COLLECTION AND PROCESSING SYSTEMS

6-32. The HUMINT team must rely on automation to achieve and maintain information dominance in a given operation. With time, effective collection planning and management at all echelons, the HUMINT team can collect a wealth of information. The sorting and analysis of this information in a timely and efficient manner is crucial to operations. Automation helps the HUMINT team to report, database, analyze, and evaluate the collected information quickly and to provide the supported unit with accurate data in the form of timely, relevant, accurate, and predictive intelligence.

6-33. Automation hardware and software must be user friendly as well as interoperable among different echelons and services. They must interface with the communications equipment of the HUMINT team as well as facilitate the interface of audiovisual devices. Technical support for hardware and software must be available and responsive.

6-34. The demand for accurate and timely HUMINT reporting, DOCEX, and open-source information has grown tremendously. Biometric (physiological, neurological, thermal analysis, facial and fingerprint recognition) technologies will allow rapid identification; coding and tracking of adversaries and human sources; and cataloging of information concerning enemy prisoners of war (EPWs), detainees, and civilians of HUMINT interest on the battlefield. Biometrics will also provide secure authentication of individuals seeking network or facility access.

6-35. HUMINT teams work with multinational forces, and other foreign nationals, and require the ability to communicate in their respective languages. Often HUMINT personnel have little or no training in the target language, and lack of skilled interpreters can hinder HUMINT activities. HUMINT teams require textual and voice translation devices, source verification, and deception detection machines (biometrics) to improve collection capability and accuracy.

6-36. HUMINT teams require dynamic machine language translation (MLT) tools that provide both non-linguists and those with limited linguist skills a comprehensive, accurate means to conduct initial screenings and basic interviews in a variety of situations. HUMINT elements will focus on in-depth interviews and communications with persons of higher priority. MLT tools will minimize reliance on contract linguists and will allow soldiers to concentrate on mission accomplishment.

MISSION SPECIFIC

6-37. The HUMINT team may conduct night operations and must be equipped with NVDs for its members and photographic and weapons systems. The HUMINT team also may operate in urban and rural areas, where the threat level can vary from semi-hostile to hostile. The safety of the HUMINT team can be enhanced with equipment that can detect, locate, suppress, illuminate, and designate hostile optical and E-O devices. In addition, high power, gyro-stabilized binoculars, which can be used from a moving vehicle, also increase the survivability of the HUMINT team. It also gives the team another surveillance and collection device.

6-38. Some of the HUMINT team missions may require the documentation of incidents. The teams can use the following equipment in their open-source collection efforts.

- Small, rugged, battery-operated digital camcorders and cameras which are able to interface with the collection and processing systems as well as communication devices.
- GPSs that can be mounted and dismounted to move in the AO efficiently.
- Short-range multichannel RF scanning devices that can also identify frequencies which enhance their security.

6-39. In some cases HUMINT teams require a stand-off, high resolution optical surveillance and recording capability that can provide target identification at extended ranges to protect the intelligence collector while avoiding detection by the adversary target. An advanced optical capability provides intelligence collectors the ability to locate and track adversarial targets (passive and hostile) for identification, collection, and target exploitations.

INTEGRATION OF LINGUISTS

6-40. Integrating linguists into the HUMINT team should take place as soon as possible. Security clearances and contractual agreements will help the team determine the level of integration.

6-41. Along with the basic briefing of what is expected of the civilian linguists as interpreters, HUMINT teams should be informed about the civilians' chain of command and the scope of their duties beyond interpreting. The HUMINT team leader must ensure that linguists are trained and capable of completing all tasks expected of them.

BATTLE HAND-OFF

6-42. HUMINT teams are always engaged. A good battle hand-off is critical to smooth transition and mission success. The battle hand-off can directly contribute to mission success or failure of the outgoing team, but especially of the incoming team. The battle hand-off begins the first day the HUMINT team begins to operate in an AO. Regardless of how long the team believes it will operate within the AO, it must ensure there is a seamless transition to an incoming team, other US unit, or agency. The HUMINT team accomplishes this transition by establishing procedures for source administration, database maintenance, and report files.

6-43. Teams must plan and implement a logical and systematic sequence of tasks to enable an incoming team to assume the operations in the AO. Adequate time must be allotted for an effective battle hand-off. In some environments, a few weeks may be necessary to accomplish an effective battle hand-off. Introductions to sources of information, especially HUMINT contact operations sources, are critical and teams must prioritize their time. During this time the outgoing HUMINT team must familiarize the new HUMINT team with all aspects of the operation, which includes past, present, and planned activities within the AO. Area orientation is critical. These include major routes, population centers, potential hot spots, and other points of interest (such as police stations, political centers, and social centers).

ORGANIZATION

6-44. HUMINT activities require a complex C2 relationship to ensure that the requirements of the supported commander are fulfilled while balancing the need for strict integrity and legality of HUMINT operations. This complex relationship balances the role of the SIO as the requirements manager and the *2X as the mission manager with the MI commander as the asset manager. ("*2X" indicates 2X functions at all levels.)

COMMAND VERSUS CONTROL

6-45. ARFOR will normally deploy as part of a joint and/or multinational operation. In all cases, commanders at each echelon will exercise command over the forces assigned to their organization. Command includes the authority and responsibility for effectively using resources, planning for and employment of forces, and ensuring that forces accomplish assigned missions. Leaders and staffs exercise control to facilitate mission accomplishment.

6-46. While the MI commander supervises subordinates and produces reports, the *2X synchronizes activities between intelligence units and provides single-source processing and limited analysis. While the MI commander takes care of the operators executing missions, the *2X obtains the data and reports from higher echelons required to execute the missions.

STAFF RESPONSIBILITIES AND FUNCTIONS

*2X Staff

6-47. The *2X staff is responsible for the integration, correlation, and fusion of all Human Sensor information into the Intelligence BOS within the *2X

AOIR. The *2X is also responsible for analyzing adversary collection, terrorist and sabotage activities, developing countermeasures to defeat threat collection activities, identifying and submitting collection requirements to fill collection gaps, and providing input to the all-source picture regarding adversary intelligence activities.

*2X Staff Officer

6-48. The *2X Staff Officer provides CI and HUMINT collection expertise. The *2X—

- Is the single focal point for all matters associated with CI and HUMINT in the AOIR.
- Is the CI and HUMINT advisor to the G2 and commander.
- Is an extension of the collection manager and ensures that the best asset or combinations of assets are used to satisfy information requirements.
- Along with his subordinate elements (CICA, HOC, OSC, CI Analysis Cell [CIAC], and HAC), exercises technical control over his assigned Army CI and HUMINT elements in the designated AOIR.
- Is the principal representative of the G2 and the commander when coordinating and deconflicting CI and HUMINT activities with national or theater agencies operating in the AOIR.
- Supports specific RM efforts in conjunction with the requirements manager through—
  - Planning and coordinating CI and HUMINT operations.
  - Reviewing and validating HUMINT requirements.
  - Recommending assignment of tasks to specific collectors.
  - Conducting liaison with non-organic HUMINT collection. This liaison includes national level and multinational force assets for source deconfliction and special activities outside the *2X AOIR.
- Will provide OMTs with capability to reach back to current database information, technical information and guidance, and source deconfliction necessary to monitor the collection activities of the HUMINT teams.

The HUMINT Analysis Cell (HAC)

6-49. The HAC is the single-source fusion point for all HUMINT reporting and operational analysis. It determines gaps in reporting and coordinates with other analysis teams and technical controllers to cross-cue other collection sensor systems. The HAC—

- Uses analytical tools to develop long-term collection plans and provides reporting feedback that supports all HUMINT and CI entities in the supported command's AOIR.
- Produces and disseminates HUMINT products and provides input to intelligence summaries.
- Uses analytical tools found at the ACE or JISE to develop long-term analyses and provides reporting feedback that supports the J/G/S2X, HUMINT operations section, OMTs, and HUMINT teams.
- Produces country and regional studies tailored to HUMINT collection.

- Compiles target folders to assist J/G/S2X assets in focusing collection efforts.
- Analyzes and reports on trends and patterns found in HUMINT reporting.
- Analyzes source reliability and credibility as reflected in reporting and communicating that analysis to the collector.
- Develops and maintains databases specific to HUMINT collection activities that directly support the collection efforts of HUMINT teams and are directly accessible by HUMINT teams.
- Provides collection requirements input to the HOC.
- Supports RM through the development of HUMINT SIRs based on command PIRs.
- Answers HUMINT-related RFIs.

6-50. For intelligence reach operations, HUMINT products are available and disseminated in a variety of forms. It is incumbent on the requestor to ensure that the HUMINT product can be transmitted over the available communications systems. This includes verifying the appropriate security level of the communications systems.

HUMINT Operations Cell (HOC)

6-51. The HOC in the *2X coordinates and synchronizes all HUMINT activities in the AOIR. The HOC exercises technical control over all HUMINT entities in the designated AOIR and deconflicts HUMINT activities with higher, lower, and adjacent HUMINT elements. The HOC accomplishes all responsibilities through coordination with the operational units and the CICA and operations support cell (OSC). The HOC tracks all HUMINT activities in the AOR. The J/G2X uses this information to advise the SIO on all HUMINT activities conducted within the AO. The HOC—

- Exercises technical control of all HUMINT assets and coordinates and deconflicts HUMINT activities in the deployed AO.
- Establishes and maintains a HUMINT source database.
- Coordinates and supervises HUMINT FP source operations conducted by all services and components in the AO.
- Develops and manages collection requirements for HUMINT in coordination with the requirements manager.
- Develops and provides the HUMINT portion of the intelligence synchronization plan to the J/G/S2X and requirements manager for inclusion in the intelligence synchronization plan.
- Coordinates the activities of HUMINT collectors assigned or attached to interrogations and debriefing facilities.
- Expedites preparation of intelligence information reports and their distribution to consumers at all levels.
- Performs liaison with HN and US national HUMINT organizations.

Operations Support Cell

6-52. The OSC in the *2X staff maintains the source registry for all HUMINT activities in the designated AOIR. The OSC provides management of intelligence property book operations, source incentive programs, and ICFs.

6-53. Integrating linguists and DOD emergency-essential civilians, such as technical support contractors, into the HUMINT team should take place as soon as possible. Security clearances and contractual agreements will help the team determine the level of integration.

6-54. Along with the basic briefing of what is expected of the civilian linguists as interpreters and the emergency-essential civilians as support personnel, HUMINT teams should be informed about the civilians' chain of command and the scope of their duties beyond interpreting and technical support. The HUMINT team leader must ensure that linguists and emergency-essential civilians are trained and capable of completing all tasks expected of them.

Counterintelligence Coordinating Authority

6-55. The CICA coordinates all CI activities for a deployed force. There can be only one CICA in a theater of operations. When multiple echelons exist, the highest echelon has the CICA and subordinate G2X offices have a CICA. Depending on the size and scope of the operation, the CICA could be the unified command's CI staff officer; the CI or HUMINT staff officer from corps or division; or a senior warrant officer or branch CI officer designated by the unified command CI staff officer or task force commander. For more information about the CICA functions, see Chapter 11.

HUMINT TEAM STRUCTURE

Operational Management Team

6-56. The OMT is a four-person team consisting of a warrant officer (WO), two noncommissioned officers (NCOs), and a junior enlisted soldier. Civilians may be inserted into this structure when appropriate. Rank structure and standards of grade for OMTs vary depending upon the skill sets required and mission focus. HUMINT OMTs provide operational guidance for two to four HUMINT teams, depending on mission focus and operational tempo. When two or more HUMINT teams are deployed in DS of a maneuver element, an OMT also deploys to provide technical control. The OMT works closely with the supported S2 and analysis and control team (ACT) to furnish current threat information and answer the supported commander's PIRs and IRs. OMTs coordinate with the supported 2X and manage subordinate HUMINT team echelons to—

- Provide guidance and technical control of operational activity.
- Provide the collection and operational focus for HUMINT teams.
- Provide quality control and dissemination of reports for subordinate HUMINT teams.
- Conduct single-discipline HUMINT analysis, and assist in mission analysis for the supported commander.

- Act as a conduit between subordinate HUMINT teams, the HOC, and supported unit headquarters.
- Provide administrative support for subordinate HUMINT teams, to include reporting mission and equipment status to the HOC and the supported unit headquarters.
- Educate the supported commander on the capabilities of the HUMINT teams.
- Integrate the HUMINT teams directly into the maneuver commander's ISR planning.

HUMINT Team

6-57. The HUMINT team is a four-person team consisting of two NCOs and two junior enlisted personnel. Civilians may be inserted into this structure when appropriate. Rank structure and standards of grade for HUMINT teams vary depending upon the skill sets required and the mission focus. HUMINT teams are trained to execute the full range of HUMINT functions as defined in **HUMINT FUNCTIONS** section. However, they may be assigned to mission-focused elements such as DOCEX, interrogation, debriefing, or contact operations.

Chapter 7

Imagery Intelligence

DEFINITION

7-1. IMINT is intelligence derived from the exploitation of imagery collected by visual photography, infrared, lasers, multi-spectral sensors, and radar. These sensors produce images of objects optically, electronically, or digitally on film, electronic display devices, or other media.

ROLE

7-2. The role of imagery is to assist the commander in focusing and protecting his combat power. Imagery often enhances the commander's situational understanding of the battlespace. Other than direct human observation, imagery is the only intelligence discipline that allows the commander to see the battlefield in real time as the operation progresses. In those cases where maps are not available, digital imagery in hardcopy or softcopy can be used as a substitute. Imagery can also be used to update maps or produce grid-referenced graphics. Detailed mission planning often requires imagery, to include three-dimensional stereo images, in order to provide the degree of resolution necessary to support such specialized planning.

FUNDAMENTALS

7-3. Some imagery assets are very responsive to the individual commander's intelligence requirements. Some imagery systems can directly transmit imagery into the tactical operations center (TOC); examples include imagery from UAVs and the Joint Surveillance Target Attack Radar System (JSTARS). This direct downlink enables the G2/S2 to use the imagery as soon as possible instead of having to wait for finished imagery products. A note of caution is required, however, if not a trained imagery analyst because the G2/S2 could incorrectly interpret the imagery.

7-4. Imagery-related equipment has undergone a reduction in size as well as a reduction in the time it takes to provide products, particularly softcopy imagery. The modularity and size reduction of imagery analysis, processing, and display systems make transport easier; they also allow the commander to bring lesser amounts than in the past, while still retaining those systems (or subsystems) required to complete the mission. Additionally, data compression allows faster transmission of imagery products directly to the warfighter.

SOURCES OF IMAGERY

7-5. There are three general sources of imagery: national, civil, and commercial. National imagery traditionally refers to imagery collected by DOD imagery systems. However, there are other sources of imagery provided by non-national sources such as handheld photography (film, digital, and video), UAV imagery, and gun-camera images.

National

7-6. National systems are developed specifically for supporting the President of the United States, the Secretary of Defense, other national agencies, and US military forces. These systems respond to the needs of the nation and those of the combatant commands.

Civil

7-7. Civil imagery systems are usually government funded in terms of building, launching, and operating the system. In many, but not all, cases the agencies operating these civil imagery systems also process, distribute, and archive the imagery data or images.

Commercial

7-8. Commercial companies build, launch, and operate imagery systems for profit. In times of crises, license agreements with the US Government obligate US commercial satellite imaging systems to provide data only to the US Government at the market value. This protects information concerning US operations from adversarial exploitation from commercial systems such as the ACE Imagery Company. However, foreign commercial imagery systems are not bound to this arrangement, and thus may be used by our nation's adversaries. Commercial imagery has become increasingly valuable for many reasons:

- Due to its unclassified nature, civil and commercial imagery is useful in an open environment, may be released to multinational partners, and can be made available to the press. They are especially useful for geospatial products. Their use allows national systems more time to focus on other intelligence functions.
- Civil and commercial imagery offer radar and multi-spectral. Some offer large area collection useful for broad area coverage purposes.
- Commercial satellite imagery resolution varies from less than one meter to several kilometers.

7-9. The Central Imagery Tasking Office (CITO) is responsible for ordering commercial imagery. The Commercial Satellite Imagery Library is available to research DOD purchased commercial imagery. The G2/S2 should consult the CITO when forming commercial imagery requests. NGA will deliver the imagery primarily on CD-ROM media via courier or mail service. Limited digital or electronic delivery is available as well.

TYPES OF IMAGERY SENSORS

7-10. There are four types of imagery sensors: visible (optical), infrared, radar, and multi-spectral. Each sensor has a unique capability, with distinct advantages and disadvantages. The G2/S2 must understand each sensor's capability in order to select the best sensor for the mission and thus enable the user to better understand the intelligence received. Certain sensors are better suited for military operations than others. (See Table 7-1.)

Table 7-1. Sensor Characteristics Matrix.

SENSORS	ADVANTAGES	DISADVANTAGES
Visible (Optical) Best tool for daytime, clear weather, detailed analysis. Includes video and electro-optical.	• Affords a familiar view of a scene. • Offers system resolution that cannot be achieved in other optical systems or in thermal images and radars. • Preferred for detailed analysis and mensuration. • Offers stereoscopic viewing.	• Restricted by terrain and vegetation. • Limited to daytime use only. • Reduced picture size.
Infrared Best tool for nighttime, clear weather, detailed analysis. Includes Overhead Non-Imaging Infrared (ONIR).	• A passive sensor and is impossible to jam. • Offers camouflage penetration. • Provides good resolution. • Nighttime imaging capability.	• Not effective during thermal crossover periods. • Product not easily interpretable. • Requires skilled analysis. • Cannot penetrate clouds.
Radar Useful for detecting presence of objects at night and in bad weather. Includes synthetic aperture radar (SAR), coherent change detection (CCD), and MTI.	• All weather; can penetrate fog, haze, clouds, smoke. • Day or night use. • Does not rely on visible light nor thermal radiation. • Good standoff capability. • Large area coverage. • Allows moving target detection. • Foliage and ground penetration.	• Product not easily interpretable. • Requires skilled analysis. • Terrain masking inhibits use.
Multi-Spectral Imagery (MSI) Best tool for mapping purposes and terrain analysis.	• Large database available. • Band combinations can be manipulated to display desired requirements. • Images can be merged with other digital data to provide higher resolution.	• Product not easily interpretable. • Requires skilled analysis. • Computer manipulation requires large amounts of memory and storage; requires large processing capabilities.

PLAN

7-11. The first step in planning for IMINT is determining the need for IMINT products based on the PIRs and the initial IPB. The G2/S2 should research targets using online imagery databases early and request those imagery products that are not perishable for contingency planning. National and COCOM imagery databases may hold recently imaged targets that could meet the commander's immediate needs instead of requesting new imagery. The staff must clearly articulate their intelligence requirements to include communicating what the mission is and how the requested product will aid in mission accomplishment. The G2/S2 should submit the imagery requirement using established procedures such as those in the unit's SOP or as established by the COCOM.

7-12. The G2/S2 must also determine the specific imagery requirements so as not to burden the system with unnecessary requests. The desire for imagery products often exceeds the capabilities of the IMINT system. Therefore, it is imperative that the G2/S2 consider what type of analysis they require, and request only that which they require. The specifications of the request for IMINT products often affect the timeliness of the response. For example, determining if vehicles are tanks takes less time and requires less resolution than determining if a tank is a T-64 or a T-72.

7-13. Here are some of the roles IMINT can perform that the G2/S2 may consider when determining IMINT requirements:

- Imagery can detect and/or identify and locate specific unit types, equipment, obstacles, potential field fortifications, etc., from which intelligence analysts are able to analyze enemy capabilities and develop possible COAs.
- Imagery can also update maps and enhance the interpretation of information from maps. Detailed mission planning uses imagery to include stereo images for three-dimensional viewing of the terrain and many other geospatial uses.
- Imagery can be used as a substitute for maps when maps are not available. The most common application of this technique is by constructing imagery mosaics: a combination of two or more overlapping photographic prints that form a single picture.
- MTI displays or products can provide an NRT picture of an entity's movement by indicating its speed, location, and direction of travel. MTI systems do not differentiate friendly from enemy. Imagery assets, particularly MTI systems, are useful in cueing other ISR systems.
- Imagery can be used to support protection of the force by helping the commander visualize how his forces look—including their disposition, composition, and vulnerabilities—as exploited by enemy IMINT systems.
- Imagery analysts use combat assessment imagery to confirm destruction, determine the percentage of destruction, or whether the target was unaffected.

PREPARE

7-14. The G2/S2 IMINT-related actions during the prepare function of the intelligence process include establishing or verifying the portion of the intelligence communications architecture that supports IMINT display and analysis functions properly. Additionally, the G2/S2 must ensure that required IMINT analytical assets and resources are prepared to provide support or are available through intelligence reach. Lastly, the G2/S2 must also ensure IMINT reporting and dissemination channels and procedures are in place and rehearsals are conducted with all pertinent IMINT elements to ensure interoperability.

COLLECT

7-15. As previously mentioned, there are four types of imagery sensors. Depending on the type of sensor, it can record hardcopy or softcopy single frame or continuous (video). A given imagery target will not necessarily receive continuous coverage due to the possible conflict between the number and priority of targets and the number and availability of imagery assets. However, a commander may decide to have continuous surveillance of certain targets, for specified periods of time, usually using his own imagery assets (for example, UAV) even though this detracts from the commander's ability to use these assets for other imagery targets within his AOI.

PROCESS

7-16. The process function regarding IMINT involves converting imagery data into a form that is suitable for performing analysis and producing intelligence. Examples of IMINT processing include developing film, enhancing imagery, converting electronic data into visual displays or graphics, and constructing electronic images from IMINT data.

PRODUCE

7-17. The IMINT producer must ensure the IMINT product satisfies the associated intelligence requirements and that the product is in the required format. The quality and resolution of the product is highly dependent upon the type of sensor, the time of day, and the weather conditions, as well as the imagery analyst's ability to identify items, vehicles, equipment, and personnel within the images. Specific IMINT products are discussed in the **ANALYZE** section below.

ANALYZE

7-18. Timeliness is critical not only to IMINT collection but also to IMINT analysis and reporting. It is difficult to separate IMINT reporting from IMINT analysis in this discussion. This is demonstrated by the three phases of IMINT reporting presented below; all are dependent upon the timeliness requirements. Each phase represents a different degree of analysis and period of time available to accomplish the exploitation of the imagery.

- First Phase imagery analysis is the rapid exploitation of newly acquired imagery and reporting of imagery-derived information within a specified time from receipt of imagery. This phase satisfies priority requirements of immediate need and/or identifies changes or activity of immediate significance. First Phase imagery analysis results in an Initial Phase Imagery Report (IPIR).
- Second Phase imagery analysis is the detailed exploitation of newly acquired imagery and the reporting of imagery-derived intelligence and information while meeting the production and timeliness requirements. Other intelligence discipline source material may support phase two imagery as appropriate. Second phase imagery analysis results in a Supplemental Imagery Report (SUPR).
- Third Phase imagery analysis is the detailed analysis of all available imagery pertinent to an SIR, and the subsequent production and reporting resulting from this analysis within a specified time. This

phase provides an organized detailed analysis of an imagery target or topic, using imagery as the primary data source but incorporating data from other sources as appropriate.

7-19. The two types of imagery exploitation are national and direct support (DS).

- National exploitation is imagery exploitation that supports presidential requirements, National Security Council (NSC) requirements, congressional requirements, or requirements of a common concern to the intelligence community.
- DS exploitation is imagery exploitation that supports assigned missions of a single agency, department, or command (the warfighter).

7-20. IMINT assets will complete DS exploitation in order to satisfy (First Phase) requirements and report the results as soon as possible, but not later than 24 hours after receipt of the imagery. Collectors will complete national exploitation in order to satisfy (Second and Third Phases) requirements and report the results within the time specifications of each individual requirement.

DISSEMINATE

7-21. IMINT products are distributed or disseminated in hardcopy, softcopy, or direct viewing such as through remote terminals. The distribution of hardcopy products will be via couriers or other type of mail system. The dissemination of softcopy products will be either as hardcopy products (for example, CD-ROM and 3.5-inch disks) or electronically. The requestor must ensure that the requested product is transmittable over the available communications systems.

ASSESS

7-22. The requestor should immediately assess the imagery product upon receipt for accuracy and relevance to the original request. The requestor must then notify the producer and inform him of the extent to which the product answered the PIR. Providing feedback to the producer regarding the product helps ensure the producer will provide the required information in the correct format. The following are some of the questions which the requestor should consider when providing feedback to the producer.

- Is the format of the product acceptable?
- Is additional information needed on the product or future products?
- Is excess information included on the product?

Chapter 8

Signals Intelligence

DEFINITION

8-1. SIGINT is a category of intelligence comprising either individually or in combination all COMINT, ELINT, and FISINT, however transmitted; intelligence is derived from communications, electronics, and foreign instrumentation signals. SIGINT has three subcategories:

- COMINT – The intelligence derived from foreign communications by other than the intended recipients.
- ELINT – The technical and geo-location intelligence derived from foreign non-communications electromagnetic radiations emanating from other than nuclear detonations or radioactive sources.
- FISINT – Technical information and intelligence derived from the intercept of foreign electromagnetic emissions associated with the testing and operational deployment of non-US aerospace, surface, and subsurface systems. Foreign instrumentation signals include but are not limited to telemetry, beaconry, electronic interrogators, and video data links. (See JP 1-02)

8-2. SIGINT provides intelligence to the commander based upon intercepted communications and provides transmitter location data.

ROLE

8-3. SIGINT provides intelligence on threat capabilities, disposition, composition, and intentions. In addition, SIGINT provides targeting information for the delivery of lethal and non-lethal effects.

FUNDAMENTALS

8-4. It is important that the G2/S2 understand how SIGINT assets are organized not only within the Army but also throughout the DOD. The majority of SIGINT assets are located at EAC. SIGINT assets from all the armed services, combined with national SIGINT assets, work together to support commanders from the tactical to the strategic level. Only by understanding the SIGINT structure that transcends traditional service component boundaries can the G2/S2 understand how to use SIGINT effectively.

TECHNICAL CONTROL AND ANALYSIS ELEMENT (TCAE) ARCHITECTURE

8-5. The TCAE architecture supports Army SIGINT collectors and analysts by providing SIGINT-specific intelligence and guidance to SIGINT personnel. The SIGINT technical architecture complements existing C2 relationships; it does not replace the commander's authority or chain of command.

8-6. There is a TCAE at each Army operational echelon. The TCAE at each echelon provides a single POC and resource dedicated to supporting the commander at that level. TCAEs or SIGINT elements at varying echelons draw information from each other in order to provide a more complete and detailed intelligence picture for their respective commanders.

ARMY TCAE (ATCAE)

8-7. The ATCAE, established at the national level, plays a significant role in TCAE operations by providing technical support oversight and providing access to national databases—in coordination with the NSA and other intelligence organizations. Additionally, the ATCAE works closely with NSA's IO element assisting in operations, monitoring technical capabilities, and providing liaison with Army's 1st Information Operations Command (Land). (This unit was previously called Land Information Warfare Activity.)

8-8. The ATCAE supports the Army's Quick Reaction Capability (QRC) SIGINT and signals research and target development (SRTD) operations. The QRC systems are responsive to the ground force commander's requirements and allow the ability to conduct SIGINT operations against modern communications systems.

REGIONAL TCAE (RTCAE)

8-9. The Army established RTCAEs to allow supported units at all echelons access to SIGINT regarding their respective AOI. The Army placed an RTCAE within each Regional Security Operations Center (RSOC). An RSOC is a joint service facility established by the Director, NSA, to conduct continuous security operations on selected targets in support of national and warfighter intelligence requirements using remoting technologies. The RSOC embeds a focus on the tactical commander's SIGINT requirements at a national level. It also creates a framework for enhanced interoperability among service SIGINT activities, especially within the context of a deployed JTF.

8-10. In addition to each RSOC, the RTCAE works closely with Theater TCAEs as well as Corps and Division SIGINT elements with the same regional focus. It provides analytic support and tailored products and answers RFIs. During peacetime and real-world operations RTCAEs facilitate intelligence reach support. They provide tailored technical support packages, support for surge and survey operations, analytic expertise, and signal data access and digital file transfers of live collection for Army tactical units.

THEATER TCAE

8-11. The Theater TCAE performs SIGINT and EW technical control and analysis and management. It provides SIGINT technical support for assigned, OPCON, and lower echelon SIGINT resources deployed in the theater. This includes mission tasking, processing, analyzing, and reporting of SIGINT data, information, and intelligence. The TCAE provides direction for the Theater Collection and Exploitation Battalion's SIGINT mission and for other theater tactical SIGINT assets.

SIGINT THEATER INTELLIGENCE BRIGADE/GROUP (TIB/TIG)

8-12. The TIB/TIG conducts operational level multidiscipline SIGINT operations. It provides timely intelligence to the commander throughout the full spectrum of operations. The TIB/TIG is structured to provide ground and aerial SIGINT support. It executes the full range of SIGINT missions or tasks.

SIGINT AT ECHELONS CORPS AND BELOW

8-13. There is a varying mixture of SIGINT assets within each corps. A reference listing the types of SIGINT assets found at echelons corps and below is listed in ST 2-50.

PLAN

8-14. An important SIGINT planning consideration is that if at all possible, SIGINT collection should be employed in conjunction with another intelligence discipline collection system. SIGINT is often used to cue, and be cued by, other ISR assets.

8-15. During planning, the G2/S2 should retrieve, update, or develop any required SIGINT databases. This includes effecting coordination with other SIGINT assets or elements that can support the operation.

8-16. SIGINT ground collection assets are usually placed in proximity to enemy signal sources due to the limited height of collector antennas, the low power output of threat or enemy emitters, and the line-of-sight (LOS) constraints imposed by terrain. Ground-based SIGINT teams are most effective when positioned to—

- Maximize threat emitter interception. This allows teams to overcome the constraints of threat emitter characteristics usually allowed by proximity to the threat.
- Minimize system receiver interference. This increases the potential capability of the team to acquire threat emitters in a timely manner.
- Optimize overlapping areas of intercept coverage. This ensures coverage of the AOIR and allows targets to be handed off from team to team if necessary.

8-17. Aerial SIGINT assets have additional planning requirements. Coordination must be effected with Army Airspace Command and Control (A2C2) for ingress and egress routes, restricted operating zones (ROZs) where the aircraft can travel on an intercept track, as well as with the supporting air operations center to determine the availability of asset types and times.

PREPARE

8-18. The G2/S2 ensures the SIGINT unit and asset leaders have effected all necessary coordination and conducted rehearsals. This includes establishing or verifying the operation of the SIGINT technical architecture. The G2/S2 also ensures all required SIGINT assets and resources are available, SIGINT reporting and dissemination channels and procedures are in place, and connectivity and interoperability exist with all pertinent SIGINT elements.

8-19. SIGINT OPCON is the authoritative direction of SIGINT activities, including tasking and allocation of effort, and the authoritative prescription of those uniform techniques and standards by which SIGINT information is collected, processed, and reported.

8-20. SIGINT operational tasking is the authoritative operational direction of and direct levying of SIGINT information needs by a military commander on designated SIGINT resources. These requirements are directive, regardless of other priorities, and are conditioned only by the capability of those resources to produce such information. Operational tasking includes authority to deploy all or part of the SIGINT resources for which SIGINT operational tasking authority (SOTA) has been delegated.

8-21. SOTA is the military commander's authority to operationally direct and levy SIGINT requirements on designated SIGINT resources; it includes authority to deploy and redeploy all or part of the SIGINT resources for which SOTA has been delegated.

COLLECT

8-22. SIGINT performs two major collection activities: signals intercept and direction finding (DF).

Signals Intercept

8-23. Signals intercept are those SIGINT actions used to search for, intercept, and identify threat electromagnetic signals for the purpose of immediate threat recognition. Signals intercept provides information required to answer PIRs, and other intelligence requirements in support of the ISR effort.

Direction Finding

8-24. Even when threat radio operators use COMSEC procedures, SIGINT teams can often intercept and approximate the location of the threat's signals. Specifically, SIGINT teams can use DF to determine—

- Movement of threat personnel or equipment.
- Locations of emitters associated with weapon systems and units.
- New emitter locations and confirm known emitter locations.
- Possible friendly targets the enemy intends to attack (lethal and non-lethal).

8-25. In addition to finding threat forces, DF operations can assist the (radio-equipped) friendly force by—

- Locating and vectoring assets or units during limited visibility.
- Locating downed aircraft and personnel radio beacons.
- Conducting signal security assessments.
- Locating sources of communication interference and jamming.

PROCESS

8-26. SIGINT processing involves converting intercepts of SIGINT into written and verbal reports, automated message, graphic displays, recordings, and other forms suitable for analysis and intelligence production. Since US forces routinely conduct operations against adversaries who speak languages other than English, SIGINT processing often also includes translation of these intercepts.

8-27. Due to the complexity of many SIGINT systems, automated processing may occur several times before SIGINT data or information receives any human interaction.

PRODUCE

8-28. The SIGINT producer must ensure the SIGINT product satisfies the associated intelligence requirements and that the product is in the required format. The quality, fidelity, and timeliness of SIGINT products are highly dependent upon the type of intercept, the collection system, the system's position in relation to the threat emitter, the weather (including space-based weather), as well as the SIGINT operator's ability to identify the appropriate threat signal activity.

8-29. SIGINT production results in some of the reports and formats mentioned in the process section above; however, the objective for SIGINT is to be used in an all-source analytical approach.

ANALYZE

8-30. The intelligence staff analyzes intelligence and information about the enemy's communications capabilities to determine appropriate SIGINT collection strategies. Conversely, a corresponding analysis of the friendly forces' SIGINT capabilities must be conducted to ensure the continued effectiveness of, or to improve upon, SIGINT collection.

8-31. SIGINT analysts also sort through large amounts of SIGINT and information and intelligence to identify and use only that which is pertinent to the CCIRs (PIRs and FFIRs).

DISSEMINATE

8-32. SIGINT of critical importance to the force, including answers to the CCIRs (PIRs and FFIRs), is disseminated via the most expeditious means possible. Due to the usually highly perishable nature of SIGINT, the most expeditious reporting means is often immediately augmented with a follow-up report or augmented by a report transmitted through additional means, enhancing the probability of receipt. Sometimes the most expeditious means of reporting critical SIGINT information to the commander is face to face.

8-33. For intelligence reach operations, SIGINT products are available and disseminated in a variety of forms: hardcopy, softcopy, direct viewing, or listening (television or radio). It is incumbent on the requestor to ensure that the SIGINT product can be transmitted over the available communications systems. This includes verifying the appropriate security level of the communications system.

ASSESS

8-34. The primary goal of the assess function when applied to SIGINT is to determine whether the results of SIGINT collection meet the requirements of the unit's ISR effort. SIGINT producers must assess all facets of SIGINT operations, from receipt of the ISR task to the dissemination of SIGINT, in an effort to determine their effectiveness. This assessment is not only directed at each SIGINT asset individually but also throughout the supporting SIGINT architecture and the unit's entire ISR effort.

8-35. The G2/S2 immediately assesses SIGINT products upon receipt for accuracy and relevance. He must inform the SIGINT producer of the extent to which the product answered the PIR or intelligence requirement. Providing feedback to the SIGINT producer—and collector—helps improve the effectiveness and efficiency of SIGINT.

Chapter 9

Measurement and Signatures Intelligence

DEFINITION

9-1. MASINT is technically derived intelligence that detects, locates, tracks, identifies, and/or describes the specific characteristics of fixed and dynamic target objects and sources. It also includes the additional advanced processing and exploitation of data derived from IMINT and SIGINT collection.

9-2. MASINT collection systems include but are not limited to radar, spectroradiometric, E-O, acoustic, RF, nuclear detection, and seismic sensors, as well as techniques for gathering NBC and other material samples.

9-3. It requires the translation of technical data into recognizable and useful target features and performance characteristics. Computer, communication, data, and display processing technologies now provide MASINT in support of commanders throughout the full spectrum of operations.

9-4. The subdisciplines within MASINT include, but are not limited to, the following:

- **Radar Intelligence (RADINT).** The active or passive collection of energy reflected from a target or object by LOS, bistatic, or over-the-horizon radar systems. RADINT collection provides information on radar cross-sections, tracking, precise spatial measurements of components, motion and radar reflectance, and absorption characteristics for dynamic targets and objectives. A SAR system, coupled with advanced MASINT processing techniques—
  - Provides a high resolution, day and night collection capability.
  - Can produce a variety of intelligence products that identify or provide change detection, terrain mapping, underwater obstacles, dynamic sensing of targets in clutter, and radar cross-section signature measurements.
- **Frequency Intelligence.** The collection, processing, and exploitation of electromagnetic emissions from a radio frequency weapon (RFW), an RFW precursor, or an RFW simulator; collateral signals from other weapons, weapon precursors, or weapon simulators (for example, electromagnetic pulse signals associated with nuclear bursts); and spurious or unintentional signals.
  - *Electromagnetic Pulses.* Measurable bursts of energy that result from a rapid change in a material or medium, resulting in an explosive force, produces RF emissions. The RF pulse emissions associated with nuclear testing, advanced technology devices, power and propulsion systems, or other impulsive events can be used to detect, locate, identify, characterize, and target threats.

  - *Unintentional Radiation Intelligence (RINT).* The integration and specialized application of MASINT techniques against unintentional radiation sources that are incidental to the RF propagation and operating characteristics of military and civil engines, power sources, weapons systems, electronic systems, machinery, equipment, or instruments. These techniques may be valuable in detecting, tracking, and monitoring a variety of activities of interest.
- **E-O Intelligence.** The collection, processing, exploitation, and analysis of emitted or reflected energy across the optical portion (ultraviolet, visible, and infrared) of the EMS. MASINT E-O provides detailed information on the radiant intensities, dynamic motion, spectral and spatial characteristics, and the materials composition of a target. E-O data collection has broad application to a variety of military, civil, economic, and environmental targets. E-O sensor devices include radiometers, spectrometers, non-literal imaging systems, lasers, or laser radar (LIDAR).
  - *Infrared Intelligence (IRINT).* A subcategory of E-O that includes data collection across the infrared portion of the EMS where spectral and thermal properties are measured.
  - *LASER Intelligence (LASINT).* Integration and specialized application of MASINT E-O and other collection to gather data on laser systems. The focus of the collection is on laser detection, laser threat warning, and precise measurement of the frequencies, power levels, wave propagation, determination of power source, and other technical and operating characteristics associated with laser systems—strategic and tactical weapons, range finders, and illuminators.
  - *Hyperspectral Imagery (HSI).* A subcategory of E-O intelligence produced from reflected or emitted energy in the visible and near infrared spectrum used to improve target detection, discrimination, and recognition. HSI can detect specific types of foliage—supporting drug-crop identification; disturbed soil—supporting the identification of mass graves, minefields, caches, underground facilities or cut foliage; and variances in soil, foliage, and hydrologic features—often supporting NBC contaminant detection.
- **Spectroradiometric Products.** Include E-O spectral (frequency) and radiometric (energy) measurements. A spectral plot represents radiant intensity versus wavelength at an instant in time. The number of spectral bands in a sensor system determines the amount of detail that can be obtained about the source of the object being viewed. Sensor systems range from multispectral (2 to 100 bands) to hyperspectral (100 to 1,000 bands) to ultraspectral (1,000+ bands). More bands provide more discrete information, or greater resolution. The characteristic emission and absorption spectra serve to fingerprint or define the makeup of the feature that was observed. A radiometric plot represents the radiant intensity versus time. An example is the radiant intensity plot of a missile exhaust plume as the missile is in flight. The intensity or brightness of the object is a function of several conditions including its temperature, surface properties or material, and how fast it is moving. For each point along a time-intensity

radiometric plot, a spectral plot can be generated based on the number of spectral bands in the collector.

- **Geophysical Intelligence.** Geophysical MASINT involves phenomena transmitted through the earth (ground, water, atmosphere) and manmade structures including emitted or reflected sounds, pressure waves, vibrations, and magnetic field or ionosphere disturbances.
  - *Seismic Intelligence.* The passive collection and measurement of seismic waves or vibrations in the earth surface.
  - *Acoustic Intelligence.* The collection of passive or active emitted or reflected sounds, pressure waves or vibrations in the atmosphere (ACOUSTINT) or in the water (ACINT). ACINT systems detect, identify, and track ships and submarines operating in the ocean.
  - *Magnetic Intelligence.* The collection of detectable magnetic field anomalies in the earth's magnetic field (land and sea). An example is a Remotely Emplaced Battlefield Surveillance System (REMBASS) sensor detection indicating the presence and direction of travel of a ferrous object.
- **Nuclear Intelligence (NUCINT).** The information derived from nuclear radiation and other physical phenomena associated with nuclear weapons, reactors, processes, materials, devices, and facilities. Nuclear monitoring can be done remotely or during onsite inspections of nuclear facilities. Data exploitation results in characterization of nuclear weapons, reactors, and materials. A number of systems detect and monitor the world for nuclear explosions, as well as nuclear materials production.
- **Materials Intelligence.** The collection, processing, and analysis of gas, liquid, or solid samples. Materials intelligence is critical to collection against NBC warfare threats. It is also important to analyzing military and civil manufacturing activities, public health concerns, and environmental problems. Samples are both collected by automatic equipment, such as air samplers, and directly by humans. Samples, once collected, may be rapidly characterized or undergo extensive forensic laboratory analysis to determine the identity and characteristics of the sources of the samples.

ROLE

9-5. MASINT provides intelligence to the commander throughout the full spectrum of operations to facilitate situational understanding. MASINT can thwart many of the camouflage, concealment, and deception techniques currently used to deceive ISR systems.

9-6. MASINT is perceived as a "strategic" discipline with limited "tactical" support capabilities. But, by application of real-time analysis and dissemination, MASINT has a potential ability to provide real-time situation awareness and targeting not necessarily available to the classic disciplines. Specifically, MASINT "sensors" have unique capabilities to detect missile launch, detect and track aircraft, ships, and vehicles; do non-cooperative target identification (NCTI), combat assessment, and BDA; and detect and track fallout from nuclear detonations. Often, these contributions are the first indicators of hostile activities. For example, two EXOCET-equipped

Mirage F-1s were shot down during the Operation DESERT STORM due to MASINT collection and analysis. As evidenced by Operation IRAQI FREEDOM (OIF), MASINT will play a decisive role in the targeting of smart munitions with the signatures (fingerprint) of the targets they are seeking (for example, infrared signatures).

9-7. The MASINT systems most familiar on today's battlefield are employed by ground surveillance and NBC reconnaissance elements.

9-8. MASINT spans the entire EMS and its capabilities complement, rather than compete with, the other intelligence disciplines. MASINT provides, to varying degrees, the capability to—

- Use automatic target recognition (ATR) and aided target recognition (AiTR).
- Penetrate manmade and/or natural camouflage.
- Penetrate manmade and/or natural cover, including the ability to detect subterranean anomalies or targets.
- Counter stealth technology.
- Detect recently placed mines.
- Detect natural or manmade environmental disturbances in the earth's surface not discernible through other intelligence means.
- Provide signatures (target identification) to munitions and sensors.
- Enhance passive identification of friend or foe.
- Detect the presence of NBC agents to include prior to, during, or after employment.
- Detect signature anomalies that may affect target-sensing systems.

FUNDAMENTALS

9-9. Before discussing the functions of the intelligence process within a MASINT, the following paragraph provides an overview of organizational structure of MASINT.

9-10. Within DOD, the DIA provides central coordination for MASINT collection efforts through the Central MASINT Office. Each service, in turn, has a primary command or staff activity to develop requirements and coordinate MASINT effort. Army responsibility currently resides with INSCOM. Army weapons systems programs that require MASINT information to support system design or operations submit requests through the Army Reprogramming Analysis Team (ARAT) or INSCOM channels for data collection and processing. The S&TI community also performs MASINT collection and processing primarily to support R&D programs and signature development. Every S&TI center has some involvement in MASINT collection or production that reflects that center's overall mission (for example, NGIC does work on armored vehicles and artillery). Service R&D centers such as the Communications-Electronics Command (CECOM) Research, Development, and Engineering Center (RDEC), and Night Vision and Electronic Systems Laboratory are also involved in developing sensor systems for collecting and processing MASINT.

9-11. In addition to supporting the S&TI mission, INSCOM units also execute limited ground-based operational collection to support Theater and ASCC PIRs. This capability will expand upon the standup of INSCOM TIBs and TIGs.

PLAN

9-12. Some MASINT sensors can provide extremely specific information about detected targets, whereas other sensors may only be capable of providing an indication that an entity was detected. Additionally, there are varying capabilities of detection, identification, and classification among MASINT sensors. It is these varying capabilities that require synchronizing the employment of MASINT sensors both within the MASINT discipline and within the ISR effort as a whole. See FM 2-01 for more specific information on ISR synchronization.

9-13. As previously mentioned, there are many types of MASINT sensors. Depending on the type of sensor employed, a given MASINT collection target or NAI may not necessarily receive continuous coverage due to the possible conflict between the number and priority of targets and the number and availability of MASINT assets. However, a commander may decide to have continuous surveillance of certain targets by using his own MASINT assets (for example, REMBASS or Improved-REMBASS).

9-14. Another consideration when planning MASINT missions is whether to use active, passive, or a combination of both when planning MASINT coverage.

PREPARE

9-15. The G2/S2 MASINT related actions during the prepare function of the intelligence process include properly establishing or verifying the MASINT portion of the intelligence communications architecture functions. Additionally, the G2/S2 must ensure that required MASINT analytical assets and resources are prepared to provide support or are available through intelligence reach. Since the products of MASINT are not as well known as products from other intelligence disciplines, the G2/S2 must be aware of the types of MASINT products available to support the operation, and then educate the rest of his unit's staff on the use of these MASINT products. Lastly, the G2/S2 must also ensure MASINT reporting and dissemination channels and procedures are in place and rehearsals are conducted with all pertinent MASINT elements to ensure interoperability.

COLLECT

9-16. MASINT provides information required to answer PIRs and other intelligence requirements in support of the ISR effort. As stated earlier in this chapter, MASINT collection must not only be synchronized within its own discipline but also be synchronized and integrated into the unit's overall ISR effort in order to be effective.

9-17. MASINT sensors are employed throughout the full spectrum of operations from a variety of platforms—sub-surface, ground, marine, and aerospace.

PROCESS

9-18. Just as in the other intelligence disciplines, MASINT involves dealing with huge volumes of data that have to be processed before beginning analysis and production. The process function regarding MASINT involves converting esoteric data into a form that is suitable for performing analysis and producing intelligence. MASINT processing can include relatively simple actions such as converting a REMBASS sensor activation into a report, to a complex task such as processing HSI into a report identifying the composition and concentrations of carcinogenic chemicals present in the emissions from a factory upwind from a US forces encampment.

PRODUCE

9-19. Effective and timely MASINT requires personnel with diverse skill sets. The MASINT producer must ensure the MASINT product satisfies the associated intelligence requirements and that the product is in the required format. The quality, fidelity, and timeliness of MASINT products are highly dependent upon the type of target, the collection system, the system's position in relation to the target or NAI, and the weather, as well as the MASINT system operator's ability to identify the appropriate threat activity.

9-20. The objective of MASINT production is to be used in an all-source analytical approach.

ANALYZE

9-21. The intelligence staff analyzes intelligence and information about the enemy's equipment, doctrine, and TTP to determine appropriate MASINT collection strategies. Conversely, a corresponding analysis of the friendly force's MASINT capabilities must be conducted to ensure the continued effectiveness of, or to improve upon, MASINT collection.

DISSEMINATE

9-22. MASINT of critical importance to the force, including answers to the PIRs, is disseminated via the most expeditious means possible.

9-23. For intelligence reach operations, MASINT products are available and disseminated in a variety of forms. The requestor must ensure that the MASINT product can be transmitted over the available communications systems. This includes verifying the appropriate security level of the communications system.

ASSESS

9-24. The primary goal of the MASINT assess function is to determine whether the results of MASINT collection and production meet the requirements of the unit's ISR effort. MASINT producers must assess all facets of MASINT operations, from receipt of the ISR task to the dissemination of MASINT, in an effort to determine the effectiveness of MASINT. This assessment is not only directed at each MASINT asset individually but also throughout the supporting intelligence communications architecture, to include intelligence reach and the unit's entire ISR effort.

9-25. Also, the G2/S2 immediately assesses MASINT products upon receipt for accuracy and relevance. He must inform the MASINT producer of the extent to which the product answered the PIR or intelligence requirement. Providing feedback to the MASINT producer—and collector—helps improve the effectiveness and efficiency of MASINT.

Chapter 10

Technical Intelligence

DEFINITION

10-1. TECHINT is intelligence derived from the collection and analysis of threat and foreign military equipment and associated materiel.

ROLE

10-2. The strength of the US military lies, in part, to the diversity and extent of its technology base. While the US aspires to be the leader in integrating technology, the threat can achieve temporary technological advantage in certain areas by acquiring modern systems or capabilities. The world arms market is willing to provide these advanced systems to countries or individuals with the resources to pay for them. A concerted TECHINT program is vital to providing precise direction and purpose within the US R&D process to ensure quick and efficient neutralization of this advantage.

10-3. The role of TECHINT is to ensure that the warfighter understands the full technological capabilities of the threat. With this understanding, the warfighter can adopt appropriate countermeasures, operations, and tactics.

10-4. TECHINT has two goals within its role:

- To ensure the US armed forces maintain technological advantage against any adversary.
- To provide tailored, timely, and accurate TECHINT support to the warfighter throughout the entire range of military operations. This includes providing US forces intelligence, information, and training on foreign weapons systems to an extent that allows their use of CEE.

FUNDAMENTALS

10-5. The G2/S2 must understand how TECHINT assets are organized in order to properly apply the intelligence process, as the majority of TECHINT assets are located at EAC.

DEFENSE INTELLIGENCE AGENCY

10-6. DIA manages and reviews overall TECHINT activities. The S&TI Directorate within DIA is the action element for TECHINT. This directorate coordinates with external TECHINT agencies on non-policy matters concerning the production of S&TI. The following organizations provide TECHINT support under the control of DIA:

- **Armed Forces Medical Intelligence Center (AFMIC).** AFMIC, based at Fort Detrick, MD, is a DOD intelligence production center under DIA control. AFMIC is responsible for exploiting foreign medical materiel. The director supports the Army Foreign Materiel Exploi-tation

Program (FMEP) and Army medical R&D requirements. The director coordinates planning, programming, and budgeting with the Army Deputy Chief of Staff, Intelligence (DCS, G2).

- **Missile and Space Intelligence Center (MSIC).** MSIC, based at Redstone Arsenal, AL, is a DOD intelligence production center under DIA control and supports the FMEP. The MSIC acquires, produces, maintains, and disseminates S&TI pertaining to missile and space weapons systems, subsystems, components, and activities. The S&TI produced at MSIC also covers foreign state-of-the-art technology and research applicable to missiles.
- **Defense HUMINT Service.** DHS conducts worldwide HUMINT operations in support of foreign materiel acquisition (FMA) and foreign materiel exploitation (FME).

10-7. The organizations and agencies discussed below constitute the Army TECHINT structure.

ARMY DCS, G2

10-8. The Army DCS, G2 exercises general staff responsibility for all Army TECHINT activities. The Army DCS, G2 forms policies and procedures for S&TI activities, supervises and carries out the Army S&TI program, coordinates DA staff and MSC requirements for TECHINT, and is responsible for the Army Foreign Materiel Program (FMP).

INSCOM

10-9. Under the direction of Headquarters, Department of the Army (HQDA), INSCOM is responsible for peacetime TECHINT operations. HQ, INSCOM, fulfills its responsibilities through its TECHINT oversight function and manages the Army's Foreign Materiel for Training (FMT) Program and FMEP. It provides the interface with strategic S&TI agencies in support of FME and organizes, trains, and equips EAC TECHINT organizations during peacetime. TECHINT exploitation within INSCOM is performed by the following elements:

- **National Ground Intelligence Center.** HQ, INSCOM, exercises direct OPCON over the NGIC. NGIC produces and maintains intelligence on foreign scientific developments, ground force weapons systems, and associated technologies. NGIC analysis includes but is not limited to military communications electronics systems, types of aircraft used by foreign ground forces, NBC systems, and basic research in civilian technologies with possible military applications.
- **203d Military Intelligence Battalion.** The 203d MI Battalion is a multi-component unit headquartered at Aberdeen Proving Ground, MD, and is the Army's sole TECHINT battalion. It performs the following functions:
  - Conducts TECHINT collection and reporting in support of validated S&TI objectives.
  - Acts as the HQDA executive agent for foreign materiel used for training purposes.

- Conducts TECHINT training for DOD analysts and RC TECHINT personnel.
- Supports INSCOM's FMA and FME operations as directed.
- Analyzes and exploits foreign captured enemy documents (CEDs), equipment, weapon systems, and other war materiel.
- Reports on the capabilities and limitations of enemy combat materiel.
- Provides reports alerting the command to the tactical threat posed by technical advances in new or recently discovered foreign or enemy materiel.
- Provides countermeasures to any enemy technical advantage.
- Provides foreign or enemy equipment for troop familiarization and training.
- Provides recommendations on the reuse of CEM.
- Supervises evacuating items of TECHINT interest.
- Provides task-organized battlefield TECHINT teams to support a subordinate command's TECHINT effort.

US ARMY MATERIEL COMMAND (AMC)

10-10. AMC plays a significant support role in TECHINT. Among AMC elements are a series of RDECs, the Army Research Laboratory System, and the Test and Evaluation Command (TECOM). Each conducts highly technical evaluations of foreign equipment. In peacetime, the AMC conducts FME on equipment purchased by each laboratory and RDEC for the intelligence community and for DOD as part of the International Materiel Evaluation Program (IMEP). AMC elements include—

- The Foreign Ordnance Exploitation Team, which is located at the Fire Support Armaments Center (FSAC) in the Picatinny Arsenal. This team exploits foreign ground ordnance and develops render safe procedures (RSPs) for foreign ordnance. It also prepares detailed intelligence reports to support explosive ordnance disposal (EOD), MI, and US munitions developers.
- The Science and Technology Center Europe and the Science and Technology Center Far East, which have the responsibility of collecting information on foreign technical developments by attending arms shows and technology exhibitions.
- The Soldier Biological Chemical Defense Command, which is the headquarters for the US Army Technical Escort Unit (TEU). The mission of the TEU is to collect and escort (transport) chemical, biological, radiological, and nuclear (CBRN) samples for testing and evaluation.

10-11. There are many other agencies with TECHINT responsibilities within the DOD. Refer to FM 34-54 for more information on TECHINT.

PLAN

10-12. TECHINT collection usually begins when an organization or individual reports the recovery or acquisition of unusual, new, or newly employed threat materiel. However, there are often indications that the threat may be using materiel not yet associated with the threat among the myriad intelligence products available. Conversely, it may be known that the threat is using a particular item, the capabilities of which are unknown to US forces. It is in these cases that a unit may receive a TECHINT related ISR task. It is conceivable that such a task may be reflected in the PIR. An example of such a task is the identification or verification of suspected external modifications on a particular model of an enemy's main battle tank; the results of which are linked to a commander's decision.

10-13. TECHINT related ISR tasks should include a mission or target folder. At a minimum, this folder should include a description of the item, with its associated major combat systems, as well as handling instructions, reporting instructions, and a photograph or sketch of the item if available.

10-14. Additionally, the G2/S2 is responsible for ensuring the staff coordinates and establishes a plan for evacuating the desired materiel.

PREPARE

10-15. The G2/S2 must ensure that required TECHINT analytical assets, resources, and evacuation means are prepared to provide support. This includes verifying coordination effected with the task-organized battlefield TECHINT teams from the 203d MI Battalion. The G2/S2 must also ensure the means to report and disseminate TECHINT results to the unit and its soldiers are in place so that they can immediately adopt appropriate countermeasures, operations, or tactics in order to enhance their survival and mission accomplishment.

COLLECT

10-16. TECHINT collection includes capturing, reporting, and evacuating CEM. TECHINT collection begins when an organization or individual reports the recovery or acquisition of threat materiel. An item of materiel is exploited at each level, and continues on through succeeding higher levels until an appropriate countermeasure to neutralize the item's capabilities is identified or developed.

10-17. Army personnel (soldiers and civilian) and units will normally safeguard CEM and report it through intelligence channels to the first TECHINT element in the reporting chain. The location of this TECHINT element will be in accordance with the METT-TC factors; however, there will usually be TECHINT representation at the Corps G2 or the COCOM J2. The TECHINT representative or element will verify if the type of materiel is of intelligence value and determine its further disposition in conjunction with the unit's staff.

PROCESS

10-18. TECHINT processing starts (simultaneously with collection) with the capture of a piece of equipment of TECHINT value. This confirms that the enemy is indeed employing this materiel. In accordance with METT-TC factors, a TECHINT team may move to the location of the item at the capture site or wait until the item is evacuated before conducting a hasty exploitation. After hasty exploitation, the team decides if further processing is required. If it is, the items are sent to the first (or nearest) Captured Materiel Exploitation Center (CMEC). If the item is deemed to yield no immediate tactical intelligence value, it may still be evacuated to the S&TI centers in CONUS for further analysis if the systems represent a change in the technological posture of an enemy.

PRODUCE

10-19. Battlefield TECHINT teams normally report initial and secondary examinations of CEM using either a preliminary technical report or a complementary technical report.

- A preliminary technical report—
  - Includes a general description of the item reported and recommended RSP.
  - Alerts others to information that can be used immediately by tactical units.
- A complementary technical report is more in-depth and—
  - Follows a secondary or an in-depth initial examination.
  - Allows the CMEC to compare new information with intelligence holdings.

10-20. At each successive echelon of exploitation, TECHINT analysts add to the overall body of information on an item by either adding to previous reports or by preparing new reports. The CMEC or other national level S&TI activities prepare more advanced technical reports and analyses. These reports include—

- Detailed technical reports.
- Translation reports.
- Special technical reports.

10-21. Other TECHINT products include—

- CMEC publications such as operator manuals, maintenance manuals, TECHINT bulletins, and tactical user bulletins.
- S&TI analysis bulletins.
- Foreign materiel exploitation reports.

ANALYZE

10-22. TECHINT analysts use checklists established by S&TI agencies and the CMECs to analyze each type of the adversary's equipment for which requirements exist. Analysis always begins with what is, and what is not, known about the piece of equipment. TECHINT units maintain procedures and plans for sampling, analyzing, and handling materiel.

DISSEMINATE

10-23. TECHINT of critical importance to the force, including answers to the PIR, is disseminated via the most expeditious means possible.

10-24. Routine TECHINT reports and products are usually transmitted through the unit's existing intelligence communications architecture in the format of an intelligence information report (IIR) format. For intelligence reach operations, TECHINT products are available and disseminated in a variety of forms. The requestor must ensure that the TECHINT product can be transmitted over the available communications systems. This includes verifying the appropriate security level of the communications systems.

ASSESS

10-25. The primary goal of the TECHINT assess function is to determine whether the results of TECHINT production meet the unit's PIR or intelligence requirements. The G2/S2 immediately assesses TECHINT products upon receipt for accuracy and relevance. He must inform the TECHINT producer of the extent to which the product answered the PIR or intelligence requirement. Providing feedback to TECHINT analysts helps improve the effectiveness and efficiency of TECHINT.

10-26. The G2/S2 also assesses the success of the unit's ISR effort in accomplishing any TECHINT associated ISR task and shares his assessment with the staff and the pertinent units or personnel.

Chapter 11

Counterintelligence

DEFINITION

11-1. CI counters or neutralizes intelligence collection efforts through collection, CI investigations, operations, analysis and production, and functional and technical services. CI includes all actions taken to detect, identify, exploit, and neutralize the multidiscipline intelligence activities of friends, competitors, opponents, adversaries, and enemies. It is the key intelligence community contributor to protect US interests and equities. Figure 11-1 shows the CI overview.

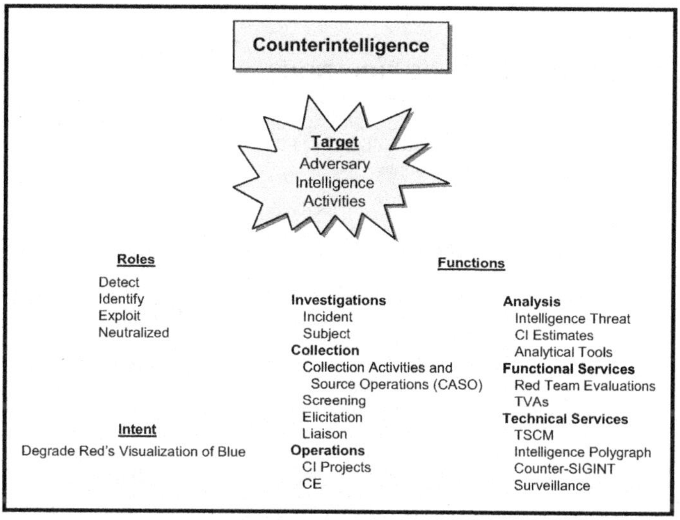

Figure 11-1. Counterintelligence Overview.

ROLE

11-2. The role of CI is to detect, identify, exploit, and neutralize all adversary intelligence entities targeting US and multinational interests. CI will focus on countering adversary intelligence collection activities targeting information or material concerning US personnel, activities, operations, plans, equipment, facilities, publications, technology, or documents—either classified or unclassified—without official consent of designated US release authorities, for any purpose that could cause damage or otherwise adversely impact the interests of national security of the US ability to fulfill national policy and objectives. Adversary intelligence threats include but are not limited to any (US, multinational, friendly, competitor, opponent, adversary, or recognized enemy) government or NGOs, companies, businesses, corporations, consortiums, groups, agencies, cells or persons, terrorists, insurgents, guerrilla entities, and persons whose demonstrated actions, views, or opinions are inimical to US interests.

11-3. CI elements are instrumental in contributing to situational awareness in the AOI. CI elements may corroborate other intelligence discipline information as well as cue other intelligence assets through the CI core competencies and through CI technical services. The CI core competencies are collection, investigations of national security crimes within the purview of CI, operations, and analysis and production. CI technical services include computer network operations, technical surveillance countermeasures (TSCM), and polygraph. CI focuses on combating adversary intelligence activities targeting Army personnel, plans, operations, activities, technologies, and other critical information and infrastructure.

COUNTERINTELLIGENCE FUNCTIONS

11-4. CI functions are interrelated, mutually supporting, and can be derived from one another. No single function or technical capability can defeat adversary intelligence efforts to target US interests. CI functions are discussed below.

CI INVESTIGATIONS

11-5. Investigative activity is essential to countering the adversary intelligence threat to Army interests. CI places emphasis on investigative activity to support force, infrastructure and technology protection, homeland defense, information assurance, and security programs. CI investigations focus on resolving allegations of known or suspected acts that may constitute National Security Crimes under US law. The primary objective in any CI investigation is the detection, identification, exploitation, and/or neutralization of adversary intelligence threats directed against the US Army. CI investigations are also conducted to identify systemic security problems that may have damaging repercussions to Army operations and national security interests. All CI investigations are conducted within guidelines established in AR 381-10, AR 381-12, AR 381-20, applicable DOD policy and directives, and US laws.

CI OPERATIONS

11-6. CI operations are characterized as those activities that are not solely associated with investigative, collection, analysis, or production functions. CI operations can be either offensive or defensive in nature; they are derived from or transition to a collection or investigative activity depending on the scope, objective, or continued possibility for operational exploitation. CI operations fall into two categories: CI support operations and CI sensitive operations.

- **CI Support Operations**. These are defensive operations used to support ARFOR and technology protection, security projects, and programs. They include technical services support, support to acquisition, FP, special access, international security, foreign visitor or contact, treaty verification, information assurance, homeland defense, and other approved projects and programs.
- **CI Sensitive Operations**. These operations are generally offensive in nature and involve direct or indirect operations against a known adversary intelligence threat. These operations include counter-espionage (CE) and CI projects and are conducted by designated units.

CI COLLECTION

11-7. **Collection Activities**. CI elements conduct collection activities focused on adversary intelligence threats that target US and multinational interests. CI collection is conducted through the use of sources, assets, official contacts, and other human or multimedia sources to obtain information that impacts the supported unit. CI will not be used as a substitute for Army HUMINT collection. These activities are designed to collect specific information or develop leads concerning adversary intelligence collection requirements, capabilities, efforts, operations, structure, personalities, and methods of operation targeting US and multinational interests. CI collection can result from ongoing CI investigations or operations or serve to initiate CI investigations and/or operations.

11-8. **Liaison**. CI elements conduct liaison with US, multinational, and HN military and civilian agencies, to include NGO, for the purpose of obtaining information of CI interest and coordinating or deconflicting CI activities. Liaison activities are designed to ensure a cooperative operating environment for CI elements and to develop CI leads for further exploitation.

11-9. **CI Collection Activities and Source Operations (CASO)**. CASO is used to collect information on direct threats to US Army installations, organizations, activities, and personnel. The CASO program is not intended to be used as a substitute for tactical HUMINT contact operations. CASO can be used to initiate CI investigations, identify potential leads for offensive operations, or develop additional CASO leads. **Only designated units, in accordance with applicable policy, will be authorized to pursue investigative and offensive operational leads from ongoing CASO operations.**

11-10. **Screening**. CI Special Agents work jointly with HUMINT Collectors during screening operations to identify civilians on the battlefield, EPWs, detainees, and other noncombatants who may have information of CI interest or to develop CI leads. CI screening is also conducted during the process of hiring HN citizens for Army and DOD employment. Information obtained during screening operations may be used to initiate CI investigations and operations or to cue other intelligence collection disciplines such as HUMINT, IMINT, SIGINT, and MASINT.

11-11. **Debriefing**. CI Special Agents conduct debriefings of friendly force, HN, or the local population who may have information of CI interest regarding adversary intelligence collection or targeting efforts focused on US and multinational interests.

11-12. **Functional Services**. CI elements conduct functional services to assist traditional CI activities of investigations, collection, operations, analysis, and production; they also provide tailored support to US, DOD, and Army protection and security programs for commanders at all echelons. CI elements may use one or several of the functional services simultaneously to provide tailored support to a particular CI mission or supported program. CI functional services consist of—

- CI threat vulnerability assessments (TVAs).
- Adversary intelligence simulation (Red Team Evaluation).
- Covering agent support.

11-13. **CI Technical Services**. CI technical services are used to assist the CI functions of investigations, collections, and operations or to support functional services conducted by CI elements. For additional information on CI technical services, refer to AR 381-20 (SECRET/NOFORN) and FM 34-5. CI technical services consist of—

- Surveillance.
- Intelligence polygraphs.
- TSCM.
- Computer Network Operations (CNO).
- IO.
- Counter-Signals Intelligence (C-SIGINT).

ANALYSIS

11-14. IPB, all-source, and single-source analysis are used to template adversary intelligence activities. Analysis is also used to make recommendations to the commander on how to counter adversary intelligence efforts and to refine CI activities to potentially neutralize and/or exploit those efforts and to continually focus the efforts of CI teams. CI analysis focuses on the multidiscipline adversary intelligence collection threat targeting information on US multinational personnel, operations, activities, technology, and intentions. Raw information, open-source material, and finished intelligence products are analyzed in response to local and national requirements. Analysis occurs at all levels from tactical to strategic.

- At the tactical level, CI teams focus their efforts on supporting mission requirements and contributing to the all-source COP.
- Operational analysis is used to assess how adversary intelligence views and targets US interests; identify US vulnerabilities that could be exploited or targeted; and to determine adversary intelligence targeting methods of operation (MOs).
- CI collection priorities; assessing adversary intelligence technical options for countering US weapons and intelligence systems; and

assessing the impact of technology transfer activities on the US military's technological overmatch.

PRODUCTION

11-15. CI products consist of, but are not limited to, target nomination, CI input to TVAs, CI estimates, and investigative and intelligence information reports. Finalized intelligence derived from CI activities are incorporated into joint and national intelligence databases, assessments, and analysis products. CI products are also incorporated into the COP to support battlefield situational awareness. CI production takes place at all levels.

- Operational and tactical production includes tactical alerts, spot reports, and current intelligence; CI threat and/or vulnerability assessments tailored to specific activities, units, installations, programs, or geographic areas; CI studies to support contingency planning and major exercises; studies of adversary intelligence organization, MO, personnel, activities, and intentions that pose a current or potential threat to the supported command.
- Strategic products include assessments supporting national and Army programs including SAPs and acquisition programs; worldwide assessments of the organization, location, funding, training, operations capabilities, and intentions of terrorist organizations; global trends in adversary intelligence MO; after-action studies of individual espionage cases; analyses of the intelligence collection capabilities of international narcotics trafficking organizations; and multimedia threat products to support Army CI awareness programs.

TECHNICAL SERVICES

11-16. CI organizations with technically trained CI Special Agents are chartered with providing unique technical capabilities to augment CI investigations, collection, and operations. These technical capabilities are not used as substitutes for CI activities, but support traditional CI techniques employed to counter and neutralize adversary intelligence activities targeting US interests.

11-17. Selected CI Special Agents are trained to identify human deception indicators during the course of investigations, operations, and collection missions. Required training and skills include the use of polygraph and emerging biometric technologies that can recognize indications of deception by human sources, contacts, and subjects of investigations, as well as analysis and reporting of results.

11-18. TSCM CI Special Agents are trained in sophisticated electronic and sensing equipment to identify technical collection activities carried out by adversary intelligence entities. The use of TSCM is critical to ensuring that sensitive and/or restricted areas are clear of any adversary intelligence-placed active or passive electronic sensing, or eavesdropping or collection devices. These areas include SCIF, secure working and planning areas, and C2 facilities.

11-19. CNO consist of computer network attack (CNA), computer network defense (CND), and computer network exploitation (CNE). CI Special Agents are specially trained in the areas of computer operation, network theory and administration, and forensics, along with IO in order to ensure US information dominance. The reliance on networked systems will result in greater emphasis being placed on information assurance. Specially selected CI agents will be trained in CNO in order to assist in protecting US information and information systems while exploiting and/or attacking adversary information and information systems.

11-20. For more information on technical services, refer AR 381-20 and FM 34-5.

OPERATIONAL EMPLOYMENT

11-21. Army CI supports the full spectrum of military operations. The Army requires a well-trained CI force consisting of AC, RC, civilian government employee, and contractor personnel. CI elements are focused on and dedicated to detecting, identifying, neutralizing, and/or exploiting adversary intelligence elements attempting to collect information on US forces. Effective employment of Army CI elements in all phases of operations and at all levels from tactical to strategic is paramount to countering any adversary intelligence threat to US interests and resources. Moreover, Army CI protects special programs that provide R&D, acquisition, and integration of technologies leading to future technological overmatch. CI elements have the capability and authority to conduct complex, nontraditional operations in this Information Age to CE, to both protect critical technologies and satisfy Army, DOD, and national level CI objectives.

LEVELS OF EMPLOYMENT

Strategic and Departmental

11-22. Strategic and departmental operations will be conducted by CI elements supporting national, DOD, and DA required missions (for example, support to NATO and special operations and missions). Strategic and departmental CI will conduct compartmented investigations and operations to affect the knowledge of adversary intelligence regarding CONOPS and defense information. Army CI will execute the full range of CI functions and missions at the strategic and departmental level including CI investigations and operations, CE, technology protection, SAP support, treaty verification, and technical CI services (polygraph, TSCM, and computer forensics). Strategic and departmental CI will also support SOF and special mission units (SMUs) within the scope of applicable national, DOD, and DA security policies and regulations.

Operational

11-23. Operational missions of CI elements will support combatant commanders, generally in geographic AOIR. Operational CI elements will focus on threat identification and countering regional adversary intelligence threats. Operational level CI activities and functions include CI investigations and operations, CE, technology protection, SAP support, treaty verification, and technical CI services (polygraph, TSCM, and computer forensics). CI elements must be capable of quickly transitioning from a peacetime mission to crisis operations to support combatant commander requirements. Theater CI assets will conduct unilateral to multinational operations in designated theaters. Operational elements may also be deployed to support or reinforce tactical forces in CONOPS.

Tactical

11-24. CI teams will conduct operations throughout the battlespace during CONOPS. CONOPS support activities include conduct of CASO, limited CI investigative capability, personnel security investigations, screenings, and debriefings. CI activities in CONOPS focus on countering the adversary intelligence threat and assisting in conducting TVAs. TVAs will be conducted in conjunction with MPs, Engineers, and Medical Service personnel to provide the commander with a comprehensive FP assessment. During peacetime, organic tactical CI teams conduct activities in accordance with approved regulations and command guidance.

SUPPORT TO CONTINGENCY OPERATIONS

11-25. The initial phase of operations from PME to MTW lays the foundation of future team operations. In general, the priority of effort focuses inward on security of operating bases, areas of troop concentration, and C2 nodes to identify the collection threat to US forces that could be used by adversary elements to plan hostile acts against US activities and locations.

11-26. Once security of the operating bases has been established, the operational focus of CI teams shifts outside the operating base to continue to detect, identify, and neutralize the collection threat to US forces as well as to provide I&W of hostile acts targeting US activities. The CI team uses several collection methods, to include CASO, elicitation, and liaison, to answer the supported commander's requirements. This is referred to as the continuation phase. The CI team conducts CI investigations to identify, neutralize, and exploit reported threat intelligence collection efforts.

11-27. A key element to the CI team's success is the opportunity to spot, assess, and develop relationships with potential sources of information. Operating as independent teams, without being tied to ISR or combat assets, enables the CI team's maximum interaction with the local population, thereby maximizing the pool of potential sources of information. Along with the opportunity to spot, assess, and interact with potential sources of information, a second key element of a CI team's success is its approachability to the local population. A soft posture enables a CI team to appear as non-threatening as possible. Experience has shown that the local population in general is apprehensive of fully and openly armed patrols and soldiers moving around population centers.

11-28. During some operations, civilian attire or nontactical vehicles may be used to lower the CI team profile. In some special situations, these measures are taken to make the operation less visible to the casual observer. Also, in some cultures, sharing food and beverages among friends is expected; exceptions to restrictions or general orders should be considered to facilitate successful CI team operations, many of which are geared towards developing relationships with potential sources of information.

SUPPORT TO INSTALLATIONS AND OPERATING BASES

11-29. CI teams, as part of a multi-agency team consisting of MPs, CA, medical, and EOD, support the conduct of TVAs of installations and operating bases to identify the intelligence threat to the operating locations. Detailed TVAs identify weaknesses in operational and physical security procedures and recommend countermeasures to mitigate intelligence collection on friendly forces limiting the ability to plan hostile acts on US activities and locations. CI activities supporting installations and operating bases include—

- Interviewing walk-in sources and locally employed personnel.
- Screening local national (LN) hires. Commanders, staff planners, and SIOs should always provide input to personnel assigned to establish and negotiate contracts using local national (LN) hires. This requirement ensures that LN hires can be screened, interviewed, and in some instances used as CI sources or assets in order to provide intelligence information that impacts the security of the base camp.
- Debriefing friendly force personnel who are in contact with the local population, such as—
  - ISR patrols.
  - MP patrols.
  - Combat patrols.
  - Liaison personnel.
  - CA and PSYOP teams.
- Conducting limited local open-source information collection.
- Providing support to TVAs of the base camp.

TACTICS, TECHNIQUES, AND PROCEDURES

11-30. At the CI team level, team members conduct mission analysis and planning specific to their AO. Backwards planning and source profiling are used extensively to choose CI targets. To verify adequate area coverage, the CI team may periodically develop and use CI target overlays and other CI analytical tools that illustrate the CI situation, identify CI gaps, and help refocus the collection effort.

11-31. The CI team is also in constant contact with the supported S2 and the other ISR assets (Scouts, PSYOP, CA, and MP) in order to coordinate and deconflict operations and to cross-check collected information. The supported unit S2, with the help of the CI team, regularly and systematically debriefs all ISR assets.

11-32. The CI team must be integrated into the supported unit's ISR plan. The CI OMT chief will advise the supported unit on the specific capabilities and requirements of the team to maximize mission success.

OPERATIONAL RISK MITIGATION

11-33. The employment of CI teams includes varying degrees of contact with the local population. As the degree of contact with the population increases, both the quantity and quality of CI collection increases. In many instances, however, there is a risk to the CI team inherent with increased exposure to the local population. The decision at what level to employ a CI team is METT-TC dependent. The risk to the CI assets must be balanced with the need to collect priority information and to protect the force as a whole. ROE, SOFA, direction from higher headquarters, and the overall threat level may also restrict the deployment and use of CI teams. The commander should consider exceptions to the ROE to facilitate CI collection.

11-34. Risks are minimized through the situational awareness of CI team members. They plan and rehearse to readily react to any situation and carry the necessary firepower to disengage from difficult situations. If it becomes necessary to call for assistance, adequate and redundant communications equipment is critical. These scenarios and actions should be trained prior to deployment into a contingency area and rehearsed continuously throughout the deployment.

11-35. A supported unit commander is often tempted to keep the CI team "inside the wire" when the THREATCON level increases. The supported commander must weigh the risk versus potential information gain when establishing operational parameters of supporting CI teams. This is necessary especially during high THREATCON levels when the supported unit commander needs as complete a picture as possible of the threat arrayed against US and multinational forces.

11-36. When it is not expedient to deploy the CI team independently due to threat levels or other restrictions, the team can be integrated into other ongoing operations. The CI team may be employed as part of a combat, ISR, or MP patrol or used to support CA, PSYOP, engineer, or other operations. This method reduces the risk to the team while allowing a limited ability to collect information. It has the advantage of placing the team in contact with the local population and allowing it to spot, assess, and interact with potential sources of information. However, this deployment method restricts collection by subordinating the team's efforts to the requirements, locations, and timetables of the unit or operation into which it is integrated and does not allow for the conduct of sensitive source operations. **This method of employment should be considered a last resort.**

COUNTERINTELLIGENCE EQUIPMENT

11-37. Basic C2, transportation, and weapons requirements do not differ significantly from most soldier requirements and are available as unit issue items. However, CI teams have unique communications, collection, processing, and mission-specific requirements.

COMMUNICATIONS

11-38. **Dedicated and Secure Long-Range Communications.** These are keys to the success of the CI team mission. CI team operations require a secure, three-tiered communications architecture consisting of inter/intra-team radios, vehicle-based communications, and a CI and HUMINT base station.

11-39. **Communications Network.** The CI team must have access to existing communications networks such as the tactical LAN. The CI team must also be equipped with its own COMSEC devices. It is imperative that the CI team acquire access to the public communication system of the HN. This can be in the form of either landlines or cellular telephones. Such access enables the CI team to develop leads which can provide early indicators to US forces.

11-40. **Interoperability.** Communications systems must be equipped with an open-ended architecture to allow for expansion and compatibility with other service elements, government organizations, NGOs, and multinational elements to effectively communicate during CONOPS. All ISR systems must be vertically and horizontally integrated to be compatible across all BOSs and with Legacy and Interim Force elements.

11-41. **SOTM.** To provide real-time and NRT information reporting, CI elements must have the capability to transmit voice, data, imagery, and video while on the move. CI teams must be able to transmit while geographically separated from their parent unit while operating remotely. This broadband requirement can only be achieved through a SATCOM capability and must be achievable while mobile.

CI COLLECTION AND PROCESSING SYSTEMS

11-42. The CI team must rely on automation to achieve and maintain information dominance in a given operation. With time, effective collection planning and management at all echelons, the CI team can collect a wealth of information. The sorting and analysis of this information in a timely and efficient manner is crucial to operations. Automation helps the CI team to report, database, analyze, and evaluate the collected information quickly and to provide the supported unit with accurate data in the form of timely, relevant, accurate, and predictive intelligence.

11-43. Automation hardware and software must be user friendly as well as interoperable among different echelons and services. They must interface with the communications equipment of the CI team as well as facilitate the interface of audiovisual devices. Technical support for hardware and software must be available and responsive.

11-44. The demand for accurate and timely CI reporting, DOCEX, and open-source information has grown tremendously. Biometric (physiological, neurological, thermal analysis, facial and fingerprint recognition) technologies will allow rapid identification, coding, and tracking of adversaries and human sources; as well as cataloging of information concerning EPWs, detainees, and civilians of CI interest on the battlefield. Biometrics will also provide secure authentication of individuals seeking network or facility access.

11-45. CI teams work with multinational forces and other foreign nationals and require the ability to communicate in their respective languages. Often CI personnel have little or no training in the target language, and lack of skilled interpreters can hinder CI activities. CI teams require textual and voice translation devices, source verification, and deception detection machines (biometrics) to improve collection capability and accuracy.

11-46. CI teams require dynamic MLT tools that provide both non-linguists and those with limited linguist skills a comprehensive, accurate means to conduct initial CI screenings and basic interviews in a variety of situations. CI elements will focus on in-depth interviews and communications with persons of higher priority. MLT tools minimize reliance on contract linguists and allow soldiers to concentrate on mission accomplishment.

MISSION SPECIFIC

11-47. The CI team may conduct night operations and must be equipped with NVDs for its members, and photographic and weapons systems. The CI team also may operate in urban and rural areas, where the threat level can vary from semi-hostile to hostile. The safety of the CI team can be enhanced with equipment that can detect, locate, suppress, illuminate, and designate hostile optical and E-O devices. In addition, high power, gyro-stabilized binoculars, which can be used from a moving vehicle, increases the survivability of the CI team and also gives the team another surveillance and collection device.

11-48. Some of the CI team missions may require the documentation of incidents. The CI teams can use the following equipment in their open-source collection efforts.

- Small, rugged, battery-operated digital camcorders and cameras which are able to interface with the collection and processing systems as well as communication devices.
- GPSs that can be mounted and dismounted to move in the AO efficiently.
- Short-range multichannel RF scanning devices that can also identify frequencies which enhance their security.
- In some cases CI teams require a stand-off, high resolution optical surveillance and recording capability that can provide target identification at extended ranges to protect the intelligence collector while avoiding detection by the adversary target. An advanced optical capability provides intelligence collectors the ability to locate and track adversary targets (passive and hostile) for identification, collection, and target exploitations.

INTEGRATION OF LINGUISTS

11-49. Integrating linguists into the CI team should take place as soon as possible. Security clearances and contractual agreements will help the team determine the level of integration.

11-50. Along with the basic briefing of what is expected of the civilian linguists as interpreters, CI teams should be informed about the civilians' chain of command and the scope of their duties beyond interpreting. The CI team leader must ensure that linguists are trained and capable of completing all tasks expected of them.

BATTLE HAND-OFF

11-51. CI teams are always engaged. A good battle hand-off is critical to smooth transition and mission success. The battle hand-off can directly contribute to mission success or failure of the outgoing team, but especially of the incoming team. The battle hand-off begins the first day the CI team begins to operate in an AO. Regardless of how long the team believes it will operate within the AO, it must ensure there is a seamless transition to an incoming team, other US unit, or agency. The CI team accomplishes this transition by establishing procedures for source administration, database maintenance, and report files.

11-52. Teams must plan and implement a logical and systematic sequence of tasks that enables an incoming team to assume the operations in the AO. Adequate time must be allotted for an effective battle hand-off. In some environments, a few weeks may be necessary to accomplish an effective battle hand-off. Introductions to sources of information, especially CASO sources, are critical, and teams must prioritize their time. During this time the outgoing CI team must familiarize the new CI team with all aspects of the operation, which include past, present, and planned activities within the AO. Area orientation is critical. These include major routes, population centers, potential hot spots, and other points of interest (such as police stations, political centers, and social centers).

ORGANIZATION

11-53. CI activities require a complex C2 relationship to ensure that the requirements of the supported commander are fulfilled while balancing the need for strict integrity and legality of CI operations. This complex relationship balances the role of the SIO as the requirements manager and the 2X as the mission manager with the MI commander as the asset manager.

COMMAND VERSUS CONTROL

11-54. ARFOR will normally deploy as part of a joint, multinational, and/or combined operation. In all cases, commanders at each echelon will exercise command over the forces assigned to their organization. Command includes the authority and responsibility for effectively using resources, planning for and employment of forces, and ensuring that forces accomplish assigned missions. Leaders and staffs exercise control to facilitate mission accomplishment.

11-55. While the MI commander supervises subordinates and produces reports, the *2X synchronizes activities between intelligence units and provides single-source processing and limited analysis. **(*2X " indicates 2X functions at all levels.)** While the MI commander takes care of the operators executing missions, the *2X obtains the data and reports from higher echelons required to execute the missions.

STAFF RESPONSIBILITIES AND FUNCTIONS

11-56. The *2X staff is responsible for the integration, correlation, and fusion of all Human Sensor information into the Intelligence BOS within the *2X AOIR. The *2X is also responsible for analyzing adversary intelligence collection, terrorist and sabotage activities, developing countermeasures to defeat threat collection activities, identifying and submitting collection requirements to fill CI collection gaps, and providing input to the all-source picture regarding adversary intelligence activities.

11-57. The *2X Staff Officer provides CI and HUMINT collection expertise. The *2X—

- Is the single focal point for all matters associated with CI and HUMINT in the AOIR.
- Is the CI and HUMINT advisor to the G2 and commander.
- Is an extension of the collection manager and ensures that the best asset or combinations of assets are used to satisfy information requirements.
- Along with his subordinate elements—CICA, HOC, OSC, CIAC, and HAC—exercises technical control over his assigned Army CI and HUMINT elements in the designated AOIR.
- Is the principal representative of the G2 and the commander when coordinating and deconflicting CI and HUMINT activities with national or theater agencies operating in the AOIR.
- Supports specific RM efforts in conjunction with the requirements manager through the planning and coordination of CI and HUMINT operations; the review and validation of CI requirements; the recommendation for assignment of tasks to specific collectors; and the conduct of liaison with non-organic HUMINT collection. This liaison includes national level and multinational force assets for source deconfliction and special activities outside the *2X AOIR.
- Will provide OMTs with capability to reach back to current database information, technical information and guidance, and source deconfliction necessary to monitor the collection activities of the CI teams.
  - *CICA.* The CICA is responsible for coordinating and synchronizing all CI activities in the designated AOIR. The CICA exercises technical control over all CI entities in the designated AOIR and deconflicts CI activities with higher, lower, and adjacent CI elements. The CICA accomplishes all responsibilities through coordination with the operational units and other *2X staff elements.
  - *OSC.* The OSC in the *2X staff maintains the source registry for all CI activities in the designated AOIR. The OSC provides management of intelligence property book operations, source incentive programs, and ICFs.
  - *CIAC.* The CIAC analyzes adversary intelligence collection capabilities. The CIAC leverages all intelligence discipline reporting and analysis to counter threat collection capabilities against the deployed force. CIAC analysis provides information and analysis to the COP.

11-58. The ACE and JISE CI analysis team analyzes threat intelligence collection and the intelligence collection efforts of foreign organizations involved in terrorism and sabotage in order to develop countermeasures against them. CI analysis cross-cues IMINT, SIGINT, MASINT, and TECHINT resources in addition to CI-related HUMINT reporting and analysis to counter threat collection capabilities against the deployed force. While the HAC supports the positive collection efforts of the force, the CI analysis team supports the "defend" aspects of the commander's FP program.

11-59. CI analysis is the analysis of the adversary's HUMINT, IMINT, SIGINT, and MASINT capabilities in support of intelligence collection, terrorism, and sabotage in order to develop countermeasures against them. It involves a reverse IPB process in which the analyst looks at US forces and operations from the threat's perspective. CI analytical products are an important tool in the COA development in the MDMP. This analytical tool supports the commander's FP program and facilitates the nomination of CI targets for neutralization or exploitation. (See FM 2-01.2 (FM 34-60) for more information on CI analysis.) Specifically, CI analysis—

- Produces and disseminates CI products and provides input to INTSUMs.
- Provides collection requirements input to the CICA.
- Analyzes source reliability and credibility as reflected in reporting and communicating that analysis to the collector.
- Nominates CI targets for neutralization or exploitation.
- Identifies and submits CI-related requirements to fill collection gaps.
- Assists HAC personnel in focusing the CI aspects of the HUMINT collection program.
- Presents CI analysis products such as CI estimates, target lists, reports, and graphics that support the commander.

11-60. For intelligence reach operations, CI products are available and disseminated in a variety of forms. It is incumbent on the requestor to ensure the CI product can be transmitted over the available communications systems. This includes verifying the appropriate security level of the communications systems.

CI TEAM STRUCTURE

11-61. **OMT.** The OMT is a four-person team consisting of a WO, two NCOs, and a junior enlisted soldier. (Civilians may be inserted into this structure as appropriate.) Rank structure and standards of grade for OMTs will vary depending upon the skill sets required and mission focus. CI OMTs will provide operational guidance for 1 to 4 CI teams, depending on mission focus and operational tempo. When two or more CI teams are deployed in a DS role, an OMT is also deployed to provide technical control. The OMT works closely with the supported S2 and ACT to furnish current threat information and to answer the supported commander's PIRs and IRs. OMTs coordinate with the supported 2X and manage subordinate CI teams to—

- Provide guidance and technical control of operational activity.
- Provide the collection and operational focus for CI teams.

- Provide quality control and dissemination of reports for subordinate CI teams.
- Conduct single-discipline CI analysis and assist in mission analysis for the supported commander.
- Act as a conduit between subordinate CI teams, the CICA, and supported unit headquarters.
- Provide administrative support for subordinate CI teams to include reporting mission and equipment status to the CICA and the supported unit headquarters.
- Educate the supported commander on the capabilities of the CI teams.
- Integrate the CI teams directly into the maneuver commander's ISR planning.

11-62. **CI Team.** The CI team is a four-person team consisting of two NCOs and two junior enlisted personnel. Rank structure and standards of grade for CI teams will vary depending upon the skill sets required and mission focus. CI teams are trained to execute the full range of CI functions; however, they may be assigned to mission-focused elements (for example, CE, CI projects). Assignment to a TSCM, polygraph, or information warfare team requires additional, specialized technical training.

Appendix A

Intelligence and Information Operations

THE INFORMATION ENVIRONMENT

A-1. The information environment is the aggregate of individuals, organizations, and systems that collect, process, store, display, and disseminate information; also included is the information itself. (FM 3-0) The information environment, it should be noted, is not an exclusively military one; in fact, the military applications of information are almost obscured in today's universal usage of the information spectrum by national, international, and non-state players.

THE COMMANDER AND INFORMATION

A-2. Information is facts, data, or instructions in any medium or form; the meaning that a human assigns to data by means of the known conventions used in their representation. (JP 1-02) Information provides the key to battlefield success in the 21st century. Commanders must have detailed information to command. Information is the medium that allows the commander's decisionmaking and execution cycle to function. Information gives direction to actions by the force, identifies the enemy's centers of gravity, provides COAs for force activity, and enables the force to accomplish its operational mission.

INFORMATION SUPERIORITY

A-3. Information superiority is the operational advantage derived from the ability to collect, process and disseminate an uninterrupted flow of information while exploiting or denying an adversary's ability to do the same. (FM 3-0) Relevant information drawn from intelligence supports the creation and development of situational understanding that contributes directly to information superiority during decisive operations. The requirement for information superiority is not new; what is new is that today's information technologies are creating a base of knowledge for military planning and execution that is unprecedented in scope, volume, accuracy, and timeliness. This means commanders receive accurate, timely information that enables them to make better decisions and act faster than their adversaries.

A-4. Information superiority, however, is neither a staple in today's battlespace nor is it necessary in military operations as a constant condition; rather, at specific times during operations, information superiority becomes a key enabler for assuring military success. At the operational level of predominant interest to the land component commander (LCC), information superiority is realized through the integration of operational level, interdependent ISR, information management, and IO to gain and maintain operational initiative and to achieve an operational advantage.

INFORMATION OPERATIONS

A-5. IO are the employment of core capabilities of electronic warfare, computer network operations, PSYOP, military deception, and operations security, in concert with specified supporting and related capabilities to affect or defend information and information systems, and to influence decisionmaking.... Information systems are...the equipment and facilities that collect, process, store, display, disseminate information. These include computers, hardware, software, and communications, as well as policies and procedures for their use. (FM 3-0). Offensive IO are capable of degrading an adversary's will to resist and ability to fight. Defensive IO measures passively and actively protect friendly information and C2 systems and limit their vulnerability.

A-6. Offensive and defensive IO are conducted as a fully coordinated effort to ensure the complementary, asymmetric, and reinforcing effects to attack enemy forces, influence others, and protect friendly forces. Relevant information assures that the right person has the right information at the right time for decisionmaking and execution.

THE ELEMENTS OF IO

A-7. Full-spectrum IO incorporates, integrates, and synchronizes traditionally independent capabilities and activities in support of the commander's mission. (See FM 3-0 and FM 3-13.) The Army's doctrinal view of full-spectrum IO as evolving core capabilities are—

- PSYOP.
- OPSEC.
- EW.
- Military deception.
- CNO.
  - CNA.
  - CND.
  - CNE.

A-8. Supporting IO capabilities are—

- Physical destruction.
- Physical security.
- Information assurance.
- Counterpropaganda.
- Counterdeception.
- Counterintelligence.

IO-RELATED ELEMENTS

A-9. The IO-related activities of PA and CMO are closely associated and integrated with the elements of IO as key contributors to information superiority.

INTELLIGENCE SUPPORT TO IO

A-10. All-source intelligence support, encompassing all of its forms, is the principal enabler for successful IO. The commander's capability to synchronize military operations can be heavily influenced by the ability to identify the threat and understand the adversary's capabilities and intentions in the information environment.

A-11. It is essential that the elements of the command's IO capabilities and vulnerabilities are integrated into the command intelligence plan. This leads to COAs that synchronize the elements of IO and IO-related activities into the commander's warfighting plans. Integration of the full-spectrum aspects of IO ensures that the relative importance of information is recognized in the development of a synchronized OPLAN. Further, integration of IO into the planning process provides a methodology for analyzing the threat from a knowledge base that enhances protection of friendly systems and assets while exposing windows of opportunity for attack or exploitation. IPB is the best process we have for understanding the battlefield and the options it presents to friendly and threat forces.

A-12. Provide Intelligence support to IO includes, but is not limited to, the following:

- **Provide Intelligence Support to Offensive IO**. The Intelligence BOS supports offensive IO by providing information to identify critical enemy C2 nodes. Intelligence also helps to identify enemy systems and procedures that may be vulnerable for offensive IO. Additionally, intelligence plays a key role in evaluating and assessing the effectiveness of offensive IO.
  - *Provide Intelligence Support to PSYOP.* This task identifies the cultural, social, economic, and political environment of the AO. It identifies target groups and subgroups and their location, conditions, vulnerabilities, susceptibilities, cultures, attitudes, and behaviors. PSYOP influence foreign target audiences in the AO to support achieving the commander's goals in the AO.
    - Identify the cultural, social, economic, and political environment of the AOI; for example, adversary mechanisms for political control, adversary communication and broadcast systems used to elicit support from the populace, and current and past adversary propaganda activities and their effectiveness.
    - Identify target groups and subgroups and their location, conditions, vulnerabilities, susceptibilities, cultures, attitudes, and behaviors. This includes determining the audience demographics of popular radio and television programs and periodicals; groups influenced by media personalities and political cartoons.
    - Identify impact of planned PSYOP on individuals outside the targeted group (for example, multinational partners, neighboring populations).
  - *Provide Intelligence Support to Military Deception.* This task identifies the capabilities and limitations of the adversary's

intelligence-gathering systems and identifies adversary biases and perceptions.

- Profiles of key adversary leaders.
- Cultural, religious, social, and political characteristics of the country and region.
- Sources of military, economic, or political support.
- Adversary decisionmaking processes, patterns, and biases.
- Adversary perceptions of the military situation in the AO.
- Capabilities and limitations of adversary CI and security services.

▪ *Provide Intelligence Support to Electronic Attack.* This task supports electronic attack employing jamming, electromagnetic energy, or directed energy against personnel, facilities, or equipment. It identifies critical adversary information systems and C2 nodes. It includes determining and presenting the adversary's electronic OB and their information system infrastructure. The enemy's C2 system vulnerabilities and the means they use to protect their C2 systems is part of the electronic OB.

- **Provide Intelligence Support to Defensive IO**. The Intelligence BOS supports defensive IO by providing information to identify threat IO capabilities and tactics. Intelligence provides information relating to CND, physical security, OPSEC, counterdeception, and counterpropaganda. The Intelligence BOS supports defensive IO by providing information to identify threat IO capabilities and tactics. Intelligence provides information relating to CND. Provide Intelligence Support to OPSEC—identify capabilities and limitations of the adversary's intelligence system to include adversary intelligence objectives and the means, methods, and facilities used by the threat to collect, process, and analyze information—supports the identification of indicators that could be interpreted or pieced together to penetrate EEFI in time to be useful to adversaries.
- **Provide Intelligence Support to Activities Related to IO**. The Intelligence BOS when operating outside US territories supports activities related to IO under some circumstances.

▪ *Provide Intelligence Support to CMO.* This task allows military intelligence organizations to collect and provide information and intelligence products concerning foreign cultural, social, economic, and political elements within an IO in support of CMO. Identify cultural, social, economic, and political environment of the AOI, including—

- Population demographics.
- Civilian populace attitudes, alliances, and behavior.
- Availability of basic necessities (food, clothing, water, shelter, medical care) and the ability of the populace to care for itself.
- Access to medical care.
- Locations and potential routes, destinations, and assembly areas or sites of displaced persons.
- Local government type, status, organization, and capabilities.

  - Availability of local material and personnel to support military operations.
  - NGOs in the AOI, their agenda, resources, and capabilities.
- *Provide Intelligence Support to Public Affairs.* This task identifies the multinational and foreign public physical and social environment, as well as world, HN national, and HN local public opinion, in addition to the propaganda and misinformation capabilities, activities, targets, themes, and dissemination means of the adversary. Identify world, national, and local public opinion (location, biases or predispositions, and agenda of national and international media representatives in the AOI, and trends reflected by the national and international media).

Appendix B

Linguist Support

ROLE OF LINGUISTS

B-1. Military operations are highly dependent on foreign language support. The requirement to communicate with and serve on multinational staffs, communicate with local populations, and exploit enemy forces necessitates the use of linguists. The growing focus on multinational operations increases the competition for limited linguist resources that are vital for mission success. This appendix establishes the framework and process to access, prioritize, and employ the Army's limited organic linguist resources.

LINGUISTIC SUPPORT CATEGORIES

B-2. Foreign language support requirements of US Armed Forces typically fall into one of four broad categories:

- Intelligence and Information Gathering. This category includes the traditional SIGINT and HUMINT disciplines, as well as foreign language support to FP and exploitation of open-source information.
- CMO. This category encompasses all functions relating to military interaction with the civilian population. Foreign language support is critical to CMO in areas such as government liaison, legal agreements, medical support and operations, law enforcement, engineering projects, public safety, security and population control, CA, and PSYOP.
- Logistics. This category consists of foreign language support to sustainment or transportation functions. These include logistical contracting, port, railhead, airhead, or transshipment operations and convoy operations.
- Multinational Operations and Liaison. This category includes the coordination of military operations and liaison with multinational partners, previously unaffiliated nations, and at times adversary or former adversary nations. Multinational operations are becoming more common and increasingly important.

DETERMINING LINGUIST REQUIREMENTS

B-3. To identify linguist requirements, the staff conducts mission analysis and identifies specified or implied tasks requiring foreign language support. Other critical factors are the organization or echelon of command and the location of the mission. The staff uses these criteria to determine the allocation of linguists, such as one linguist team per echelon of command, one linguist per piece of equipment, or one linguist team per location where the function is to be performed. The staff then applies task organization and scheme of maneuver to determine the number of linguists needed for an operation.

B-4. The staff must analyze each linguist assignment to determine the minimum level of foreign language proficiency needed. While interpretation for a peace negotiation requires not only outstanding linguistic capability but also cultural acumen, the translation of routine documents (with the aid of a dictionary) requires a much different skill set. Poor identification of linguist proficiency requirements can tie up the best linguists in less effective roles, creating linguist shortfalls in other areas.

B-5. The relative importance of each of the four linguist support categories is mission dependent. For example, during a NEO civil and military coordination would probably not be as critical as intelligence and information gathering. However, the situation is reversed for a humanitarian assistance mission in which CMOs have a significant impact on mission success. Identifying these "dynamics" helps the commander and staff prioritize linguist requirements.

B-6. Determining linguist requirements for any operation can be difficult because each operation is unique. However, commanders and staffs with a basic knowledge of organic Army linguistic assets, foreign language resource alternatives, and MI skills can successfully assess, prioritize, and employ linguists in support of their military operations.

PLANNING AND MANAGING LINGUIST SUPPORT

B-7. Commanders must consider the linguist requirements as part of their MDMP for every CONPLAN and OPLAN assigned to their commands. Prior staff planning and identification of linguist requirements should prompt commanders to initiate linguist support requests and identify command relationships prior to actual operations. If the mission analysis reveals requirements for linguistic support, the commander must identify what foreign languages are needed, the foreign language proficiency levels needed for each assignment, and the best source of linguists. In addition, if the mission includes intelligence and information collection, the commander must identify MI collection skills required. During mission analysis, the commander should consider linguist requirements for every CONPLAN and OPLAN assigned to his command.

LINGUIST CATEGORIES

B-8. The commander and staff must identify linguist requirements by category:

- Category I – Have native proficiency in the target language (level 4-5) and an advanced working proficiency (Interagency Language Round Table [ILRT] level 2+) in English. They may be locally hired or from a region outside the AO. They do not require a security clearance. They must be screened by the Army CI support team.
- Category II – Are US citizens screened by Army CI personnel and are granted access to SECRET by the designated US government personnel security authority. Have native proficiency in the target language (level 4-5) and an advanced working proficiency (ILRT 2+) in English.
- Category III – Are US citizens screened by Army CI personnel and are granted either TS/SCI clearance or an interim TS/SCI clearance by the

designated US government personnel security authority. Meet a minimum requirement of ILRT level 3. They are capable of understanding the essentials of all speech in a standard dialect. They must be able to follow accurately the essentials of conversation, make and answer phone calls, understand radio broadcasts and news stories, and oral reports (both of a technical and non-technical nature).

PRIMARY STAFF RESPONSIBILITIES

B-9. Primary staff at each echelon has responsibilities for evaluating requirements and managing linguist support. The responsibilities include but are not limited to those discussed below. In addition, each staff section is responsible for determining its linguist support required to meet its operational missions.

Assistant Chief of Staff, G1 (S1):

- Identify linguist requirements needed to support G1/S1 functions in all contingency areas. G1/S1 requirements for linguist support include but are not limited to the following:
  - Coordinate with local authorities on matters of civilian hire, finance, and recordkeeping.
  - Contract for local hire personnel.
  - Coordinate for local morale support and community activities.
  - Coordinate with local authorities for postal operations.
  - Support for administration, counseling, personal affairs, and leave for LN and third-country national (TCN) personnel.
  - Coordinate for local medical support.
  - Liaison with multinational counterparts.
- Linguist staffing and linguist replacement management.
- Identify foreign language skill identifiers for all assigned, attached, or OPCON Army linguists.
- Identify all Army foreign language skilled soldiers not identified on electronic Military Personnel Office System (eMILPO) and Defense Integrated Management Human Resource System (DIMHRS). The Standard Installation Division Personnel System (SIDPERS) was replaced by eMILPO.
- Deploy and provide administrative support of DA and DOD civilian linguists.
- Hire, contract for, and provide administrative support of LN linguists.
- Procure Army foreign language support personnel for screening local labor resources.

Assistant Chief of Staff, G2 (S2):

- Identify linguist requirements needed to support G2/S2 functions in all contingency areas. G2/S2 requirements for linguist support include but are not limited to—
  - Evaluate and/or use local maps and terrain products in operations.
  - Process for MI purposes material taken from EPWs or civilian internees.

  - At lower echelons, conduct tactical questioning of refugees, detainees, and EPWs.
  - Assess local open-source information for intelligence value.
  - Coordinate intelligence and liaison with multinational and HN counterpart.
- Determine, during the initial IPB, all foreign languages (spoken and written) and dialects needed for mission accomplishment.
- Collect, process, produce, and disseminate information derived from linguist sources.
- Provide intelligence training for MI linguists employed in AOs.
- Coordinate for security investigations, as necessary, for local hire linguists.
- Provide support to CI screening of contracted linguists and LN labor force.

Assistant Chief of Staff, G3 (S3):

- Identify linguist requirements needed to support G3/S3 functions in all contingency areas. G3/S3 requirements for linguist support include but are not limited to—
  - Operational coordination and liaison with multinational and HN counterparts.
  - Translate OPORDs and OPLANs for use by multinational counterparts.
- Consolidate unit linguistic requirements and establish priorities.
- Develop linguist deployment and employment plans.
- Develop plans to train linguists and to use linguists for training the force in AO's foreign language survival skills. In addition to global language skills, linguists must have training in specific vocabulary used in the AO; for example, terms used for military, paramilitary, civilian or terrorist organizations, and ethnic groups within the area, nomenclatures of equipment used, and other military or technical vocabulary. Training in the specific dialect used in the AO would also be beneficial.
- Assign, attach, and detach linguists and linguist teams.
- Integrate additional or replacement linguists through operational channels.
- Recommend modernization and development of linguist systems and methods.
- Coordinate mobilization and demobilization of RC linguist support.
- Plan linguist usage for deception operations.
- Plan linguist support for movement of EPWs, detainees, and refugees.
- Coordinate evaluation of linguist support by all staff elements.

Assistant Chief of Staff, G4 (S4):

- Identify linguist requirements needed to support G4/S4 functions in all contingency areas. G4/S4 linguist requirements for linguist support include but are not limited to—

    - Procure local supply, maintenance, transportation, and services.
    - Coordinate logistics at air and seaports of debarkation.
    - Contract with local governments, agencies, and individuals for sites and storage.
- Provide logistical, supply, maintenance, and transportation support to attached linguists.

Assistant Chief of Staff, G5 (S5):

- Identify linguist requirements needed to support G5/S5 functions in all contingency areas. G5/S5 linguist requirements for linguist support include but are not limited to—
    - Determine civilian impact on military operations.
    - Minimize civilian interference with combat operations.
    - Inform civilians of curfews, movement restrictions, and relocations.
    - Provide assistance to liaison with HN and multinational agencies, dignitaries, and authorities.
    - Promote positive community programs to win over support.
    - Determine if multinational operations PSYOP efforts are mutually planned and synchronized.
    - Interpret support to assist resolution of civilian claims against the US Government.
    - Solicit linguistic and cultural knowledge support to protect culturally significant sites.
    - Use linguistic and cultural support to identify cultural and religious customs.
- Assist the G1 in the contracting of local hire linguists.
- Identify foreign language requirements for CMOs.

Assistant Chief of Staff, G6 (S6):

- Identify linguist requirements needed to support G6/S6 functions in all contingency areas. G6/S6 linguist requirements for linguist support include but are not limited to—
    - Coordinate suitable commercial information systems and services.
    - Coordinate with multinational forces on command frequency lists.
    - Coordinate signal support interfaces with HN and multinational forces.
- Manage RF assignments for supporting SIGINT linguist elements.
- Support linguist operations with internal document reproduction, distribution, and message services.
- Integrate automation management systems of linguist units.

SPECIAL STAFF OFFICER RESPONSIBILITIES

B-10. If no special staff officer is assigned the duties below, the corresponding coordinating staff officer should assume those responsibilities. Linguist requirements for special staff officers include but are not limited to the following staff officers.

Liaison Officer:

- Should speak the required foreign language. If not, he requires a translator or interpreter for all aspects of his duties.
- Request interpreters to assist when representing the multinational operations.
- Translate orders, maps, traces, overlays, and documents into multinational foreign languages.

Civilian Personnel Officer:

- Recruit, interview for suitability, and hire civilian local labor force if required.
- Negotiate host country on labor agreements.

Dental Surgeon:

- Administer dental care to support humanitarian mission requirements.
- Rehabilitate, construct, and gain usage of existing dental facilities as required.

Finance Officer:

- Support the procurement process of local goods and services not readily available through normal logistical channels.
- Ensure limited non-US and US pay functions to foreign national, HN, civilian internees, and EPWs are provided.
- Ensure all necessary banking functions are performed in theater.

Surgeon:

- Support medical humanitarian assistance and disaster relief operations.
- Provide medical care of EPWs and civilians within the command's AO.
- Coordinate medical laboratory access in AO.
- Determine the nature of local health threats to the force through populace interviews.
- Determine the identity of local or captured medical supplies.

Veterinary Officer:

- Determine source and suitability of local foods.
- Assist the local population with veterinary service needs.

Chemical Officer:

- Identify enemy force chemical weapons and equipment.
- Communicate NBC risks to supported populations.

Engineer Coordinator:

- Procure proper local materials to support engineering missions.
- Communicate engineering project requirements to contracted local work force.

- Communicate engineering project impact on local landowners and other affected parties.
- Determine, in coordination with G2/S2, suitability of local topographic maps and terrain products.
- Assess environmental concerns of HN and local populations in combined operations.

Provost Marshal:

- Support dislocated and civilian straggler control activities.
- Support internment and resettlement operations, to include displaced civilians.
- Support weapons buy-back programs, as required, and work closely with civil-military liaisons for payments to local officials.
- Support counter-drug and customs activities.
- When authorized, help foreign civil authorities maintain control.
- Conduct liaison with local LEAs.

PSYOP Officer:

- Produce approved PSYOP propaganda and counter-propaganda media.
- Evaluate PSYOP impact on target audience.

Air Defense Coordinator:

- Identify enemy air defense artillery (ADA) weapons and radars.
- Communicate air defense warnings to supported populations.
- Communicate air defense project requirements to contracted local work force.

Safety Officer:

- Provide safety training to local labor force.
- Communicate warnings of dangerous military operations and other hazards to local populace.

Transportation Officer:

- Coordinate commercial and local transportation needs.
- Coordinate movement scheduling and routes with multinational forces and/or HN.

PERSONAL STAFF OFFICER RESPONSIBILITIES

B-11. Personal staff officers are under immediate control of the commander and have direct access to the commander. Most personal staff officers also perform special staff officer duties, working with a coordinating staff officer. These assignments are on a case-by-case basis, depending on the commander's guidance and the nature of the mission; they are very common in stability operations and support operations. Linguist requirements for special staff officers include but are not limited to the following staff officers.

Chaplain:

- Coordinate religious support with multinational partners.
- Determine the impact of local population religious group faiths and practices on military operations.
- Provide religious support to the community to include hospital patients, EPWs, refugees, and civilian detainees.
- Conduct liaison with local population religious leaders in close coordination with the G5.

Public Affairs Officer:

- Act as the commander's spokesman for all communication with external media.
- Assess the accuracy of foreign media interpretation of Public Affairs Office (PAO) releases.
- Assess and recommend news, entertainment, and other information (assisting G5) for contracted services foreign nationals.

Staff Judge Advocate:

- Translate and interpret foreign legal codes, SOFAs, and international laws.
- Determine local environmental laws and treaties through translation services.
- Assess the treatment of EPWs and civilian internees.
- Translate documents to support G4 in local contracts.

SOURCES OF LINGUISTS

B-12. There are various sources that a commander can use to obtain the linguists necessary to support operations. It is vital to know the advantages and disadvantages of each type of linguist and to carefully match the available linguists to the various aspects of the operation.

ARMY LANGUAGE-QUALIFIED MOSs

B-13. The AC MI language-dependent military occupational specialities (MOSs) are 98G with a skill qualification identifier (SQI) of L/352G (Cryptologic Communications Interceptor/ Locator), and their related WO fields. Some soldiers in MOS 96B (All-Source Intelligence Analyst), MOS 97B (CI Agent), MOS 97E/351E (HUMINT Collector) and MOS 98C (SIGINT Analyst), and their related WO fields are trained in foreign languages. Using soldiers in the MOSs mentioned above has many advantages. They are already trained in the military system, are not subject to deployment restrictions (a limiting factor with civilian linguists), have a security clearance and, as US personnel, support the command's interests. The major disadvantage to utilizing these individuals for general foreign language support is that in doing so, they are removed from their primary MI functions. They should be used only in linguistic duties that include intelligence potential. For example, a HUMINT collector (97E) provides linguist support to a medical assistance team as a method to provide access to the local population to determine their attitudes toward US Forces.

B-14. Non-MI Army language qualified MOSs include some enlisted and WOs in career management fields 18 (Special Forces), 37 (PSYOP), 180A (Special Forces); and commissioned officers with a branch code 18 (Special Forces); and functional areas 39 (PSYOP and CA) and 48 (Foreign Area Officer). Particular attention must be paid to the recorded language proficiency and test date of these individuals since the standards vary by field. The same advantages and disadvantages apply as with the AC MI linguists.

B-15. RC language-dependent MOSs include those listed above in the AC. RC linguists have the same set of advantages and disadvantages as listed above for AC language-dependent MOSs. The RC also includes linguists in MOS 97L (translator/interpreter). The 97Ls are specifically trained to be a translator and interpreter. They have the same advantages as the AC linguists. An added advantage is that since their sole job is translation and interpretation, they do not have to be removed from another job in order to be used as a linguist. Their major disadvantage is that they have no additional skill that gives them dual functionality.

Army Linguists Not DOD Trained

B-16. The Army also includes numerous soldiers of all grades who are proficient in a foreign language and are receiving Foreign Language Proficiency Pay (FLPP) but whose primary duties do not require foreign language proficiency. They may have attended a civilian school to learn a foreign language, or they may have acquired proficiency through their heritage. They have the advantage of being trained soldiers and are therefore readily deployable to all areas of the battlefield.

B-17. These soldiers may have the specific vocabulary and military skill knowledge for certain linguist support missions. For example, a supply sergeant who speaks the local language would be an invaluable asset to the G4. There are disadvantages in that they already have another job and units are reluctant to give up personnel especially if they are in key positions. Their capabilities are difficult to assess. Since they are not required to take the Defense Language Proficiency Test (DLPT) if they are not receiving FLPP, it is often difficult for the G1/S1 to identify them as a linguist or for a non-linguist to judge the level of their foreign language capability.

Other Service Linguists

B-18. Other service linguists have the advantage of deployability, loyalty, and clearance, but must often learn the Army system and specific Army vocabulary. They are also difficult to obtain since their parent service probably also lacks a sufficient number of trained linguists. Other service linguists, however, will be valuable in joint operation centers and joint activities. When serving the JTF headquarters, Army commanders and staffs must be aware of the linguists in the other services in order to plan for the participation and optimize their employment.

US Contract Linguists

B-19. US civilians can be contracted to provide linguist support. They have an advantage over LN hires in that their loyalty to the US is more readily evaluated, and it is easier for them to be granted the necessary security clearance. However, there are usually severe limitations on the deployment and use of civilians. A careful assessment of their language ability is important because, in many cases, they use "old fashioned" terms, or interject US idioms. If the linguists are recent émigrés, the use of the language in their country of origin could be dangerous to them, or their loyalty may reside with their own country when at odds with US interests.

Multinational Linguists

B-20. Multinational linguists have their own set of advantages and disadvantages. These linguists may be unfamiliar with the US military system unless they have previously participated in a multinational operation with US forces. They may have a security clearance, but clearances are not necessarily equal or reciprocal, automatically guaranteeing access to classified or sensitive information between nations. They support the command's interest but may have differing priorities or responsibilities within their assigned AOs. These linguists also are already fulfilling specific duties for their own nation, which may also have a shortage of linguists. The major disadvantage to acquiring and maintaining multinational linguist support is that they are outside the C2 (via military authority or military contract) of the US forces. These linguists will be valuable in multinational operations centers and activities.

Local National Contract Linguists

B-21. LN hires will provide the bulk of your linguist support. They are usually less expensive to hire than US civilians and will know the local dialect, idioms, and culture. The expertise of these linguists in particular areas or subject matters can be an asset. However, there are several potential problems with using LN hires, to include limited English skills and loyalty considerations. Therefore, a screening interview or test is necessary to determine their proficiency in English. These individuals must also be carefully selected and screened by CI personnel (with US linguist support) initially and periodically throughout their employment. Their loyalty is always questionable. Local prejudices may influence them, and they may place their own interests above those of the US.

EVALUATING LINGUIST PROFICIENCY

B-22. Commanders and staffs must understand the Army linguist proficiency evaluation system in order to effectively plan for and employ linguists. Evaluation and reevaluation of linguist proficiency is covered in detail in AR 611-6, Section III. Language testing is required for all Army personnel in a language-dependent MOS, who have received foreign language training at government expense, who are receiving FLPP, or who are in a language-required position regardless of MOS. Other Army personnel who have knowledge of a foreign language are encouraged to take the proficiency test and may work as linguists.

B-23. The Army uses the DLPT to determine foreign language proficiency levels. DLPTs are listed by foreign language in DA Pam 611-16. In foreign languages where no printed or recorded test exists, oral interview tests are arranged. The DLPT is an indication of foreign language capability, but it is not the definitive evaluation of an individual's ability to perform linguist support.

B-24. AR 611-6, Appendix D, Sections 1 through 4, describes the proficiency levels for the skills of speaking, listening, reading, and writing a foreign language based on the interagency roundtable descriptions. The plus-level designators, shown as a "+" symbol, are used to designate when a linguist is above a base level, but not yet to the capability of the next level. For example, 2+ would indicate a better than limited working proficiency in the foreign language. The six "base levels" of proficiency, as established by DLPT and/or oral exam, are—

- Level 0 (No proficiency). The soldier has no functional foreign language ability. Level 0+. The minimum standard for Special Forces personnel indicates a memorized proficiency only.
- Level 1 (Elementary proficiency). The soldier has limited control of the foreign language skill area to meet limited practical needs and elementary foreign language requirements.
- Level 2 (Limited working proficiency). The linguist is sufficiently skilled to be able to satisfy routine foreign language demands and limited work requirements.
- Level 3 (General professional proficiency). The linguist is capable of performing most general, technical, formal, and informal foreign language tasks on a practical, social, and professional level.
- Level 4 (Advanced professional proficiency). The linguist is capable of performing advanced professional foreign language tasks fluently and accurately on all levels.
- Level 5 (Functionally native proficiency). The linguist is functionally equivalent to an articulate and well-educated native in all foreign language skills; and reflects the cultural standards of the country where the foreign language is natively spoken.

B-25. The above proficiency base levels designate proficiency in any of the four language skills: listening, reading, speaking, and writing. The most evaluated skills on the DLPT are reading and listening. These tests are not for use in evaluating linguists above the 3 proficiency level. Most Army linguist DLPT scores show only two skill levels: listening and reading (for example, 2+/3, or 3/1+).

SUSTAINING MILITARY LINGUIST PROFICIENCY

B-26. Language proficiency diminishes with lack of use and absence of exposure to the foreign language. To ensure combat readiness, commanders should require all military linguists receive periodic language training. In-country language immersion training, in-garrison contracted language instructors, on-line foreign newspapers, and foreign radio broadcasts are all examples of language training resources. Funding for language training is available through MACOM language training program funds.

Glossary

The glossary lists acronyms and terms with Army or joint definitions, and other selected terms. Where Army and joint definitions are different, (Army) follows the term. Terms for which FM 2-0 is the proponent FM (authority) are marked with an asterisk (*) and followed by the number of the paragraph (¶) where they are discussed. For other terms, the number of the proponent FM follows the definition. JP 1-02 and FM 1-02 are posted in the Joint Electronic Library, which is available online and on CD-ROM.

- Use this URL to access JP 1-02 online: http://www.dtic.mil/doctrine/jel/doddict/.
- Use this URL to access FM 1-02 online: http://www.dtic.mil/doctrine/jel/service_pubs/101_5_1.pdf.
- Follow this path to access JP 1-02 on the Joint Electronic Library CD-ROM: Mainmenu>Joint Electronic Library>DOD Dictionary.
- Follow this path to access FM 1-02 on the Joint Electronic Library CD-ROM: Mainmenu>Joint Electronic Library>Service Publications>Multiservice Pubs> FM 101-5-1.

A2C2	Army Airspace Command and Control
AA	avenue of approach
AAMDC	Area Air and Missile Defense Command
ABCS	Army Battle Command System
AC	Active Component
ACE	analysis and control element
ACINT	acoustic intelligence (water)
ACOUSTINT	acoustic intelligence (atmosphere)
ACR	armored cavalry regiment
ACT	analysis and control team
ADA	air defense artillery
ADCON	administrative control
ADDO	assistant deputy director for operations
ADO	air defense officer
ADP	automated data processing
AFACSI	Air Force Assistant Chief of Staff for Intelligence
AFIWC	Air Force Information Warfare Center
AFMIC	Armed Forces Medical Intelligence Center

AFOSI	Air Force Office of Special Investigation
AGM	attack guidance matrix
AIA	Air Force Intelligence Agency
AITR	aided target recognition
ALOC	administrative and logistics operations center
AMC	US Army Materiel Command
AMD	Air Missile Defense
AMHS	Automated Message Handling System
AMIB	Allied Military Intelligence Battalion
***analysis**	Determination of the significance of the information, relative to information and intelligence already known, and drawing deductions about the probable meaning of the evaluated information. (¶1-58)
AO	area of operations
AOC	area of concentration
AOI	area of interest
AOIR	area of intelligence responsibility
AOR	area of responsibility
AR	Army Regulation
ARAT	Army Reprogramming Analysis Team
ARCENT	US Army Central Command
ARFOR	Army forces
ARISC	Army Reserve Intelligence Support Center
ARL	airborne reconnaissance low
ARNG	US Army National Guard
ARSOF	Army Special Operations Forces
ASCC	Army Service Component Commander
ASPO	Army Space Program Office
AT	antiterrorism
ATCAE	Army Technical Control Analysis Element
ATCCS	Army Tactical Command and Control System
ATR	automatic target recognition
ART	Army tactical task
BDA	battle damage assessment
BDU	battle-dress uniform

BFACS	Battlefield Functional Area Control System
C/J2	combined J2
C2	command and control
CA	civil affairs
CASO	Collection Activities and Source Operations
CAT	crisis action team
CBRN	chemical, biological, radiological, and nuclear
CBRNE	chemical, biological, radiological, nuclear, and high-yield explosive
CCD	coherent change detection
CCIR	commander's critical information requirement
CD-ROM	compact disc-read only memory
CE	counterespionage
CECOM	Communications-Electronics Command
CED	captured enemy document
CEE	captured enemy equipment
CEM	captured enemy materiel
CFSO	counterintelligence force protection source operations
CGS	common ground station
CI	counterintelligence
CIA	Central Intelligence Agency
CIAC	counterintelligence analysis cell
CICA	Counterintelligence Coordinating Authority
CITO	Central Imagery Tasking Office
CJCS	Chairman, Joint Chiefs of Staff
CMEC	Captured Materiel Exploitation Center
CMF	career management field
CMISE	Corps Military Intelligence Support Element
CMO	civil-military operations
CNA	computer network attack
CND	computer network defense
CNE	computer network exploitation
CNN	Cable News Network
CNO	computer network operations

COA course of action

COCOM combatant command (command authority)

COG center of gravity

COMINT communications intelligence

COMSEC communications security

CONOPS contingency operations

CONPLAN contingency plan

CONUS continental United States

COP common operational picture

COSCOM Corps Support Command

***counterintelligence** **1.** Information gathered and activities conducted to protect against espionage, other intelligence activities, sabotage, or assassinations conducted by or on behalf of foreign governments or elements thereof, foreign organizations, or foreign persons, or international terrorist activities. (JP 1-02) **2.** (Army) Counterintelligence counters or neutralizes intelligence collection efforts through collection, CI investigations, operations, analysis and production, and functional and technical services. CI includes all actions taken to detect, identify, exploit, and neutralize the multidiscipline intelligence activities of friends, competitors, opponents, adversaries, and enemies; and is the key intelligence community contributor to protect US interests and equities. (FM 2-0) (¶1-111)

CP command post

CRN command radio net

CS combat support

CSG cryptologic support group

CSS combat service support

CSSCS Combat Service Support Control System

CT Counterterrorism

***cueing** The use of one or more sensor systems to provide data that directs collection by other systems. (¶2-53)

CWT combat weather team

DA Department of the Army

DAC Department of the Army Civilian

DAPDM Department of the Army Production and Dissemination Management

DCI Director of Central Intelligence

DCS, G2 Deputy Chief of Staff, Intelligence

***debriefing** The systematic questioning of individuals to procure information to answer specific collection requirements by direct and indirect questioning techniques. (¶6-4)

DF direction finding

DHS Defense HUMINT Service

DIA Defense Intelligence Agency

DIAC Defense Intelligence Analysis Center

DIMHRS Defense Integrated Management Human Resource System

DIO Defensive Information Operations

DISCOM Division Support Command

DISUM daily intelligence summary

DLPT Defense Language Proficiency Test

DMS Director of Military Support

DOCEX document exploitation

***document exploitation** The systematic extraction of information from all media formats in response to collection requirements. (¶6-4)

DOD Department of Defense

DOJ Department of Justice

DOT Department of Transportation

DP decision point

DPM dissemination program manager

DS direct support

DSM decision support matrix

DST decision support template

EA electronic attack

EAC echelons above corps

ECOA enemy course of action

EEFI essential elements of friendly information

ELINT electronic intelligence

EM electromagnetic

eMILPO electronic Military Personnel Office System

EMS electromagnetic spectrum

E-O electro-optical

EOB electronic order of battle

EOD explosive ordinance disposal

EP electronic protection

EPW enemy prisoner of war

ES electronic warfare support (Army)

***evaluating** **1.** Comparing relevant information (RI) on the situation or operation against criteria to determine success or progress. (FM 6-0) **2.** In intelligence usage, appraisal of an item of information in terms of credibility, reliability, pertinence, and accuracy. (FM 2-0) (¶1-26 and 1-98)

EW electronic warfare

EWS electronic warfare support (joint)

FBI Federal Bureau of Investigation

FDA functional damage assessment

FDU Force Design Update

FEMA Federal Emergency Management Agency

FFIR friendly force information requirement

FISINT foreign instrumentation signals intelligence

FISS Foreign Intelligence Security Service

FLPP Foreign Language Proficiency Pay

FM Field Manual

FMA foreign materiel acquisition

FME foreign materiel exploitation

FMEP Foreign Materiel Exploitation Program

FMP Foreign Materiel Program

FMT Foreign Materiel for Training

FP force protection

FRAGO fragmentary order

FS fire support

FSAC Fire Support Armaments Center

FSE fire support element

FSO fire support officer

G1 Assistant Chief of Staff, Personnel

G2 Assistant Chief of Staff, Intelligence

G3 Assistant Chief of Staff, Operations

G4 Assistant Chief of Staff, Logistics

G5 Assistant Chief of Staff, Civil/Military Affairs

G6	Assistant Chief of Staff, Command, Control, Communications, and Computer Operations
G7	Assistant Chief of Staff, Information Operations
GEOINT	geospatial intelligence
GMI	general military intelligence
GPS	Global Positioning System
GS	general support
GS-R	general support-reinforcing
HN	host nation
HOC	HUMINT operations cell
HPT	high-payoff target
HQ	headquarters
HQDA	Headquarters, Department of the Army
HSI	hyperspectral imaging
HUMINT	human intelligence
***human intelligence**	Collection by a trained HUMINT collector of foreign information from people and multimedia to identify elements, intentions, composition, strength, dispositions, tactics, equipment, personnel, and capabilities. It uses human sources and a variety of collection methods, both passively and actively, to gather information to satisfy the commander's intelligence requirements and cross-cue other intelligence disciplines. (¶1-106 and ¶6-1)
HVT	high-value target
I/R	internment and resettlement (operations)
I&W	indications and warnings
ICC	Intelligence Coordination Center (US Coast Guard)
ICF	intelligence contingency fund
ICL	intelligence coordination line
IDC	Information Dominance Center
IIR	intelligence information report
IM	information management
IMA	individual mobilization augmentee
IMEP	International Materiel Evaluation Program
IMETS	Integrated Meteorological System
IMINT	imagery intelligence

***indications and warning** (joint) Those intelligence activities intended to detect and report time-sensitive intelligence information on foreign developments that could involve a threat to the United States or allied and/or coalition military, political, or economic interests or to US citizens abroad. It includes forewarning of enemy actions or intentions; the imminence of hostilities; insurgency; nuclear or non-nuclear attack on the United States, its overseas forces, or allied and/or coalition nations; hostile reactions to US reconnaissance activities; terrorists attacks; and other similar events. (¶2-19)

***indicator** (Army) Positive or negative evidence of threat activity or any characteristic of the AO which points toward threat vulnerabilities or the adoption or rejection by the threat of a particular capability, or which may influence the commander's selection of a COA. Indicators may result from previous actions or from threat failure to take action. (FM 2-0) (¶5-14)

INFOSEC information security

***infrastructure** **1.** (joint) All building and permanent installations necessary for support, redeployment, and military forces operations (e.g., barracks, headquarters, airfields, communications facilities, stores, port installations, and maintenance stations). (JP 4-01.8) **2.** (Army) In intelligence usage, the basic underlying framework or feature of a thing; in economics, basic resources, communications, industries, and so forth, upon which others depend; in insurgency, the organization (usually hidden) of insurgent leadership. (FM 2-0) (¶5-8)

INR Bureau of Intelligence and Research, Department of State

INS Immigration and Naturalization Service

INSCOM US Army Intelligence and Security Command

intelligence **1.** (joint) The product resulting from the collection, processing, integration, analysis, evaluation, and interpretation of available information concerning foreign countries or areas. **2.** Information and knowledge about an adversary obtained through observation, investigation, analysis, or understanding. (¶1-3)

***intelligence coordination line** A line that designates the boundary between AOIRs. The G2/S2 establishes ICLs to facilitate coordination between higher, lateral, and subordinate units; coordinates with the G3/S3 to direct subordinates to track enemy units and HPTs in their areas; and hands over intelligence responsibility for areas of the battlefield. The establishment of ICLs ensures that there are no gaps in the collection effort; that all echelons are aware of the location, mission, and capabilities of other assets; facilitates asset cueing, and provides timely exchange of information between assets. The G2/S2 keeps abreast of collection activities in progress (all echelons) and battlefield developments through the ICLs. (¶3-13)

intelligence discipline (joint) A well-defined area of intelligence collection, processing, exploitation, and reporting using a specific category of technical or human resources. There are seven major disciplines: human intelligence, imagery intelligence, measurement and signature intelligence, signals intelligence (communications intelligence, electronic intelligence, and foreign instrumentation signals intelligence), all-source intelligence, technical intelligence, and counterintelligence. (¶1-104)

***intelligence process** A theoretical model used to describe intelligence operations. The intelligence process is not a framework for actual operations because the functions and common tasks occur in parallel as opposed to sequentially. (¶1-91)

***intelligence products** Intelligence products are generally placed in one of six categories: I&W, current, general military, target, S&T, and CI. The categories are distinguished from each other primarily by the purpose for which the intelligence was produced. The categories often overlap, and the same intelligence data can be used in each of the categories. (¶1-5)

INTREP intelligence report

INTSUM intelligence summary

***intelligence requirements** (Army) Those requirements generated from the staff's IRs regarding the enemy and environment that are not a part of the CCIR (PIR and FFIR). Intelligence requirements require collection and can pro-vide answers in order to identify indicators of enemy actions or intent, which reduce the uncertainties associated with an operation. Significant changes (i.e., branches and sequels) with an operation usually lead to changes in intelligence requirements. (¶2-7)

***intelligence running estimate** The intelligence running estimate is a continuous flow and presentation of relevant information and predictive intelligence that, when combined with the other staff running estimates, enable the decisionmaker's visualization and situational understanding of the AOI in order to achieve information superiority. The intelligence running estimate requires constant verification to support situational understanding of the current situation as well as predictive assessments for future operations. (¶5-30)

***intelligence synchronization** The task that ensures ISR operations are linked to the commander's requirements and respond in time to influence decisions and operations. The intelligence officer, with staff participation, synchronizes the entire collection effort to include all assets the commander controls, assets of lateral units and higher echelons units and organizations, and intelligence reach to answer the commander's CCIR (PIR and FFIR). (¶1-24)

***intelligence synchronization plan**	The plan the intelligence officer uses, with staff input, to synchronize the entire collection effort to include all assets the commander controls, assets of lateral units, and higher echelon units and organizations, and intelligence reach to answer the commander's CCIR (PIR and FFIR). (¶1-38)
intelligence warfighting function	The related tasks and systems that facilitate understanding of the operational environment
IO	information operations
IPB	intelligence preparation of the battlefield
IPIR	Initial Phase Imagery Report
IPTF	international police task force
IR	information requirement
IRINT	infrared intelligence
IRR	Individual Ready Reserve
ISM	intelligence synchronization matrix
ISR	intelligence, surveillance, and reconnaissance
J2	Joint Staff Directorate, Intelligence
J2A	Joint Staff Directorate, Intelligence (Administrative)
J2J	Joint Staff Directorate, Intelligence (Joint)
J2M	Joint Staff Directorate, Intelligence (Management)
J2O	Joint Staff Directorate, Intelligence (Operations)
J2P	Joint Staff Directorate, Intelligence (Assessments, Doctrine)
J2T	Joint Staff Directorate, Intelligence (Targeting)
J2X	Joint Staff Directorate, Intelligence (CI & HUMINT)
J3	Joint Staff Directorate, Operations
J5	Joint Staff Directorate, Civil Affairs
JAC	Joint Analysis Center
JCMEC	Joint Captured Materiel Exploitation Center
JCO	Joint Commission Observers
JCS	Joint Chiefs of Staff
JDEC	Joint Document Exploitation Center
JDISS	Joint Deployable Intelligence Support System
JFC	joint force commander
JIAWG	joint interagency working group
JIC	Joint Intelligence Center

JIF	Joint Interrogation Facility
JISE	joint intelligence support element
JOA	joint operations area
JP	Joint Publication
JSTARS	Joint Surveillance Target Attack Radar System
JTF	joint task force
JTMD	joint table of mobilization and distribution
JWICS	Joint Worldwide Intelligence Communications System
LAN	local area network
LASINT	laser intelligence
LCC	land component commander
LEA	law enforcement agency
LIDAR	laser radar
LN	local national
LNO	liaison officer
LOC	line of communication
LOS	line of sight
LPT	logistics preparation of the theater
LTIOV	latest time information is of value
MACOM	major Army command
MASINT	measurement and signature intelligence
***materiel**	(Army) In intelligence usage, the all-encompassing term for the weapons systems, equipment, apparatus, documents, and supplies of a foreign military force or nonmilitary organization. (¶2-24)
MCIA	Marine Corps Intelligence Agency
MCOO	modified combined obstacle overlay
MCS	Maneuver Control System
MDMP	military decisionmaking process
MEA	munitions effects assessment
METL	mission-essential task list
METT-TC	mission, enemy, terrain and weather, troops and support available, time available, civil considerations
MI	military intelligence
MLT	machine language translation

MNFC	multinational force commander
MO	method of operation
MOPP	mission-oriented protective posture
MOS	military occupational specialty
MP	military police
MSC	major subordinate command
MSE	mobile subscriber equipment
MSI	multi-spectral imagery
MSIC	Missile and Space Intelligence Center
MTI	moving target indicator
MTOE	modified table of organization and equipment
MTW	major theater war
NAI	named area of interest
NAIC	National Air Intelligence Center
NATO	North Atlantic Treaty Organization
NBC	nuclear, biological, and chemical
NCA	National Command Authorities
NCO	noncommissioned officer
NCR	national cryptologic representative
NCTI	non-cooperative target identification
NEO	noncombatant evacuation operation
NFH	National Defense Headquarters, Canada
NGA	National Geospatial-Intelligence Agency
NGO	non-governmental organization
NIC	National Intelligence Center
NIIRS	National Imagery Interpretability Rating Scale
NIMA	National Imagery and Mapping Agency
NIST	National Intelligence Support Team
NLT	not later than
NMCC	National Military Command
NMIC	National Maritime Intelligence Center
NMJIC	National Military Joint Intelligence Center
NRO	National Reconnaissance Office
NRT	near-real time

NSA	National Security Agency
NSC	National Security Council
NUCINT	Nuclear Intelligence
NVD	night vision device
OB	order of battle
OE	operational environment
OMA	Office of Military Affairs
OMB	Office of Management and Budget
OMT	operational management team
ONDCP	Office of National Drug Control Policy
ONI	Office of Naval Intelligence
ONIR	overhead non-imaging infrared
OPCON	operational control
OPLAN	operation plan
OPORD	operation order
OPSEC	operations security
OSC	operations support cell
OSCE	Organization for Security and Cooperation in Europe
OSD	Office of the Secretary of Defense
OSINT	open-source intelligence
PAO	Public Affairs Office
PDA	physical damage assessment
PIAP	Police Information Assessment Program
PIO	police intelligence operations
PIR	priority intelligence requirement
PME	peacetime military engagement
POC	point of contact
POD	port of debarkation
POE	port of embarkation
***predictive intelligence**	Intelligence analysis conclusions, assessments or products that attempt to define, describe, present, or portray the future situation or conditions of the enemy, +terrain, weather and civil considerations. (¶1-3)
***priority intelligence requirements**	(joint) Those intelligence requirements for which a commander has an anticipated and stated priority in his task of planning and decision-making. (FM 2-0) (¶1-32)

***process/ processing** — A function of the intelligence process that involves converting collected data, which is not already in a comprehensible form when it is reported, into a form that is understandable and suitable for analysis and production of intelligence. Examples of processing include developing film, enhancing imagery, translating a document from a foreign language, converting electronic data into a standardized report that can be analyzed by a system operator, and correlating dissimilar or jumbled information by assembling like elements before the information is forwarded for analysis. (FM 2-0) (¶1-96)

***production** — In intelligence usage, conversion of information into intelligence through the integration, analysis, evaluation, and interpretation of all source data and the preparation of intelligence products in support of known or anticipated user requirements. (¶4-20)

PSYOP — psychological operations

QRC — quick reaction capability

QRF — quick reaction force

R&D — research and development

RADINT — radar intelligence

RC — reserve component

RDEC — Research, Development, and Engineering Center

***reconnaissance, surveillance, and target acquisition/intelligence surveillance and reconnaisssance** — A full spectrum combined arms mission that integrates ground and air capabilities to provide effective, dynamic, timely, accurate and assured combat information and multi-discipline actionable intelligence for lethal and non-lethal effects/decisions in direct support of the ground tactical commander.

REMBASS — Remotely Monitored Battlefield Sensor System

***requirements management** — The intelligence task that develops a prioritized list of what information needs to be collected and produced into intelligence, dynamically updates and adjusts those requirements in response to mission adjustments/changes, and places a latest time intelligence is of value to ensure intelligence and information is reported to meet operational requirements. (¶2-83)

RF — radio frequency

RFI — request for information

RFW — radio frequency weapon

RI — relevant information

RM — requirements management

ROE — rules of engagement

ROK — Republic of Korea

ROZ — restricted operating zone

RSO&I	reception, staging, onward movement, and integration
RSOC	Regional Security Operations Center
RSTA	reconnaissance, surveillance, and target acquisition
RSP	render safe procedure
RTCAE	Regional Technical Control and Analysis Element
running estimate	A staff estimate, continuously updated, based on new information as the operation proceeds. (FM 6-0) (¶5-28)
S&T	scientific and technical
S&TI	scientific and technical intelligence
S1	Staff Officer, Personnel
S2	Staff Officer, Intelligence
S2X	Staff Officer, Intelligence (CI & HUMINT)
S3	Staff Officer, Operations
S4	Staff Officer, Logistics
S5	Staff Officer, Civil Affairs
S6	Staff Officer, Command, Control, Communications, and Computer Operations
SAP	special access program
SAR	synthetic aperture radar
SATCOM	satellite communications
SBCCOM	Chemical and Biological Defense Command
SCI	sensitive compartmented information
SCIF	sensitive compartmented information facility
SIDPERS	Standard Installation Division Personnel System
SIGINT	signals intelligence
SIGO	signal officer
SII	statement of intelligence interest
SIO	senior intelligence officer
SIPRNET	Secret Internet Protocol Router Network
SIR	specific information requirement
SITMAP	situation map

***situation template** (Army) A depiction of a potential adversary course of action as part of a particular adversary operation. Situation templates are developed on the adversary's current situation (for example, training and experience levels, logistic status, losses, and disposition), the environment, and adversary doctrine or patterns of operations. The commander dictates the level to depict the adversary based on the factors of METT-TC (at minimum two levels of command below the friendly force) as a part of his guidance for mission analysis. (FM 2-0) (¶1-58)

SJA staff judge advocate

SMU special mission unit

SOCOM satellite communications on-the-move

SOF special operations forces

SOFA status of forces agreement

SOP standing operating procedure

SOTA SIGINT operational tasking authority

SR special reconnaissance

SRTD signals research and target development

SSO special security office

STANAG standardization agreement (NATO)

STICEUR Scientific and Technology Intelligence Center Europe

SUPR supplemental imagery report

SWO staff weather officer

TA target acquisition

TACON tactical control

***tactical questioning** The expedient initial questioning for information of immediate tactical value. (FM 2-0) (¶6-6)

TAI target area of interest

TCAE technical control and analysis element

TCN third-country national

***technical control** The technical function to ensure adherence to existing policies or regulations and provide technical guidance for MI activities, particularly HUMINT, SIGINT, and CI operations. (FM 2-0) (¶3-23)

TECHINT technical intelligence

TECOM Test and Evaluation Command

TENCAP Tactical Exploitation of National Capabilities

TEU Technical Escort Unit

THREATCON	threat condition
TIB	Theater Intelligence Brigade
TIG	Theater Intelligence Group
TIM	toxic industrial material
TLP	troop-leading procedure
TOC	tactical operations center
TPFDD	Time Phased Force Deployment Data
TRADOC	US Army Training and Doctrine Command
TSA	target system assessment
TSCM	Technical Surveillance Countermeasures
TSE	theater support element
TTP	tactics, techniques, and procedures
TVA	threat vulnerability assessment
UAV	unmanned aerial vehicle
UN	United Nations
US	United States
USAF	United States Air Force
USAR	United States Army Reserve
USC	United States Code
USCENTCOM	United States Central Command
USCG	United States Coast Guard
USCS	United States Cryptologic System
USEUCOM	United States European Command
USJFCOM	United States Joint Forces Command
USMC	United States Marine Corps
USNORTHCOM	United States Northern Command
USPACECOM	United States Space Command
USPACOM	United States Pacific Command
USSID	United States Signal Intelligence Directive
USSOCOM	United States Special Operations Command
USSOUTHCOM	United States Southern Command
USSPACECOM	United States Space Command
USSTRATCOM	United States Strategic Command
USTRANSCOM	United States Transportation Command

VJ2	Vice J2
VTC	video teleconference
WAN	wide area network
WARNO	warning order
WMD	weapons of mass destruction
WO	warrant officer

Bibliography

The bibliography lists field manuals published with new numbers followed by old number.

DOCUMENTS NEEDED

These documents must be available to the intended users of this publication.

JP 1-02. *Department of Defense Dictionary of Military and Associated Terms.* June 1999. [Online] Available http://www.dtic.mil/doctrine/jel/doddict/

FM 1-02/MCRP 5-2A. *Operational Terms and Symbols* (Drag Draft). 21 February 2003.

READINGS RECOMMENDED

These sources contain relevant supplemental information.

ARMY PUBLICATIONS

Most Army doctrinal publications are available online: http://155.217.58.58/atdls.htm

AR 381-10. *US Army Intelligence Activities.* 1 July 1984.

AR 381-12. *Subversion and Espionage Directed Against the U.S. Army (SAEDA).* 15 January 1993.

AR 381-20. *The Army Counterintelligence Program.* 15 November 1993.

AR 611-6. *Army Linguist Management.* 16 February 1996.

AR 715-9. *Contractors Accompanying the Force.* 29 October 1999.

FM 1 (100-1). *The Army.* 14 June 2001.

FM 2-19.402. *STRYKER Brigade Combat Team Intelligence Operations.* 1 March 2003.

FM 2-19.602. *Surveillance Troop.* 1 March 2003.

FM 2-33.5/ST. *Intelligence Reach Operations.* 1 June 2001.

FM 3-0. *Operations.* 14 June 2001.

FM 3-05.102. *Army Special Operations Forces Intelligence.* 31 August 2001 (formerly FM 34-36).

FM 3-05.30. *Psychological Operations.* 19 June 2000.

FM 3-06.11. *Combined Arms Operations in Urban Terrain.* 28 February 2002.

FM 3-13. *Information Operations.* 28 November 2003 (formerly FM 100-6).

FM 3-19. *NBC Reconnaissance.* 19 November 1993.

FM 3-19.1. *Military Police Operations.* 22 March 2001.

FM 3-19.4. *Military Police Leaders' Handbook.* 4 March 2002.

FM 3-19.40. *Military Police Internment/Resettlement Operations.* 30 September 1987.

FM 3-20.98. *Reconnaissance Platoon.* May 2001.

FM 34-2. *Collection Management and Synchronization Planning.* 8 March 1994.

FM 34-52. (FM 2-22-3) *Human Intelligence.* 28 September 1992.

FM 3-90. *Tactics.* 4 July 2001.

FM 3-90.3. *The Mounted Brigade Combat Team.* 1 November 2001.

FM 3-100.4. *Environmental Considerations in Military Operations.* 15 June 2000, Change 1, 11 May 2001.

FM 3-100.21. *Contractors on the Battlefield.* 26 March 2000.

FM 5-170. *Engineer Reconnaissance.* 5 May 1998.

FM 6-0. *Mission Command: Command and Control of Army Forces.* 11 August 2003.

FM 6-20-10. *Tactics. Techniques, and Procedures for the Targeting Process.* 8 May 1996.

FM 6-121. *Tactics, Techniques, and Procedures for Field Artillery Target Acquisition.* 25 September 1990.

FM 7-0. *Training the Force.* 22 October 2002.

FM 7-10. *The Infantry Rifle Company.* 14 December 1990.

FM 7-15. *Army Universal Task List.* 31 August 2003.

FM 7-92. *The Infantry Reconnaissance Platoon and Squad (Airborne, Air Assault, Light Infantry).* 23 December 1992.

FM 7-100. *Opposing Force Doctrinal Framework and Strategy.* 1 May 2003.

FM 17-97. *Cavalry Troop.* 3 October 1995.

FM 17-98. *Scout Platoon.* 10 April 1999.

FM 19-1. *Military Police Support for the Airland Battle.* 23 May 1988.

FM 27-10. *The Law of Land Warfare.* 18 July 1956.

FM 31-20-5. *Special Reconnaissance Tactics, Techniques, and Procedures for Special Forces.* 23 March 1993.

FM 34-2-1. *Tactics, Techniques, and Procedures for Reconnaissance and Surveillance and Intelligence to Counterreconnaissance.* 19 June 1991.

FM 34-3. *Intelligence Analysis.* 15 March 1990.

FM 34-8-2. *Intelligence Officer's Handbook.* 1 May 1998.

FM 34-10. *Division Intelligence and Electronic Warfare Operations.* 25 November 1986.

FM 34-25. *Corps Intelligence and Electronic Warfare Operations.* 30 September 1987.

FM 34-37. *Echelons Above Corps (EAC) Intelligence and Electronic Warfare (IEW) Operations.* 15 January 1991.

FM 34-54. *Technical Intelligence.* 30 January 1998.

FM 34-80. *Brigade and Battalion Intelligence and Electronic Warfare Operations.* 15 April 1986.

FM 34-130. *Intelligence Preparation of the Battlefield.* 8 July 1994.

FM 4-0. *Combat Service Support.* 29 August 2003.

FM 41-10. *Civil Affairs Operations.* 14 February 2000.

FM 100-25. *Doctrine for Army Special Operations Forces.* August 1999.

FM 101-5. *Staff Organization and Operations.* 31 May 1997.

FM 101-5-1. *Operational Terms and Graphics.* 30 September 1997.

FM 101-5-2. *US Army Report and Message Formats.* 29 June 1999.

ST 2-22.7. *Tactical Human Intelligence and Counterintelligence Operations.* April 2002.

ST 2-33.5. *Intelligence Reach Operations.* 1 June 2001.

ST 2-50. *Intelligence and Electronic Warfare Assets.* June 2002.

ST 2-50.4. *Combat Commanders' Handbook on Intelligence.* September 2001 (formerly FM 34-81, 28 September 1992).

JOINT PUBLICATIONS

Most joint publications are available online: http://www.dtic.mil/doctrine/jel/

JP 0-2. *Unified Action Armed Forces (UNAAF).* 10 July 2001.

JP 1. *Joint Warfare for the US Armed Forces.* 14 November 2000.

JP 1-06. *Joint Tactics, Techniques, and Procedures for Financial Management During Joint Operations.* 22 December 1999.

JP 2-0. *Joint Doctrine for Intelligence to Operations.* January 2000.

JP 2-01. *Joint Intelligence Support to Military Operations.* 20 November 1996.

JP 2-01.1. *Joint Tactics, Techniques, and Procedures for Intelligence Support to Targeting.* 9 January 2003.

JP 2-01.3. *Joint Tactics, Techniques, and Procedures for Joint Intelligence Preparation of the Battlespace*, 24 May 2000.

JP 2-02. *National Intelligence to Joint Operations.* 28 September 1998.

JP 2-03. *Joint Tactics, Techniques, and Procedures for Geospatial Information and Services Support to Joint Operations.* 31 March 1999.

JP 3-0. *Doctrine and Joint Operations.* 1 February 1995.

JP 3-07. *Joint Doctrine for Military Operations Other Than War.* 16 June 1995.

JP 3-07.2. *JTTP for Antiterrorism.* 17 March 1998.

JP 3-08. *Interagency Coordination During Joint Operations,* Volume II. 9 October 1996.

JP 3-09. *Doctrine for Joint Fire Support.* 12 May 1998.

JP 3-09.1. *Joint Tactics, Techniques and Procedures for Laser Designation Operations.* 28 May 1999.

JP 3-11. *Joint Doctrine for Operations in Nuclear, Biological, and Chemical (NBC) Environments.* 11 July 2000.

JP 3-13. *Joint Doctrine for Information Operations.* 9 October 1998.

JP 3-18. *Joint Doctrine for Forcible Entry Operations.* 16 July 2001.

JP 3-35. *Joint Deployment and Redeployment Operations.* 7 September 1999.

JP 3-51. *Joint Doctrine for Electronic Warfare.* 7 April 2000.

JP 3-55. *Doctrine for Reconnaissance, Surveillance, and Target Acquisition Support for Joint Operations (RSTA).* 14 April 1993.

JP 3-57. *Joint Doctrine for Civil-Military Operations.* 8 February 2001.

JP 3-60. *Joint Doctrine for Targeting.* 17 January 2002.

JP 4-0. *Doctrine for Logistic Support of Joint Operations.* 6 April 2000.

JP 5-0. *Doctrine for Planning Joint Operations.* 13 April 1995.

JP 5-00.2. *Joint Task Force Planning Guidance and Procedures.* 13 January 1999.

PUBLIC LAWS AND OTHER PUBLICATIONS

The United States Code is available online: http://uscode.house.gov/usc.htm

EO 12333. *United States Intelligence Activities.* 4 December 1981.

CJCS Instruction 1301.01, Policy and Procedures to Assign Individuals to Meet Combatant Command Mission-Related Temporary Duty Requirements. 1 July 2001.

CJCSM 3500.04C. *Universal Joint Task List.* 1 July 2002.

Military Standard 2525B, *DOD Interface Standard, Common Warfighting Symbology.* 30 January 1999.

PUBLICATIONS UNDER DEVELOPMENT

FM 2-01. *Intelligence Synchronization.* When published tentatively in July 2004, FM 2-01 will supersede FM 34-2, 8 March 1994, and FM 34-2-1, 19 June 1991.

FM 2-01.2. *Counterintelligence Analysis.* When published tentatively in April 2005, FM 2-01.2 will supersede FM 34-60, 3 October 1995.

FM 2-01.3. *Intelligence Preparation of the Battlefield.* When published tentatively December 2004, FM 2-01.3 will supersede FM 34-130, 8 July 1994.

Index

Entries are by paragraph number.

J-L

Made in the USA
Monee, IL
05 November 2026

67813160R00282